Spoiled

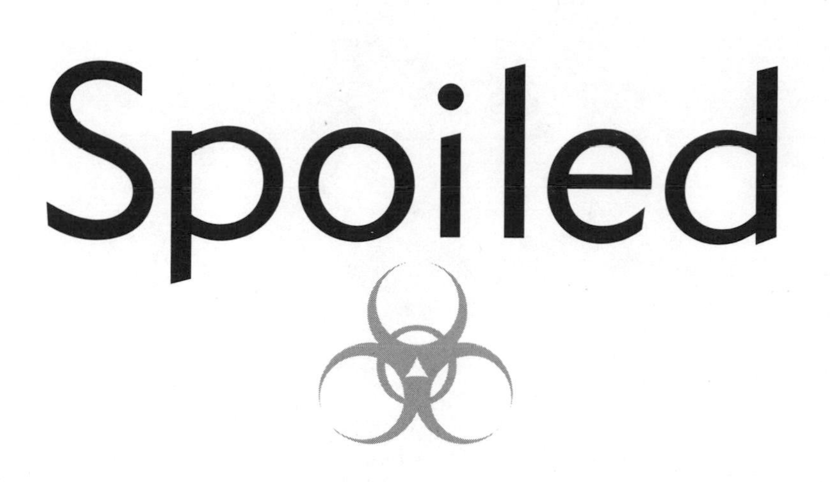

Spoiled

The Dangerous Truth About a Food Chain Gone Haywire

NICOLS FOX

BasicBooks
A Division of HarperCollins*Publishers*

Published by BasicBooks,
A Division of HarperCollins Publishers, Inc.

FIRST EDITION

Designed by Elliott Beard

Library of Congress Cataloguing-in-Publication Data

Fox, Nicols.
Spoiled : the dangerous truth about a food chain gone haywire / Nicols Fox.
p. cm.
Includes bibliographical references and index.
ISBN 0-465-01980-3
1. Foodborne diseases—Popular works. I. Title.
RA601.5.F68 1997
615.9'54—DC21 97-1239

97 98 99 00 01 ❖/RRD 10 9 8 7 6 5 4 3 2 1

For the victims of *E. coli* O157:H7,
and for Mary Heersink who asked "Why?"

Contents

Part III/Emerging Diseases and the Anatomy of Outbreak

Part IV/Our Just Deserts

Preface

Journalists are people who have found a socially acceptable way of satisfying a socially unacceptable degree of curiosity. They also tend to reject overly facile explanations. Something that doesn't make sense can set up an irritation as nagging as a sore tooth.

I felt that way in 1993 when I heard, along with the rest of the world, of the widespread outbreak of foodborne disease linked to fast-food hamburgers on the West Coast. It seemed clear at once that there was something missing to the story. The facts as they appeared in the news reports didn't add up. Hundreds were sick, and children were dying from a "urinary tract infection" linked to something they had eaten. The accounts mentioned *Escherichia coli* bacteria.

Members of the genus of *E. coli* bacteria, I thought I remembered from college biology, are extremely common and regularly found in the human gut. It seemed strange that they would be linked to a urinary tract infection, and stranger still that children should be dying of something so ordinary. My curiosity took me to a local library, and my first readings confirmed what I thought I had remembered. The bacteria are ubiquitous, and generally benign, but they have been linked to various forms of diarrhea, most of them comparatively mild—only recently. Finding out why the organism was now associated with something as dangerous as what was happening on the West Coast was more difficult, but gradually bits and pieces of a larger picture emerged. This *E. coli* went by a name: O157:H7. And it was a new strain (more properly called a serotype).

I also remembered something else about *E. coli*. Because it reproduces so quickly and because it is such a familiar microorganism, it is often used in genetic engineering experiments. Since the late 1970s, the techniques of transferring DNA from one organism to another have been widely known and are "simple" enough to be carried out in a well-equipped high school science lab. Had someone hooked something to a benign *E. coli* to make it newly dangerous? It was an interesting thought, and not at all unfeasible. Many of the early researchers in the field of genetic engineering had been (and still are) extremely concerned about the possibility. I discovered that potentially dangerous research was being undertaken despite the safeguards various governments had instituted. Genes from the human gut parasite *Giardia lamblia*, which causes debilitating persistent diarrhea, had been transferred—for reasons not clear—to *Escherichia coli*, a common resident of the human intestinal tract, by researchers at St. Bartholomew's Hospital in London in the late 1980s.[1] This experiment was in violation of both government regulations and common sense. If *E. coli* O157:H7 turned out to be a similar hookup, I would have a compelling story for any one of the several different publications I regularly wrote for.

Pursuing any story is to traverse a landscape of false trails, and this was one of them. As I learned more, the sci-fi scenario I envisioned seemed less and less likely; the organism appeared to have been around longer than genetic engineering. At the same time I was discovering a real story hardly less dramatic. A new and potentially deadly pathogen was suddenly in our food supply and was found most frequently in our favorite food: hamburger. The staff at the Centers for Disease Control and Prevention, parents of children who had been made ill or died from their infection, and experts across the country filled me in on a tale more compelling than fiction. I wrote the first of several articles.

Perhaps it was the dogged determination of women such as Mary Heersink, whose son managed to survive a horrendous bout with this pathogen, and who would then direct all her energy and time to finding out why this killer was in the meat supply; perhaps it was the dedication of Dr. Patricia Griffin at the CDC, or Dr. Marguerite Neill at the Brown University Program in Medicine, whose focus on unraveling the secrets of this pathogen was intense and compelling, and who transferred the urgency of their search for answers to me; perhaps it was just because it was a great story (which my editor at *The Economist*, Ann Wroe, almost indulgently let me pursue); but I found I could not let it go. I traveled to hearings and medical gatherings and began to identify the players in this real-life drama. I accumulated a database of my own. I collected notebook after notebook and tape after tape of interviews until my house and my life were taken over by a microbe.

I began with questions. What I discovered was that a new pathogen could find a niche and infiltrate the world's food supply, not by intentional manipulation on the part of a few malevolent or careless scientists, but by unintentional and uncoordinated shifts in our relationship to what we eat.

It was not always easy to transfer my growing interest—call it a fixation, call it compulsion—to others. Convincing editors that a vital tale of human ecology could be told through one microbe simply wasn't possible. My frustration grew.

Inevitably the experience of chasing the story becomes part of the story. I began as the ultimate generalist—I have written on everything from the arts, religion, media, and politics to granite quarrying and sardine packing. With a rudimentary background in science, I was compelled to learn more about microbiology, about epidemiology, about clinical medicine, through textbooks and the generous help of experts.

We often hear that generalists are outmoded, that this is the age of the specialist. What I found was that my broad background served me well, for this was a tale about not just science but sociology, economics, ecology, political science, and philosophy. It was a story in which my experiences abroad were a help; to which my lifelong interest in food and cooking and my passing knowledge of farming added a dimension; and to which my many friends in various places were important contributors of information and experiences.

It became clear that the troubling biography of *E. coli* O157:H7 was one that had been repeated, with variations, by a host of other emerging foodborne pathogens, and that story, inevitably, became the book I would write. I began with questions. Gradually explanations emerged from a flood of details. I've tried to tell the tale through the experiences of individuals and the work of epidemiologists as they attempt to unravel the mysteries of emerging diseases. It is a story that almost has a beginning; there is a point when one can, with some precision, say, "Here it began." But if there is an end, it is not in sight. As I wrote, outbreaks seemed to be reported more frequently, and the time in between narrowed; the pace quickened. The public learned of "mad cows" and the fatal neurological disease to which they were linked; the Japanese *E. coli* outbreak; the *Cyclospora* in raspberries, although first thought to be in strawberries; tainted apple juice in the West; poisoned oysters in the Gulf of Mexico; deadly meats in Scotland. It has sometimes seemed frighteningly like the end of food.

This has been the more difficult story to tell; a life history of the microbe that originally inspired me would have been easier. It was challenging because it could not be simplified. The microbes overlapped; they weren't simply in one dominant food but in many, not associated with one animal but with several. The causes were various and tangled. Sorting them out was the first challenge. In the end it wasn't really possible. The book mirrors the tapestry-like nature of the subject: Nothing can be separated out, and everything is interdependent.

If there is one thing that emerges from this tale it is that our apparent compulsion to simplify is at the heart of the errors we have made. We seem unable to accept that we must contend with the complex synergies and interconnections of the natural world rather than attempt to subdue and control them. Equally guilty is our drive—which since the industrial revolution has become the closest thing we now have to a universal religion—to use efficiency as the premier standard and value by which every human activity, every process, every undertaking must be judged.

This book is the story of change and the doors it opens to disease. It is a story of systems out of balance. Along the way the reader will discover, as I did, the other factors that contribute to the tale—the role played by our persistent pursuit of pleasure, novelty, and ease; the impact of politics, selfishness, and greed; the effect of ignoring traditions, of unquestioningly accepting the new, and of opting always for the cheap. And finally it confirms once again the eternal truth that in our hubris we inevitably seem to forget or choose to ignore: In the end, nature will have its way.

Acknowledgments

To repeat the cliché, that "this book wouldn't have been written without the help of many people," would be to make an understatement. I sometimes feel that I am connected by post, phone, fax and electronic mail to a vast system of unpaid researchers who have kept me supplied—without contemplating any reward other than the satisfaction of knowing they were helping—with journal articles, newspaper clippings, anecdotes, personal experiences, and, perhaps most valuable of all, encouragement. If there is one word that applies to them all, it is generosity. The circle of those who deserve and have my thanks extends around the globe and ends finally in my town, where friends and strangers alike told me stories and provided support.

I must thank first the Diarrheal and Foodborne Diseases Branch of the Centers for Disease Control and Prevention, the Minnesota Department of Health, and the Maine Bureau of Health. All three allowed me to spend time with them and offered me the benefit of their considerable wisdom on the subject of foodborne diseases. A number of other state health departments were extraordinarily cooperative as well, and deserve thanks. The assistance of the U.S. Department of Agriculture, the Food and Drug Administration, and the General Accounting Office was important and helpful, as was the information provided by groups such as Safe Tables Our Priority, the Government Accountability Project, the Safe Food Coalition, and the Council for Agricultural Science and Technology.

The names of many of the individuals who helped are scattered in the book. To single out, as I must, Mary Heersink, Rob Tauxe, Patricia Griffin, Kathy Gensheimer, Marty Blaser, Martin Skirrow, Michael Osterholm, Luise Light, Steve Bjerklie, and Craig Hedberg is to risk leaving out other major contributors, but the assistance and cooperation of these individuals went well beyond what anyone might have expected. The many others who contributed are anonymous, but they know who they are and they must know how grateful I am.

To be certain the facts were in order, various chapters or sections of the book were parceled out and reviewed by some of the individuals and experts involved. Their comments were valuable and much appreciated. I also want to thank Mary Greenlee and Ken Cochrane, who read and commented on the final manuscript. Thanks, too, go to the Jackson Laboratory Library and the Southwest Harbor Library, who helped me with research, and to my friends and family who tolerated with good nature my four-year preoccupation. Special note should be taken of those who were brave enough to invite me to dinner during this time.

But final thanks must surely go to my editor Gail Winston, who encouraged me when it was most needed, who championed the book and gave it careful attention from beginning to end, and who offered me wonderful advice. Her heart is good, her wisdom is great, and her touch deft.

Part I/News from the Front Lines

1/The Index Case

Bacteria, by any reasonable criterion, were in the beginning, are now, and ever shall be, the most successful organisms on earth.

STEPHEN JAY GOULD,
Full House: The Spread of Excellence from Plato to Darwin, 1996

Roni Rudolph lives in the La Costa area of Carlsbad, California, just north of San Diego. It is hilly, near the sea, affluent but not ostentatious, an area of tree-lined residential streets where children ride their bicycles and skateboards and mothers push strollers. In 1992 Roni had been separated from her husband, Dick, for less than two years and had settled herself and her two children, Michael and Lauren, into a comfortable townhouse about a mile from their old home. She says that she had wanted the breakup to disrupt the children's lives as little as possible. They were not far from their old neighborhood, still had the same friends, and went to the same school.[1]

The neighborhood is pleasant. It is new, but many of the same families have been there from the beginning, and there is a strong sense of attachment and a feeling of stability. Many of the children have grown up together. There are plenty of trees, palms and eucalyptus, and even the view of a hillside that has yet to be developed—the same hill, in fact, that Roni remembers pushing Lauren up in her stroller on their walks together when she was a baby. In some areas you can see the distant ocean. By the time they moved in, Lauren was five, a happy, friendly child with a mop of shiny red hair.

Roni was older than average when her children were born. She and Dick had been married thirteen years before Michael arrived; Lauren was born five years later. By that time her children seemed like an unexpected and wonderful gift. Eager for them to be comfortable and contented in their new house, she set up the family room as their joint playroom. There was a piano. In one corner was Michael's computer with its load of games; in another she created a miniature kitchen for Lauren and her friends.

Roni had run a decorating business out of her home before the separation, but she wanted something different now that she was on her own. She took a position as an assistant in the preschool Lauren attended. She wasn't a suffocating mother, far from it, she says, but she liked being close to her children, and she felt that Lauren, especially, liked knowing she wasn't far off even if they weren't in the same room. When Lauren moved on to first grade, Roni got a job as a teacher's aide in her new school.

In December 1992 things were beginning to settle down for the Rudolphs. Lauren had started first grade that fall, and Michael was in sixth. Both Roni and Dick had begun new social lives. Dick had a serious relationship, and Roni had begun to date. The children lived with Roni, but Dick saw them regularly. Often he would take them out to eat on his weekends.

This year Roni was excited about the upcoming holidays. The children would have Christmas with her, then go to their dad's, and Roni would then have a welcome week to herself. She planned to visit a friend in Colorado and was looking forward to the trip.

Friday the eighteenth was Dick's weekend to have the children. The weather was cool but beautiful, Roni remembers, standard for that time of year. San Diego's climate is generally more of a backdrop than an intrusion, and this December was no exception.

Dick picked the children up at the usual time. He had nothing special planned. Both children's report cards had been good, and as a treat he took them to a fast-food restaurant. He had a salad, Michael had a jumbo burger, and Lauren had a kid-sized cheeseburger. Then they picked up a movie from the video store next door and spent the evening together at his home. A typical, all-American kind of day.

By Sunday evening Lauren was back home but not feeling well. She complained, oddly, of a headache and seemed lethargic. Roni put her to bed with Tylenol and a hug and hoped for the best. It was getting close to the holidays. The children were off from school, and they had a big day planned for Monday. Roni had an appointment at the dentist, then they planned to drive to her mother's house in La Mesa, forty miles to the south. Together they would go to the local mall for last-minute shopping and to see Santa. Lauren wasn't feeling much better in the morning, but she still wanted to go.

"I'll be okay, Mom," she told Roni.

At the dentist she seemed to feel weak and to have trouble holding her head up. Roni was worried, but Lauren reassured her once again.

"She was just a kid. She wanted very much to see Santa," Roni remembers.

But by the time they had picked up Roni's mother and gotten to the mall, Lauren was clearly sicker. She felt so nauseated that the visit with Santa had to be abandoned. There was nothing to do but get her home. Roni took her mother back to La Mesa and headed north, back up Route 5 to La Costa. By the time they got there Lauren was throwing up.

Roni's first thought was that Lauren had stomach flu. She gave Lauren a commercial antinausea medication and put her to bed. One thing she didn't want, with Christmas just around the corner, was for Michael to catch the same thing, so she made sure Lauren used one bathroom, which she cleaned carefully with disinfectant each time Lauren was sick, and she and Michael used the other.

Lauren's condition grew worse. She would double over with painful cramps and have frequent episodes of diarrhea. In between she seemed almost lifeless, sleeping fitfully. Roni decided she had to call her pediatrician. He suspected a flu bug as well and told her to be sure Lauren's fluid intake was steady.

Tummy upsets, even severe ones such as Lauren had, usually pass within a day, or at the most two. But Lauren was getting worse, not better. By Wednesday evening her cramping was frighteningly severe and her stools were red with blood. Shortly after midnight Roni called her pediatrician again, and then Dick; together, with Michael in tow, they took Lauren to the emergency room of the nearest hospital. It was 2:30 A.M. when they got there. Christmas Eve. There was no pediatrician on duty.

"We were greeted by a few questions, endless waiting, a minimum number of perfunctory tests, and we had the overall feeling that they thought we were overreacting," remembers Roni.

They were told to take Lauren to their own pediatrician when his office opened at nine o'clock that morning, and so Dick and Roni bundled their sick little girl into the car and carried her home again, and then, a few hours later, to their doctor. He still assumed Lauren was afflicted with some fierce flu and urged them to watch her closely. Her bowel, he told them, was raw and irritated, which explained the bleeding. If she continued to be sick, she would be in danger of dehydration. If she wasn't improving by that afternoon, they were to take her to Children's Hospital in San Diego.

There was no improvement. Lauren's stools seemed more like pure blood and tissue, and her body continued to be wracked with excruciatingly painful cramps. At 2:30 that afternoon they took her to Children's Hospital, where she was admitted at once. She was quickly given intravenous fluids, and when her vital signs were monitored, Roni and Dick got the bad news.

"We were told time after time that Lauren was a very sick little girl," says Roni. Within a short time, she remembers, Lauren went from a "regular room where parents can stay with their children, to multiple test sites within the hospital, to a 'special care' unit with round-the-clock doctors and nurses monitoring her."

The pain and diarrhea became uncontrollably worse, and the powerful pain killers seemed useless. The day lapsed into night. Dick and Roni went home, dropping Michael off at his best friend's house, along with some of his presents, so that he could have a more normal Christmas in case his mother was called back to the hospital.

On Christmas Day Lauren was too weak to open the gifts her brother brought to her bedside. He opened them for her. She insisted on wearing her new Beauty and the Beast watch and keeping track of time. Every ten minutes she was allowed an ice chip to soothe her parched lips. She would consult her new watch carefully and inform the nurse.

Her brother read Christmas stories to her. He tried to interest her in *Finding Waldo*, a favorite book, but she was very weak. The day was long and troubling, but Lauren at least was talking. That night Roni went home again. She and Lauren spoke on the phone, and Lauren told her mother she loved her. When Roni arrived the next morning, Dick was already there and was visibly upset. She could see tears in his eyes, Roni remembers. He took her aside and in a horrified but hushed voice told her that Lauren had repeatedly said, "Daddy, I'm going to die. I'm going to die."

Roni took her daughter's hand and tried to comfort her. "I told her again and again that everything would be okay," she says.

It wasn't. Within the hour Lauren would suffer a massive heart attack. Dick had left for work—they were spelling each other—and Roni stood there screaming for someone to help her child as she watched her body contort and her breathing stop and her lips turn blue. The staff was there in an instant, and Roni was rushed out of the room. The minutes dragged on and on as the team worked on her daughter. They revived Lauren, but her pediatrician was deeply concerned. It had taken too long, too much effort to revive someone who should have had a young and healthy heart. Lauren was in a coma now. She'd been awake for days with the pain, unable to sleep. The doctor explained that the coma was a self-induced mechanism to rest her body, but then her kidneys ceased functioning and her body began shutting down. The only choice was to go on life support systems. After two more massive heart attacks, her tiny body could take no more. She died on December 28.

Lauren Rudolph died from foodborne illness. She died, in fact, from the cheeseburger she had eaten on December 18. It was something her parents had never suspected or even considered. They had heard the doctors refer to her illness as hemolytic uremic syndrome (HUS), but neither had heard of it or had any idea what it was; they certainly had no idea that the great majority of cases of HUS, perhaps as many as 90 percent, are caused by a bacterium found principally in ground beef.

The bacterium *E. coli* O157:H7 had been identified quickly at

Lauren's autopsy, but the public health department kept the information to itself. The pathogen is responsible for what Canadians call, with characteristic candor, the "hamburger disease." Although O157, as the medical profession calls it, has been causing outbreaks since 1982, no one at the hospital had thought to run the standard tests for it while Lauren was alive. If they had, and if it had been reported, the future for more than seven hundred other people might have been different.

The day after Lauren died, a woman from the public health department would call and leave a message, but Roni and Dick's days were consumed by the devastating task of arranging for their daughter's funeral. It would be three days before Roni returned the call. When the woman asked her, among other things, whether Lauren had eaten at a fast-food restaurant, Roni remembers thinking the question was ridiculous. Of course she had. Didn't everyone in America? She would talk to the health department four more times. Not once, she says, did they mention that *E. coli* O157:H7 had been found in Lauren's body or tell her what that might mean and what they suspected had caused her daughter's death.

Roni would discover what the health department knew only weeks later in Colorado at her friend's, where she'd gone to rest and mourn. Friends from La Costa would call her at 6:30 A.M. on January 19 to tell her that the *San Diego Union* was reporting that a six-year-old north county girl had died on December 28 from an infection caused by *E. coli* O157:H7. It had to be Lauren. Someone from the health department had leaked the story. By then it was part of a much larger story. In Seattle the hospitals were beginning to fill with O157:H7 patients. Most of them had eaten at Jack-in-the-Box, a national hamburger chain. That was where Lauren had eaten a month earlier.

One of Lauren's doctors would later speculate that the outbreak, which eventually involved five states, caused 732 illnesses, put 195 in the hospital—55 with HUS—and killed three more children, might have been identified and stopped in San Diego, where it apparently began, if only California had a reporting requirement and encouraged testing.[2] Lauren Rudolph, in fact, was probably the index case. Her sudden and unexpected death could have signaled the outbreak if enough people in the system had been paying attention. Few states were at the time, but one of the few was Washington State, where officials quickly realized what was happening.

And well they should have. Their state had experienced outbreaks from the pathogen before. In 1985 five children in two families had been infected at Fairchild Air Force Base near Spokane. Three of the children had developed HUS, and two had died. In early September 1986 several children who attended the same day-care center had developed HUS. An investigation of the center revealed that a child there had recently recovered from a bout of bloody diarrhea and apparently transferred the bacteria to his playmates. Two isolates of O157 were found, but the source was not identified. In November 1986 two older women died when food from a Mexican-style family restaurant in Walla Walla, Washington, made eighteen people ill. That outbreak, which seemed at first to be related to contaminated salad dressing, was linked to other illnesses in a nursing home not far away. An investigation by the Centers for Disease Control and Prevention (CDC) in Atlanta discovered that the hamburger meat consumed at the restaurant and in the nursing home, came from the same source, making it the likely culprit. This evidence, coupled with a survey of stool cultures from a Seattle HMO that revealed the same strain of the bacteria in patients they had seen with bloody diarrhea, made it obvious that the state had experienced a widespread outbreak probably from a single source, which, until the later study, had not been spotted, even while it was at its peak, because the cases were scattered. Washington State knew this pathogen well, and health officials there had recently pushed mandatory reporting through the legislature, making Washington the first state to do so.

Since 1982, when the first *E. coli* O157:H7 outbreak occurred in fast-food restaurants in Oregon and Michigan, scientists in Canada and the United States had been building a profile of the new pathogen. It was during the Walla Walla investigation that the CDC team traced the bacteria back through the complicated meat slaughtering and distribution system to the farm, pinpointing its reservoir in cattle and its route from there into the human food supply through the contamination of meat with fecal material. With what these observers knew about the food supply—the mass-production and widespread distribution—and the virulence of the pathogen (a term used to describe any disease-causing agent), the widespread outbreak that followed Lauren Rudolph's death was almost predictable. But it caught the rest of the world completely by surprise.

Most Americans have heard of the Jack-in-the-Box outbreak. And most probably think of it as an anomaly, a freak accident in an age of unprecedented food safety when the combined forces of government regulation, technological innovation, scientific methods of production and inspection, and industrial accountability guarantee that what we eat will not harm us. They assume that whatever went wrong has been fixed, and that such outbreaks are behind us now.

Nothing could be further from the truth. A consumer food safety group called Safe Tables Our Priority (STOP) takes calls from people around the country made sick by this microbe. They have had more than 200 reports of outbreaks from O157 since the one that killed Lauren Rudolph. There may have been more than that. The CDC's top expert in *E. coli* O157:H7, Dr. Patricia Griffin, has acknowledged that many more U.S. outbreaks have occurred "but were incompletely investigated or not reported in the medical literature."[3]

E. coli O157:H7 is just one bacterium, albeit a new and particularly vicious entry into the foodborne pathogen category, a grouping that includes not only bacteria but viruses and parasitic protozoa. All together they are referred to as microorganisms (tiny creatures) or microbes or—in the vernacular—"germs." Not all microbes are harmful. In fact, the vast majority are not. Microorganisms are everywhere in the environment, on and in anything that can support life, including our bodies, where they establish colonies. "Human skin harbors some 100,000 microbes per square centimeter," write Lynn Margulis and Dorion Sagan in *Microcosmos*. Our guts are full of harmless and even helpful bacteria. The number of *E. coli* in the gut of each human being, they say, is far greater than the total number of people who now live—or who have ever lived—on Earth. In fact, they write, an amazing "ten percent of our own dry body weight consists of bacteria, some of which, although they are not a congenital part of our bodies, we can't live without."[4] Without the help of microorganisms, we couldn't digest our foods and absorb the nutrients. These colonies are even helpful in fighting off pathogenic intruders. Therefore, a bacterium that causes disease has developed some special way around the body's defenses.

It is a common misconception that bacteria, the earliest life-forms on the planet, are simply creatures that never "developed" further. They are fully developed and adapted to their environmental niche in ways that put us to shame. Because they are simple and because the life span of some is as brief as twenty minutes, they can respond quickly through traditional means of mutation to changes going on around them. This gives them a considerable advantage. If that weren't enough, bacteria can trade genetic information with bacteria of different strains by sexual means called transduction or conjugation. By sidling up to another bacterium and extending a small tube, a bacterium can exchange the bits of genetic information contained in replicons, among which are plasmids, phages, and viruses (those tiny particles of protein and DNA that can cause our bodies so much grief). It is in this way that resistance to antibiotics is passed along so quickly, even among unrelated bacteria, if they have the appropriate receptor sites. Thus the adaptations that take complex organisms such as ourselves a million or so years may happen in days in the microbial world—as evidenced by the new "super bug" recently discovered in Great Britain that has not only become resistant to the antibiotic vancomycin but learned to use it as *food*. It actually lives and grows off the antibiotic—a horrifying development for the disease-fighting professions since vancomycin is a particularly potent antimicrobial, the weapon of last resort in dealing with infections involving some multiresistant bacterial strains.

Bacteria also live in colonies in any given niche and act as teams, coordinating the release of their complementary enzymes with the task before them.[5] And they are found in extremes of heat and cold where no other organisms could be expected to survive. When exposed to unfavorable conditions, some can protect themselves in tough cases and ride out the bad times. They have so many advantages, in fact, that one can look with incredible awe and humility at these creatures and wonder why they tolerate our presence. We can only assume that we are useful to them. It would certainly be advisable to keep on their good side, if we could figure out what that was.

Bacteria, which are one-celled creatures, are named and classified, much as plants are, into orders, families, genera, and species. Then they are further divided according to their special characteristics. In the huge *E. coli* genus the combination of letters and numbers refers to characteristics of the O (somatic) and H (flagellar) antigens (substances in the bacterium that cause an infected individual to produce certain antibodies) and is a good way to differentiate them. *E. coli* O157:H7 is a dangerous serotype of this ubiquitous and normally friendly genus because it has the ability to attach to certain human cells as well as produce a vicious toxin, which is what inflicts the extraordinary damage on the human body. The toxin destroys cells (producing the bloody diarrhea) and sometimes gets into the bloodstream, where it can shred white corpuscles until they begin to clog the renal system with the debris. This sets off, in HUS, first renal failure, then, too often, a cascade of damage that races throughout the body. Individual organs may swell to twice their normal size as they are attacked, and the invasion may end in death, as it did for Lauren Rudolph and for many others. (Adults have been made ill and died from the infection, but those at the extremes of age seem especially vulnerable to HUS.) The autopsies of several children who died from O157 infection have revealed gangrenous brains. Those who survive can be left with long-term "souvenirs" such as epilepsy, blindness, lung damage, colostomy, and kidney failure and may require organ transplants.

Shockingly for those of us who have grown to expect effective treatments from modern medicine, there is no cure for infection with this toxin. The bacteria are normally not resistant to antibiotics, but by the time the symptoms appear, the bacteria have released the toxin and thus antibiotics are not even recommended. There is some evidence, in fact, that antibiotics can make matters worse, setting the stage for complications. They may do this by destroying those colonies of friendly bacteria that, when they compete with an unwelcome intruder such as O157, may succeed in fending it off—a good argument for keeping our indigenous intestinal bacterial colonies healthy and thriving, if we could figure out precisely how to do that. Physicians treating HUS can only support the body in every way medically possible as it battles with a formidable foe.

In fact, diarrheal disease, whether caused by bacteria, viruses, or parasites, is a significant threat throughout the world. According to the World Health Organization (WHO), more than 5 million people die each year worldwide from diarrheal illness.[6] The U.S. Department of Agriculture, which has no interest in exaggerating the problem given its mandate to promote U.S. agricultural products, estimates that diarrheal disease takes nine thousand lives in the United States each year.[7] No one really knows exactly how many are affected. Diarrhea is a hidden disease, seldom cultured, often misdiagnosed as to its microbial cause, and infrequently reported. The best numbers come when surveys are done in hospital situations and the results are extended to the general population. Additionally, diarrheal illness exacts a toll not just in human misery but in economic terms. Seven pathogens alone have been estimated by the Department of Agriculture to cost the United States between $5.6 billion and $9.4 billion annually, and by a recent congressional committee to cost $22 billion annually.[8]

Over forty different pathogens cause the damage. To know that diarrhea is a common problem, you only have to watch television ads. Diarrhea medications such as Imodium A-D are advertised daily. But the humor of these advertisements—in one ad a distressed man in a car pool has the driver make frequent stops as he rushes into rest rooms—conceals the dangerous reality. What can be merely inconvenient or embarrassing can also devastate and kill.

The CDC now estimates that *E. coli* O157:H7 alone causes as many as 20,000 illnesses a year in the United States and leaves between 250 and 500 dead.[9] Unless these cases are part of a widespread outbreak that cannot be ignored, the public seldom hears of them. Despite the press coverage that accompanies the larger foodborne outbreaks, few comprehend how widespread the problem has become. The small outbreaks and isolated cases go unreported. There is sporadic coverage of major outbreaks, but no systematic tracking of the problem by the media.

We like to think of our food supply as safe. The idea that so simple an act as nourishing the human body with familiar foods can make us seriously ill or even be life-threatening is not something the mind likes to contemplate. We have become so conditioned to concerns about pesticide residues or the long-term dangers of consuming fat or coffee or salt, each scare seemingly refuted a year or two later, that warnings about real and immediate dangers are lumped into the same category. The only choice, it would seem, other than becoming neurotic, is simply to push these stories to the back of the mind. Many of us do just that.

In fact, the incidence of foodborne illness has been growing for some time. Official estimates are that cases in the United States run as high as 81 million cases a year, but recently the CDC's Dr. Morris Potter pushed that number up even higher. He suggested in the *Harvard Health Letter* that every person in the country probably has at least one episode of food poisoning a year—which, he points out, would put the number at more than 266 million cases a year.[10] Most of what doctors and the public refer to as "stomach flu" is actually food poisoning. (Influenza, as epidemiologists will tell you repeatedly, is a respiratory disease caused by a virus.)

About 25 percent of the U.S. population is especially vulnerable to foodborne pathogens because of impaired immune systems. This startling number becomes more understandable when one realizes that it includes young children and the elderly, whose immune systems are not as strong, as well as individuals with HIV, those who have had organ transplants, and cancer patients receiving treatments. Even taking an antibiotic can make one more susceptible to foodborne disease.

Foodborne illness is generally caused by bacteria (although viruses and parasites can also be culprits) that can, under the right conditions, multiply quickly to dangerous levels. The microbes can be in any food anywhere given the right circumstances. They can be in food eaten at a restaurant, a church supper, or a fair; in food bought precooked, eaten at a friend's house, or prepared right in your own kitchen. Foodborne infections can be passed on to others by what is called the fecal/oral route: by unwashed hands that prepare food or in locales, such as nursing homes and especially day-care centers, where contact with human waste (diaper changing) is frequent. One child who died in the western states outbreak had not eaten a hamburger but picked up the pathogen at a day-care facility from another infected child. Numerous outbreaks of enteric disease have been linked to day-care facilities, which are considered by the CDC to be a significant risk factor for spreading infections.

The resulting infection, however transmitted, may be a brief bout of nausea or diarrhea, or both, or it can be a life-threatening event. While children are especially vulnerable, adults are not immune, and few appreciate how serious diarrheal disease can be even in the healthy adult. A bout of infection caused by the bacteria *Salmonella* felled a friend of mine, who paid with twelve days in the hospital in intensive care and three holes in his colon. His agony, he remembers, was almost unbearable. For six months he had a colostomy until his colon could be surgically reattached.

Foodborne illness spares no one. Senator John Warner of Virginia spent five days in Bethesda Naval Hospital with *Salmonella* infection in 1995. He'd been campaigning in a tough race for the Republican nomination against a staunch conservative who was presenting him with an unusually strong challenge. Campaigns present nutritional challenges, to put it mildly. Candidates eat where and when they can, always on the run, or they consume what is prepared and presented to them at functions where it would be politically awkward to refuse or to question what it is they are about to eat. Campaign dining, as John Darnton observed in the *New York Times*, is "food as tribal bonding: I eat what you eat; vote for me."[11] Senator Warner thought his diarrhea would resolve quickly. It did not. Day after day it continued, wearing him down. Finally, out of desperation, he called his doctor and asked whether there wasn't something he could be given to stop it. "Yes, but you'll have to come to the hospital so I can take a look at you," he was told.

He was admitted at once. He was dehydrated and his electrolytes were out of kilter. He is cautious—very cautious—now about what and where he eats.[12]

Early Warnings

In the late 1980s a committee cochaired by Dr. Joshua Lederberg and Dr. Robert E. Shope was asked to look into emerging pathogens for the National Academy of Sciences. In 1992 they published a rather alarming book called *Emerging Infections: Microbial Threats to Health in the United States*. We are likely to face epidemics in the near future that will rival the 1918 influenza pandemic, during which more than 20 million died in one year, they said. The factors they identified as setting the stage for such events seemed obvious once they'd laid them out.

Lederberg and Shope were looking at all emerging infections, but the same set of factors applies equally well to those causing foodborne disease. Microbes change, for one thing. They develop methods of infecting new hosts, they pick up and pass on immunity to antibiotics, and apparently they actually change their behavior in response to a host's immunity. But we change as well. We adopt new customs and habits, and new relationships to food, some of which leave us vulnerable or increase our exposure. International travel and trade, environmental changes such as deforestation and urbanization—all these affect the prevalence of pathogens. Yet sadly, even as these changes have created hospitable new environments for emerging diseases, we have allowed our disease surveillance and public health systems to deteriorate. As if that weren't enough, more and more bacteria are becoming resistant to antibiotics even as the new pathogens emerge and the older, forgotten pathogens reemerge.

We are remarkably complacent about microbial threats. A widespread assumption in the 1950s and 1960s, encouraged by reputable medical experts and respected scientists, even the surgeon general of the United States, was that we were about to win the war against infectious diseases. It was a time of great achievement—and a confidence equal to it. Infectious diseases that had been potentially fatal one day were tamed the next as a stream of new antibiotics entered the arsenal of medical weaponry. Immunization gradually pushed what had been the common threats of childhood disease into the realm of history. If a miracle drug wasn't available to target a specific illness, the medical-scientific community could be relied on to come up with one shortly.

A host of new threats to public health in the 1970s and 1980s, from Legionnaires' and Lyme disease to AIDS, began to shake that complacency. New resistant diseases, such as multidrug-resistant tuberculosis, have begun to undermine the certainty that cures are but a test tube away. Hospitals are now confronting microbes for which they have no quick fix. Some are rethinking their transplant programs, for which these resistant bacteria present a formidable obstacle. Nosocomial diseases—ones that patients contract in the hospital—are on the rise and are probably related to the casual overuse of antibiotics in these environments. Recently, a renewed appreciation for microbes has begun to develop both within the medical community and increasingly among the public, and with it an acknowledgment that we don't really have the upper hand.

Many people are fascinated with the revelations about the changing nature of infections, but they focus on the more exotic emerging diseases such as *Ebola* or *Hanta*, names most of us had never heard of before 1994. They are missing the obvious, the serious, and often life-threatening challenges to which we can be unwittingly exposed as we go about our normal daily lives, sampling the cuisines and specialty foods that world travel and trade have made familiar and available almost everywhere, or even eating foods that we have eaten all our lives but for one reason or another are now threatening.

Food as Memory

Trying a new recipe out on guests is always an act of pure bravado. Sometimes it must be done; there are things you just don't fix for yourself. The almond–pine nut tart recipe I found in a vegetarian cookbook fell into that category. But for reasonably experienced cooks, preparing something entirely new is not as risky as it might seem to the amateur. Reading a recipe through quickly reveals whether the ingredients are compatible, and even, when the combination is unusual, whether the result is likely to be pleasing. This calls for a bit of imagination—summoning forth taste memories, sorting them out and combining them in the head, and mentally sampling the result. The almond–pine nut tart—a sweet crust spread with apricot jam, then a mixture of ground almonds, sugar, eggs, vanilla, and a few drops of almond extract, topped with pine nuts, and then baked—sounded as if it might reproduce one of those subtle creations I remembered from spending time in Europe as a child.

But in the midst of attempting to recapture that memory, there came an inevitable moment of insecurity. The taste of almond should be distinct, but not overwhelming. Had I added a sufficient number of drops of almond extract to get the proper effect?

It struck me that there was no way of knowing until the pie was baked. Tasting the egg-almond-sugar mixture, something I would have done instinctively twenty years ago, would, I knew now, put me at risk for a good case of *Salmonella* infection now that contamination is no longer just on the outside of eggs (there have long been warnings against using cracked eggs) but on the inside, because the bacterium has learned to get into the ovary of the chicken. It was certainly not worth the risk. I popped the pie into the oven hoping for the best and felt a surge of pure anger. Some very basic ingredients are now too dangerous to consume in ways we once would have considered standard.

Consider this list of foods that were happily and regularly consumed in the past but are now considered too unsafe to consume because of microbial contamination: eggs, soft-boiled or poached or sunny-side-up; mousses of all description; Caesar salad; rare hamburgers; homemade eggnog; raw shellfish; lemon meringue pie; Hollandaise sauce; homemade ice cream; beef tartar; raw cookie dough; seviche; several classic types of cake frosting. Some of these foods have quietly disappeared from restaurant menus; sometimes they are now prepared with special ingredients, such as pasteurized eggs; and sometimes they are offered to diners because those preparing the foods, even in commercial settings, have not been informed that they are dangerous. Recently, cured sausage and salsa have been called into question precisely because something about their ingredients or their potential to cause disease has changed. Decisions about food preparation and the ingredients that go into processed foods are often made without considering what effect the changes will have on microbes.

These are foods I have eaten my entire life—good foods, memorable foods, that have now, for a variety of reasons, become dangerous. Once our cookbooks and nutrition guides told us to eat the skins of vegetables because that was where the vitamins were found in abundance; now we are told to peel and throw them away because that is where the poisons are found, the ones we hope have not made it inside. And if the skins haven't been poisoned, they have certainly been made unpalatable. Now that cucumbers are covered with greasy wax to keep them appearing fresh week after week (even as their vitamin content declines), who would want to serve them unpeeled? Food seems almost an illusion now. It's there, but it isn't there. In all this magnificent abundance, with store shelves groaning from the food supply of a generous earth, it is a supreme irony that we have to worry so about what we eat. When microbial dangers are added to the dangers from pesticides and fumigants and herbicides and the concerns of eating properly for good health, the overly confident cultivate a dangerous assumption of immunity, the ignorant are in outright peril, and the educated cook is left wondering what, if anything, is safe to serve.

The microbial threat can be summed up in one sentence: Changes in the ways we produce, process, and distribute food, along with changes in our lifestyle and culture, have created niches for emerging foodborne pathogens. That is, by small, seemingly trivial changes in the way we do things, whether it is in raising animals or preparing foods, we have created custom-designed spaces in which the microbes can grow. Having slipped into these niches, these tiny opportunistic creatures spread and are now firmly established in our food chain.

Our histories are punctuated by food memories. They can mark events and celebrations as indelibly as a tattoo. It has been years since my family held what was then an annual picnic, but the Fourth of July for me will always be fried chicken, potato salad, tomato aspic, cucumbers and onions in vinegar, sliced tomatoes, ham biscuits, Grandmother's lemon tarts with their thin, crumbly shells, and mint juleps in tall frosted silver beakers. What we eat—or ate—gives us a place, a history; it pins us down, defines us absolutely as to culture and class and time.

What stands out for me when I think about food memories is that what I ate long ago still had a relationship to what was happening around me. There were certain foods that through processing—curing, drying, freezing—could transcend the dictates of the seasons and the weather, but there were many more that could not. Canning could provide something like a tomato for the winter months, but the real thing was available only briefly. And it was so different from the canned variety as to be almost a new species. Pulled from its vine—an act that released a heady, clinging smell that previewed the fruit and was itself a fulfilling experience—a freshly grown tomato was an orb of rich sensation. The lusty color, scandalously sensual texture, and overwhelming odor held much appeal for me as a child, but I had not yet identified the tomato for what it was—profoundly erotic. Asparagus, peas, lamb, shad roe, raspberries—their season was brief. Pleasant anticipation, even excitement, preceded what was not available at any other time. One waited, and the waiting was worth it.

Until recently this was the way things were. Each item of food had its time and place. The farmers around Monticello, for instance, competed with Thomas Jefferson to have the first peas each spring; the winner hosted a dinner party. Jefferson took the honors so frequently that one year he told his children not to mention his peas so that a friend could host the dinner. Any number of values were wrapped up in this ritual, not the least of which was respect for food and friends, for patience, and for immutable cycles.

If food were all that counted, my childhood was idyllic. My grandparents had fled to the country in the 1930s as pioneers in the back-to-the-land movement. They built a summer house out of old logs in the Blue Ridge Mountains of Virginia and liked it too much to go back to town. They put in heat and moved in permanently. Visiting them, which I did often, was a perpetual feast. The vegetable garden literally overflowed with produce for immediate consumption and for canning. Ducks, geese, and chickens wandered around, scratching, cackling, scurrying, producing eggs; close at hand for the occasional sacrifice on the stump my grandfather used for purposes of decapitation. Each year a steer was raised and slaughtered, its meat going into the frozen-food locker down the road, where it was ready to be retrieved as needed—ritual forays my grandfather and I often made together. Sometimes I was allowed into the huge cold lockers where massive carcasses hung and the smell of cold, clean flesh was rich and satisfying.

It's amazing now to consider the time that went into food gathering, production, and preparation just forty or fifty years ago. These ancestors of mine did not consider themselves food snobs. They were completely uninterested in "foreign" cooking; in fact, they felt absolutely no need to go beyond the tried-and-true recipes that had been handed down from their mothers and aunts. But they loved good food and would go to great lengths for it. My grandfather, who thought he hated lamb, once introduced to it had no trouble saying he'd been wrong. After that he might drive halfway across the county to pick up a freshly slaughtered one. Generally my family believed there was only one proper way to prepare anything. As a young woman, this implacable confidence profoundly irritated me. I tried to introduce new ideas—adding wine and garlic to the stew, for instance—but these were considered ruinous. Good food didn't need such things, I was told. The French had invented sauces only to hide bad meat, my great-aunt informed me authoritatively (but mistakenly).

I remember once, as a young bride, having my mother-in-law ask me, when I was staying with her, what we should have for dinner that evening. Since it was summer and the garden was only a few feet away, I suggested green beans.

"Oh, but we wouldn't have time," she said, clearly shocked.

It was then around three in the afternoon. For her, green beans could only be cooked slowly with a piece of pork fat, and that might take most of the day. They were wonderful beans, but for her they were the only version that existed. I knew better than to try and convince her otherwise. And yet even within this rigidity, which today we might consider almost pathological, there was a real respect for food. If beans, to taste the way she wanted them to taste, took half a day, then they deserved nothing less than half a day.

The important thing was the connection between food and a place and time. I knew, even as a young child, where food came from and when. I knew what it needed to thrive: The watercress loved the overflow from the springhouse; the corn preferred neat, tilled rows of well-drained soil. I understood the process by which this produce was transformed into something to eat. It had a past, a history. Consuming it linked me, whether I knew it or not, spiritually and physically to the world around me and shaped my thinking in ways I gave no thought to until recently. The beef I ate had a personality. I had played with that steer and fed it and watched it grow, and yet I accepted its death as natural and inevitable. Then it became a part of me; its flesh became my flesh, and I was united to all life in a respectful process.

This was only part of my food life, of course. With my parents, things were more typical. We bought from the store, like anyone else. My mother shook off the shackles of her heritage and tried new things. Still, her choices were, by today's standards, severely limited. During the 1950s the average grocery store in the United States had only three hundred food items on its shelves; by 1980 there were twenty-five thousand different items, an eighty-fold increase. Today it is possible to find fifty thousand different items in some supermarkets.[13] Imports bring us not just exotic ingredients from around the world but fresh fruits and vegetables throughout the year. There are no seasons in today's supermarkets, just eternal tasteless summers filled with virtual-reality vegetables. In their long treks from their exotic birthplaces, chilled and sprayed and waxed and gassed, most of these vegetables have seen their flavor flee in terror or simple exhaustion. They look like tomatoes and strawberries and potatoes and ears of corn, but they taste of nothing. Nevertheless, they are bought by many people who have no idea what they should taste like, who have never tasted anything else, and whose capacity to detect subtle flavors has been numbed by the heavily spiced and strongly flavored shock-foods we have resorted to as a replacement for genuine flavor, and so they are acceptable. My grandmother complained in the 1950s that the taste had been bred out of cantaloupes, that they had been crossed too often with pumpkins to make them sturdier and more capable of withstanding the rigors of transportation, and that this process had left them with a pumpkin taste. Familiar only with what I knew, I had no idea what she was talking about. Now, as one who can remember what corn fresh from the stalk tastes like before its sugar changes to starch, I think of her as I spurn the pre-shucked, shrink-wrapped facsimiles of corn on the cob in the produce counter. As someone who once lived where pineapples were grown and who has tasted the incredible, juicy sweetness of those that ripened in the field, the supermarket products tempt me but inevitably disappoint. I am trapped by these memories, yet there is nothing I would trade for them. I guard these memories, the scents and flavors I can bring to mind, as if I were protecting cultural icons or religious rituals.

The meat and vegetable sections are anomalies in the supermarkets of today. The space devoted to them is small compared to that devoted to processed foods. Our preferences and priorities as a culture are clear: We want convenience and speed, even to the point of mostly relegating food preparation to others. Time spent preparing food is wasted time. Almost half of every food dollar is spent eating out, and much of that is at fast-food restaurants, where it is no coincidence that the jobs are among the country's lowest paid. There is, to be sure, a culinary elite who are willing to spend large chunks of money and time to prepare excellent food, but that makes the point. Eating healthy, well-prepared, fresh, and nutritious foods should not be an activity cornered by a tiny proportion of affluent customers. Surely it is a universal right, something we all should consider basic, if only because a healthy population is in our common interest.

Most processed foods—and indeed, virtually all restaurant foods—give little hint of their origin. Most people have no idea where food comes from. In January 1995, Angus King, the newly elected and very earnest governor of Maine, said that he needed and wanted to learn more about agriculture. "I want to know how broccoli happens," he was quoted in the newspaper as telling a puzzled farmer.

Interestingly, there is even a fear of foods that are too closely identified with their source without the reassuring filter of technology. One summer not too many years ago I was visiting friends on Nantucket, and on a hike we came across a patch of ripe blueberries. We scrounged every possible container from the car, and most of the adults, and all of the children, picked all they could. That night I made them into a blueberry grunt. One of the visiting couples, the wife born and bred in New York City, declined dessert. I finally pried the reason out of them. They thought we were going to poison ourselves. We were not, of course, but the ease with which a wild blueberry can be identified and picked was simply beyond their experience—indeed, beyond their imaginations. Blueberries, for them, came in boxes. Blueberries "happened."

This worries sensible people like Dr. Mitchell Cohen, director of the Division of Bacterial and Mycotic Diseases at the CDC, who thinks there should be a relationship between what we eat and where it comes from. When his son consumes a drumstick, he tells me, he makes a point of explaining that it is, in fact, the leg of a chicken. Making this connection is a good thing to do, but it is an intellectual exercise. Seeing a chicken alive one minute and dead the next, watching it being plucked and scalded and eviscerated, cut up, floured, and fried is the way to make the point that eating is a significant act.

Food preservation, before technology, was dicey. Acid has long been known to keep foods from spoiling, as does drying and salting. Pickles, sauerkraut, corned beef, cured ham—before these were food pleasures, they were food preservation techniques. Home canning, when it became common, was known to be dangerous because the spores of *Clostridium botulinum*, which originate in the soil and are often on vegetables, can multiply in the conditions they prefer—the acid-free, oxygen-free environment of the sealed jar. Without modern food science and technology, it was only by following exactly the techniques that had proven effective in the past that fermented or salted foods could be considered safe. The other side of trial is error, and there must have been many illness-causing errors as techniques evolved over the years. They would have been remembered, even without knowing that microbes were the problem, and those memories would have become traditions. The stubborn unwillingness of my ancestors to change the way they cooked may well have been a behavioral mechanism that helped them avoid foodborne illness. Cooking beans interminably would certainly have killed off the C. *botulinum*, which is sometimes found in home-canned foods.

Still, there were no guarantees. Church suppers and class reunions, to which people brought food from home that could, in the words of modern food safety experts, suffer "temperature abuse," were frequent sources of large outbreaks in the past. Potato salad and turkeys stuffed the night before were famous for felling whole families. My father, an army officer, never forgot the Thanksgiving an entire battalion under his command was made acutely ill from turkey stuffing. The death of one soldier shocked us greatly, our confidence that such microbial threats had been "conquered" being already firmly in place, even in the 1950s. That experience, by the way, firmly established the tradition in my family of not stuffing the turkey until minutes before cooking. A food tradition based on safety was thus born.

The culprit in these outbreaks was often staphylococcal toxin, but it might have been *Salmonella*, which was found in many places, including the intestinal tracts of poultry. That is why standard warnings were issued against using cracked eggs, and today most eggs are cleaned and wiped with a disinfecting solution before being marketed. Unfortunately, microbes change and adapt. Now new microbes are appearing and old microbes are found in new places. Not understanding what has happened to the bacteria, many people think these recent warnings are exaggerated—more food hype—and unwisely dismiss them.

If you are disinclined to believe that foodborne illness is a serious matter, you only have to have lunch with an infectious-disease specialist to change your mind. Dr. Kathleen Gensheimer, a slight, intense physician in her forties, is Maine's state epidemiologist. She and her nurse assistant respond to all reports of disease outbreak in the state. Not long ago there had been several cases of hemolytic uremic syndrome (HUS) in one Maine county, apparently related to hamburger meat. One child died. Since only about 5 percent of *E. coli* O157:H7 infections end up as HUS, Dr. Gensheimer calculated that there had been at least sixty cases that she'd heard nothing about.

We had arranged to have lunch to talk about the cases, and she had picked the restaurant, one she knew and trusted. That wasn't enough. As we went over the menu and I suggested things that sounded good, she would shake her head. "Not the Caesar salad. There are raw eggs in there. Not the seafood. Who knows? Not the salad. Raw vegetables come with microbes attached—or they can be cross-contaminated."[14] I settled finally for a lentil burger, and she had a vegetable lasagna; both dishes were thoroughly cooked. I stayed healthy, but mentally I have never been the same. My assumptions and illusions were swept away in one lunch hour.

During the meal she told me how she prepares meat in her kitchen.

> I assume that poultry is contaminated, and that any package of hamburger I buy is grossly contaminated too. When I'm preparing a turkey or hamburger, I have everything out and ready, the pan on the counter, right where my meat is, so that I don't have to go in the cupboard to get it. Then I've got the soap handy, of course. After I handle the meat, I use my wrists to turn on the faucet so I'm not touching it with my dirty hands. I wash my hands thoroughly for at least thirty seconds.

Dr. Gensheimer's four children, like children everywhere, wanted to eat raw cookie dough, which, becasue it contains raw eggs, is not safe. Each day now before leaving for work she carries the vegetable peelings from the night before to the back of her house on the main street of a small town and throws them over the tall wire fence into her chicken enclosure. These hens provide her with eggs she likes to think are safe enough to allow her children that traditional pleasure. But she can't be sure. "What I do know is that they are collected quickly, refrigerated quickly, and used quickly."

When we have to treat what we eat like a biohazard, when we have to go to such extremes to secure what we hope is a safe version of something that was once assuredly safe, something is seriously wrong.

2/Policing the Pathogens

Nature seems to have intended that her creatures feed upon one another.

HANS ZINSSER,
Rats, Lice, and People, [1934]

In the summer of 1994 television news viewers were presented with a sight so horrible, so unlike anything most of them had ever experienced or even contemplated, that it seemed more like a film scene created for maximum dramatic impact. But it was not a Hollywood horror story. It was thousands upon thousands of Rwandan refugees camped in desperate conditions on a barren landscape in Zaire. As relief workers moved among them, the smoke of a thousand fires spread a hellish haze over an apparition of human misery at its unimaginable worst. Everywhere lay men, women, and children, gaunt and sunken-eyed; too weak to stand or move; their skin hanging limply on their frail bodies, grossly discolored from thousands of broken capillaries; their lips pulled back and parched, strangely silent; sprawled in rags stained with their own watery waste. The camps were in the midst of a cholera outbreak. The ill, the dying, and the dead were everywhere.

Disease had broken out because thousands of desperate people had no proper sanitation facilities and had contaminated their only source of drinking water. The bacillus (a bacterium shaped like a rod) *Vibrio cholerae* (the Greek *vibrio* refers to its curved, comma-like shape) had entered the water source, a large lake and its tributary streams, and multiplied to dangerous levels. Now it was spreading, and killing, at a terrifying rate. Refugee workers did what they could, providing simple medical care to combat dehydration, bringing in clean water and disinfectant, digging latrines to avoid further contamination, burying the infected corpses, urging that measures be taken to avoid person-to-person transmission.

In Atlanta at the Centers for Disease Control and Prevention (CDC), the nation's premier disease prevention agency, the staff had watched reports of the gathering refugees with growing alarm. Outbreaks of disease were a virtual guarantee given the mass of people and the physical conditions. Teams of CDC epidemiologists, part of the uniformed Public Health Service, are prepared to head out at a moment's notice to wherever in the world epidemic disease has raised its head if they can learn something or offer a unique level of assistance. More often than not, they do both. When cholera predictably broke out in Zaire, there were already nongovernmental organizations (NGOs) such as UNICEF and Médicines Sans Frontiers (MSF) in place. Working for the NGOs involved in the first efforts to cope with the epidemic were epidemiologists the CDC had trained. But this was an opportunity for a new crop of epidemiologists-in-training to gain firsthand experience in an outbreak setting. The agency sent a team of nine to lend assistance if they could, to set up systems to monitor the health of refugees, and to see how well efforts to combat the disease were working and how they might be improved. On another cholera outbreak a team study had discovered that many of the ill were unable to reach treatment centers and that just going among the refugees with stretchers could save lives. Learning such simple things could be invaluable in planning for the next epidemic—and there always seemed to be one.

Cholera is an old enemy, but it is instructive for our purposes because it illustrates dramatically the way a disease agent can be spread. One of many types of diarrheal disease, cholera is generally transmitted by water, but it can infect seafood, vegetables, and fruit—anything that comes into contact with contaminated water. Researchers report that food is increasingly a vehicle.[1] And today the spread of cholera is made easier by the pace of modern travel and trade.

It can exist in a population at low levels causing only sporadic infections until some change causes the microbes to multiply. The toxin these organisms produce is likely to cause disease only when there are as many as 100 million microbes in the digestive system. At that level the body's defenses are overwhelmed. The typical illness begins with painless diarrhea—pale, watery, practically odorless stools. Between the diarrhea and the vomiting, which may be severe, the victims lose a great deal of fluid—perhaps as much or more than 10 percent of body weight. This acute dehydration can lead to shock. Death can come quickly if rehydration therapy is not administered at once to replace the lost fluids and electrolytes.[2] Where treatment is excellent, about one out of every one hundred cholera victims still dies. When severe illness is not properly treated, it can cause death, within days or even hours, in 20 percent of those infected. Under the worst conditions, the death rate can be horrifying.

The cholera outbreak in Zaire lasted from July 21 to August 12, 1994. CDC teams devised ways of counting its victims and later reported that in that brief span of time more than 35,000 individuals were treated. Another 33,600, they estimated, never reached a health clinic. To compound the horror, during the final days of the outbreak refugees were also suffering from dysentery, which is infection with the bacterium *Shigella dysenteriae*. This multidrug-resistant strain, distributed through the same contaminated water and food, caused 15,500 illnesses. The two diseases together caused 85,000 illnesses and left 41,500 people dead.[3]

An outbreak of cholera in a situation such as the Rwandan refugee camps in Zaire is a reminder of how severe diarrheal illness can be, but because it occurred where there was a complete breakdown of what we consider modern sanitary and health conditions, we think of it as an aberration, not something that could happen under First World conditions. We thought we were seeing a specter from the past. And we were. But cholera still presents a threat. And it is coming closer all the time.

Cholera strikes the human population in waves that generally begin in Asia and spread to the West. Six pandemics were recorded before the twentieth century. Three of them reached the United States in the 1800s. The current pandemic, the seventh, is caused by *Vibrio cholerae* O1 biotype El Tor. It began in 1961 and has spread slowly but relentlessly to Africa, Europe, and Oceania. Unlike other pandemics, this one has not gone away. The CDC has tracked its progress; where cholera epidemics have been, its investigators could be found. For a while it seemed destined to spare the Western Hemisphere. But in January 1991, three years before Zaire, word reached the Enteric Diseases Branch at the CDC with the news of what had long been feared: Cases of cholera had begun appearing, not in Africa, where outbreaks had occurred regularly since the 1970s, but in Peru. There had been no cholera in South America for more than one hundred years.

The CDC was instantly pervaded by that curious mixture of foreboding and excitement that arises when something interesting is in the offing—a new disease outbreak, an unusual vehicle for disease, or in this case, disease where it had not been before. Chasing epidemics is, quite simply, what these doctors do. An outbreak is an opportunity to get on the trail of a potential killer, put education and experience to work, and hopefully, stop disease in its tracks. But actually stopping an outbreak is much more difficult than one might suppose when there is no clearly defined single source—a contaminated food that can be recalled, for instance. Cholera is rarely traceable to a single source. The disease, according to which strain is present, has the potential to cause widespread outbreaks in a variety of ways. It can also settle into the environment where it can stay for years, recycling through water or other vehicles contaminated with human waste.

Ironically, that had happened in the United States. There are only two places in the world where cholera is now considered, not epidemic, but endemic—where it has settled into an ecosystem apparently to stay. One is in northeastern Australia, and the other is on the U.S. Gulf Coast. The first U.S. case appeared in a fifty-one-year-old shrimp fisherman from Port Lavaca, Texas, in 1973. It was *Vibrio cholerae* O1, biotype El Tor, but the CDC was never able to determine how it had reached American shores.

Random cases continued to appear. In 1978, eleven people came down with the infection after eating cooked crabs harvested from sites widely separated along the Louisiana coastal marshes. Two more cases turned up in Texas in 1981, and that fall a worker on an oil rig barge passed his infection along to fourteen other people. In 1984 a case was identified in Maryland in a man who had eaten crabs harvested along the Texas coast, and more sporadic cases followed in Florida and Georgia from eating raw oysters. Then in the fall of 1986, eighteen people became ill from seafood, crabs, and shrimp harvested from different spots along the Louisiana coast.

Researchers have now sadly concluded that drinking water or eating foods from this reservoir presents a potential risk for acquiring cholera.[4] Called the Gulf Coast strain, it has a particular genetic profile and seems, fortunately, to cause only small outbreaks or sporadic illnesses rather than the huge outbreaks of the epidemic strain—the one some thought had now reached South America.[5]

The CDC knew that another reason the indigenous strain had not spread in the United States was the existence of an elaborate sewage system and generally reliable sources of clean water. In South America both sewage and water systems varied greatly in their reliability from place to place. If the disease was there, it would very likely spread. Controlling it would be a formidable challenge, but the opportunity simply to learn something about a disease in outbreak conditions was equally important. The more that is known about a pathogen—the conditions it prefers, how it infects, the nature of the illness it causes—the more effective the countermeasures can be when it strikes again. Medicine has been able to vanquish only one infectious disease—smallpox. Polio is the next goal. Ending cholera is not even a remote possibility, and today cholera is on the move.

The Disease Detectives

Epidemiology is the study of disease in a population. It can be traced back to the mid-1880s, when an epidemic of that old nemesis, cholera, was attacking London with a vengeance. More than five hundred people died in a bit more than a week.

"Carts groaned under the weight of corpses carried away for mass burial. Those who could, fled. Others locked themselves away in fear," wrote Peter Jaret in a recent article on epidemiology.[6] John Snow, a forty-one-year-old London physician, had taken the time to mark on a map where victims had died. Some of the deaths centered on a single public pump at the intersection of Cambridge and Broad streets. Snow did not know what in the pump might be causing disease, but he suspected it was the source of the contamination. To be certain, he interviewed families that had lost members to the disease but lived at some distance from the pump. What he discovered confirmed his suspicions. Five of those victims had regularly sent for water from the suspected Broad Street pump because they liked the taste. Three other victims were children who went to school nearby. Snow surprised members of the vestry of St. James parish, who were meeting to consider what might be causing the epidemic, by asking them to remove the handle of the pump. They agreed, and the outbreak in that area came to a halt within days. His research and his interview approach, with its basis in deductive reasoning, became the foundation of a new field of medicine.

Originally epidemiology was limited to communicable diseases, but since the 1920s it has broadened its reach. Now it attempts to understand all the factors that apply to the incidence of disease among people. Epidemiologists are interested not just in what causes *Hanta* virus, for instance, but in the factors responsible for lung cancer or heart disease. John Snow's primitive but effective interview process has been vastly elaborated to include, among other techniques, case-control studies and statistical analysis. An epidemiologist's world today is neatly divided into "ills" and "wells." It is by comparing those who have become ill with closely matched well individuals as to what they did—or in the case of foodborne disease, what they ate—that the source of an outbreak can be revealed. In the case of diarrheal disease, epidemiologists follow a pattern, teasing the disease from the mass of information: the lab report, the numbers and distribution of cases or a particular serotype in a certain geographical area; patterns of consumption; facts about the implicated food products. In the process they build pictures of infection. How do these pathogens work? What foods are they in? Is it a new vehicle or an old familiar one, contaminated in a new way? Has something gone badly wrong in a process that seemed safe? Epidemiologists are the fingers on the pulse of disease, monitoring it constantly so as to predict what it might do next. Statistical analysis is used to calculate how significant the different factors are. Simply put, if a study reveals that 95 percent of the ills in an outbreak ate cornmeal mush from Restaurant A while only 5 percent of the wells did so, most likely the problem has been identified.

The case-control method is now taken for granted, but it was introduced at the CDC only during the 1970s by Dr. Paul Blake, who would eventually serve as chief of the Foodborne and Diarrheal Diseases Branch for fourteen years. The first outbreaks to which the method was applied were of cholera.

The CDC was organized almost one hundred years after epidemiology's beginnings when it emerged from the Office of Malaria Control in War Areas after World War II to deal with the infectious diseases that had devastated soldiers. Not long afterward it took on plague control, and during the 1950s it expanded its role to surveillance, prevention, and control of venereal diseases. Today, with a vastly extended realm, one division may be tracking Lyme disease while another is analyzing the causes of urban violence—a different kind of plague. The agency works in partnership with state and local health departments, academic institutions, industry and labor, as well as professional and philanthropic foundations. It exchanges information with the National Institutes of Health (NIH), the Food and Drug Administration (FDA), the U.S. Department of Agriculture (USDA), other federal agencies, and health and disease agencies worldwide.

Public health is not something most physicians in training hear about, but if they are drawn to the field, they are likely to apply to the CDC's Epidemic Intelligence Service (EIS), a two-year training program for young epidemiologists that supplies the foot soldiers in disease investigations. After completion of their training, the graduates often take jobs in state public health departments, as Dr. Gensheimer of Maine did. Over the years these former EIS officers have established a CDC-trained network of public health officials across the nation. They supply a helpful overlay of continuity in training and orientation to what is an otherwise decentralized "system" of private and public health care in the United States.

Cholera is an archetype of the epidemic diseases that the agency is known for tackling. The disease has returned generation after generation, challenging disease specialists with subtle new mutations. If this cholera, with its epidemic potential, were to reach the Americas and establish itself, it could have frightening implications.

Into the Eye of the Storm

The younger members of the Foodborne and Diarrheal Diseases Branch took the Peruvian reports seriously, but around the table where they gathered each Friday to share what they were working on, scribbling notes in their ever-present lime-green logbooks, the more experienced hands shook their heads. If Paul Blake had been among them, the mood would have been different. Known as the cholera guru, he had long expected the disease to show up in South America, but he was in Africa. The doubters had heard stories before of cholera-like illnesses in South America, and the reports had turned out to be false. As they waited for more details, they took bets on whether or not the reports would be confirmed.

They were congregated in a deceptively ordinary-looking structure on Clifton Road in Atlanta, Georgia. Security is moderately tight but not oppressive. There are additional layers of security inside that grow increasingly impenetrable as the danger increases of accidental exposure to the disease-causing microbes in the agency's laboratories, but on the surface the atmosphere is friendly and relatively open to serious inquiry. The CDC depends, after all, on a medical profession and a public willing, ready, and able to inform investigators instantly when something unusual happens.

For years budget restraints have kept the agency in a bare-bones situation even as disease threats have increased—Lyme, Legionnaires', *Hanta* virus, *Ebola*—but recently there has been a growing awareness of the importance of infectious-disease surveillance, prevention, and control. A recent study from the National Center for Infectious Diseases concluded that the death rate from infectious diseases has increased more than 50 percent since 1980, making it the third leading cause of death in the United States in 1992, when there was a 22-percent increase in infectious disease mortality in the United States even excluding deaths due to HIV/AIDS. The increase in deaths due to septicemia alone increased 83 percent.[7] Some of this increase has been caused, no doubt, by the rapidly rising bacterial strains that have become resistant to antibiotics. If just a decade ago infectious diseases were considered a thing of the past, virtually no one feels that way anymore. Congress has taken note when parceling out the funding.

The Foodborne and Diarrheal Diseases Branch, previously in cramped quarters on the fourth floor, has moved to a suite of offices on a lower level that allows some breathing space. Still, its offices are spare and modest. The walls are cinder block, painted an innocuous cream; the gray linoleum floors are patched here and there with a newer, lighter color; and file boxes are piled at the end of the hall. When money is spent here, it is in looking for disease. One reason for the move was to be closer to the laboratories, as the most effective investigations are the result of close teamwork between epidemiologists and microbiologists.

At the head of the table, waiting for news of Peru, sat Dr. Robert V. Tauxe, acting chief of the Foodborne and Diarrheal Diseases Branch in Paul Blake's absence. Tauxe has the characteristically insatiable curiosity of a good investigator, but he also carries with him an aura of worldliness and sophistication that is unusual in a medical researcher. He studied cultural anthropology at Yale University before medical school, then returned to take Yale's course in public health. He stands out for his ability to look at human illness from a wider perspective. Nothing about humans or disease or the way they interact is simple. Those who find the answer to disease puzzles today are likely to be well versed in history, cultural anthropology, religion, basic ecology, and economics in addition to medicine and the sciences of bacteriology, virology, and parasitology. Larry Shipman, a USDA veterinarian who has worked with Tauxe over the years, describes him as someone who gathers and stores seemingly extraneous bits and pieces of the disease puzzle in his mind for five to ten years until he can say, "Ah ha, so that's where that fits."

Some of the reports from Peru were coming from the newly organized Field Epidemiology Training Program (FTEP). (Part of the CDC's efforts are spent training others around the world in the techniques it has developed, hoping to build an international network of epidemiologists just as the EIS program has developed a national one.) Other reports were coming from Peruvian physicians. Dr. Marge Pollack, a former EIS officer under contract to the Peruvian government to advise the newly organized FTEP, insisted adamantly again and again that she was seeing cholera. She urged her former colleagues at the CDC to get involved.

If the cases in Peru were indeed epidemic cholera, it was not alarmist to suggest that it could easily move through South and Central America to reach U.S. borders. The perennial challenge facing the CDC, and indeed every public health official, is to balance urgency with caution. The CDC investigators wanted proof; they wanted to see for themselves. Dr. David Swerdlow, a former EIS officer and now a CDC staff veteran, was on phone duty, a task taken in democratic rotation. He spent hours working on the logistics of getting isolates from Peru, a task that involved arranging for pilots to hand-carry the delicate cargo, then to get it to a quarantine station in the United States then on to the CDC laboratories, where it would be identified by the chief of the Epidemic Investigations Laboratory, Dr. Joy Wells.

The laboratories confirmed what the Peruvian doctors had been saying. The isolates were indeed of *Vibrio cholerae* O1, biotype El Tor, the Asian strain. Somehow it had reached the shores of this South American country. Now the cases seemed to be coming from all over Peru.

The CDC is always prepared to move quickly. Now it scrambled. This was a big one. Tauxe himself would go to Peru, taking Dr. Allen Ries with him. They began a familiar routine but an important one—packing cooler after cooler with culture medium and an assortment of other laboratory supplies that might be difficult or time-consuming to find in Peru. They didn't want to have to stop in the midst of an investigation to hunt for something it would have been just as simple to bring.

"I went first to offer immediate advice, scope out the needs, and define the possibilities for emergency investigations," remembers Tauxe. This was not his first cholera outbreak, and experience counts.

There are things one should do in an outbreak of cholera—and things one should not do, he says. The CDC has developed quick-response methods that serve as a model everywhere. Its first job, wherever the outbreak, is to share that experience. Sometimes the CDC must convince political leaders that quarantining and vaccinating, usually their first impulse, would be both wasteful and futile. The cholera vaccine is often ineffective, and the process of distributing it is time-consuming and costly; quarantining is disruptive and unnecessary if reasonable health precautions are taken. Any available resources should, in a cholera outbreak, be directed to education and sanitation.

The challenge for all the public health officials in Peru would be to stop the disease in its tracks. It wouldn't be easy—in fact, it was probably impossible—but they had to try. Unlike public health measures to control the incidence of chronic illnesses such as cancer and heart disease, where results are seen in generations, such measures can have a profound and often immediate impact on the transmission of infectious diseases.

The first task was to help the sick, then to prevent further transmission. The CDC found that the Peruvian clinics were well prepared to treat the patients, and Peruvian health officials had already begun with the basics of containment. Early on, the Peruvian Ministry of Health had issued a practical advisory. Boil drinking water. Wash fruits and vegetables in boiled water. Avoid eating seviche, a favorite national dish made of raw fish marinated in lime juice. Street vendors, a suspected source of contaminated foods, were banned from selling their wares. Despite the advisory, the number of cases was still growing. During the months of February and March there were 16,400 cases of cholera among the surrounding provincial population of 626,456. Almost 3 percent of the population of Trujillo, a coastal town north of Lima, was ill. The CDC wanted to know how the disease was being spread in Trujillo and what could be done to stop it. A CDC team led by David Swerdlow set up a case-control study.

What the team discovered was typical of how the disease is spread: It had begun, as it often does, with flaws in sanitation. The water system of Trujillo was interconnected, and it was not chlorinated. Contamination could run right through it. In many neighborhoods the taps worked for only an hour or two a day, and many families stored water in containers. These vessels would become progressively contaminated by hands as water was dipped from them. The National University of Trujillo and the municipal water service conducted tests of the water supply. They found that 60 percent of the samples contained coliform bacteria, an indication that the water was being contaminated with fecal material. Six percent of the samples yielded *V. cholerae* O1. They also discovered that water lines were being tapped, unauthorized access that could allow contamination to enter the system, and that untreated sewage was being diverted from the sewage system to irrigate crops. Cabbage, lettuce, carrots, and watermelon—whatever was growing in the fields irrigated with the sewage—was potentially being contaminated. When the activities of those who were ill were compared to those of the "wells," the cholera victims were most likely to have drunk unboiled water, attended a fiesta, or eaten cabbage.[8]

The recommendations of the investigators were simple but potentially effective. Chlorinate the water and use narrow-necked storage vessels to prevent hands from contaminating stored drinking water. (A plastic striped water container, designed by CDC officials for this purpose, and a model for use in areas remote from sanitation, now sits in David Swerdlow's office at the CDC, where it is used as a fish tank.) In Trujillo either the public health measures or the natural course of the disease slowed the epidemic; new cases declined dramatically. And happily, because of good medical facilities, the fatality rate was low. More than half the patients in the hospital study had received oral solutions to combat dehydration before arrival at the hospital, some before seeing any health care professional, indicating that public education about treating diarrhea had been effective.

But the cholera, despite the attempts to control it, had already begun to spread to other provinces. It crept person by person along routes the CDC epidemiologists and other investigators were able to trace, up through South America. It was only a matter of time. By the end of 1992, 731,312 cases and 6,323 deaths had been reported. By comparison, in the United States in 1991 there were 29 cases, all resulting from the endemic strain or from travel; the year before that, only 8 cases were reported. But in 1992, 102 cases of cholera were reported in the United States, more, according to the CDC's *Morbidity and Mortality Weekly Report (MMWR)*, than in any year since the agency began cholera surveillance in 1961.[9] Some of those affected had traveled in South America. Several had eaten foods brought in by visiting friends and relatives from South America. And 75 cases occurred among airline passengers who had eaten a seafood salad on board a plane coming from South America.[10] There were six hospitalizations, and one of those patients died. Cholera could be slowed, but the CDC never really had any illusions that cholera could be stopped. It was allied now with modern travel and trade. The CDC knew that it was fighting not only the microbes themselves but also the human behaviors that encourage them. The most they could hope for was containment.

The Relentless Move North

That there were not more cases in the United States was actually good news. In the summer of 1991 the Food and Drug Administration found *Vibrio cholerae* O1 in oysters and oyster-eating fish taken from beds in Mobile Bay. It was not the strain of *V. cholerae* endemic to the Gulf Coast, but the new, epidemic South American strain. Somehow it had been transferred to these waters, and the infection was lapping at American shores, washing over the rich beds where a thriving seafood industry harvested the sea's bounty. It was the CDC's nightmare. The beds were immediately closed to harvesting.

Growing up in Virginia, my first experience with a raw oyster came early. At its very best from the cold waters of the Chesapeake Bay, an oyster tastes like essence of sea, a delicate saltiness and a subtle, strangely appealing organic scent that is vaguely like the smell of fresh seaweed overlaid with humus. With a dash of hot sauce, it is an amazing pleasure. It tastes as if the world in all its complexity has been filtered through this animal, and in truth, it almost has, as the oyster lives by filtering its nutrients from whatever is in its watery environment.

Oysters are a traditional New Year's Day food in Virginia, and I have been to receptions where a bank of shuckers opened them and placed them on ice as fast as they could for the greedy guests, including me, who could down a dozen easily. Just a few years ago I spent a weekend on the Eastern Shore of Maryland with a famous oyster-eating friend. We were entertained by a family that owned a restaurant, and to satisfy my friend a full bushel of oysters was brought in fresh. We downed dozens each. It was a feat out of a Fielding novel, an unforgettable, extravagant orgy. At the time the thought crossed my mind that such an indulgence might never happen again, anywhere, for anybody. An oyster is only as good as its bed. And more and more beds are being contaminated. None of us became ill. But if I had known then what I know now, I would not have eaten them raw. The danger of contracting a serious illness is just too great.

With the new cholera infection, Gulf Coast residents were warned to wash their hands after handling raw seafood and to cook it well before eating it. The measures seemed to work. Isolated cases of cholera continued to appear in the United States, but they were the old strain. Of the reported cases, none came from Mobile Bay seafood.

Once cholera settles into an environment, it can be there for a long time, as the endemic strain had shown. The beds were retested a few months later, but they were still contaminated with cholera and they remained closed. Finally in August, to the relief of everyone, the tests were clear and the beds were reopened. But the FDA, the CDC, and other agencies were concerned about how they had become contaminated in the first place. They had their suspicions, and an investigation seemed to confirm their guess. Ships take on large volumes of water in harbor as ballast. When they enter a port, they often discharge it. The FDA investigative team sampled water from the ballast tanks of ships that had entered Mobile Bay from Latin American ports and found what they expected: The same strain of *V. cholerae* O1 found in the beds was in the ballast water. They did all they could to correct the situation. It wasn't much. The International Maritime Organization could only recommend that freighters empty and refill their ballast water while out at sea in the hopes that the bacteria could be flushed out before entering waters where the vulnerable shellfish beds were located.[11] There was no enforcement.

By September 1994 the number of cholera cases had increased to more than 1 million in South America; nearly ten thousand had died.[12] The epidemic had made its way from three areas of South America across the continent and into Central America. Cases were being reported from several sites in Mexico. In the United States sixty-five more cases had been reported to the CDC. Differentiating strains by subtle genetic markers can pinpoint the source of an infection, tying it, like the DNA evidence in a crime, to its origin. The CDC found that the great majority of U.S. cases were associated with travel either to South America or to Asia, where another new strain, *V. cholerae* O139, had begun causing outbreaks in 1992.

All this time most Americans were unaware that a scourge was pressing at its borders. The specter of epidemic illness is foreign to us now. Up until very recently we assumed that modern sanitation, medicine, and technology had made disease epidemics a thing of the past. The CDC had no such illusions. Its investigation teams constantly track disease as well as they can with the sketchy surveillance resources they are able to muster, looking both for the unexpected resurgence of an old enemy and for the new pathogens that are appearing more and more frequently.

The experience with cholera—although it is unlikely, given U.S. sanitation standards, ever to rise to its full epidemic fury here—illustrates how changes can occur that create a space, or a niche, for an emerging microbe. In Peru it was probably a ship discharging contaminated bilge water that began the outbreak. Then the lack of chlorination in the water, the open jugs of water in the homes, the unauthorized tapping of the water lines and the sewage pipes—all these factors allowed the disease to spread. It was not one failure but a series of failures that made the epidemic possible, and sometimes there seems to be a pattern to these failures that goes beyond coincidence. The CDC knows that little is certain about disease except for one fact: It always seems to find a way.

The Peripatetic Pathogen

Political borders mean nothing to disease-causing agents. They have moved across the landscape and through history with remarkable agility, hitching rides in the knapsacks of Mongolian horsemen across the steppes of Manchuria, in the wares of trading caravans, or on the bodies of invaders or visitors or lovers. They have crossed oceans with rats and mice and insects and then moved inland on the backs of small mammals, spreading their territory relentlessly to wherever the environment is hospitable. Today they travel even more comfortably with tourists and our extraordinary international trade in goods.

In 1900, for instance, *Pasteurella pestis*, the bacillus that causes plague, was found to have reached the United States after an outbreak in Asia, probably via ships' rats. It infected ground squirrels in California and a small outbreak of the disease occurred among the human population in San Francisco. Today, almost a century later, the microbe continues to thrive in the population of rodents and small mammals in the West, spreading from one animal community to another and occasionally infecting humans from the bites of infected fleas. Humans even facilitated this spread unwittingly by transporting sick animals in the hopes of reducing the number of prairie dogs by infecting them. That effort was unsuccessful—the prairie dogs remain—but the burrowing rodents and fleas that transmit the disease have stabilized it. Plague no longer decimates their communities and seems to exist in relative equilibrium with the human population in the infected states.[13] Today, with antibiotics, it is easily curable when correctly diagnosed, and none of the cases in humans has been of the pneumonic type, which can be passed from person to person. But most people are surprised to hear that in 1994 there were seventeen cases of the dreaded disease in the United States, all of them in the western states. The United States, in fact, is one of nine countries (Brazil, China, Madagascar, Mongolia, Myanmar, Peru, the United States, Vietnam, and Zaire) that reported human plague cases in 1992.[14] In 1996 the United States had a plague fatality. What was once limited to the western side of the continent from Canada to Mexico is now moving eastward, hitching rides wherever it can in the animal population. As technology advances, so do microbes. Now they travel by air. The faster we move, the faster they move.

We think of illness as the ultimate personal experience; misery is indeed a private language, untranslatable. But illness is never truly a private affair. Others are always affected. And when an illness moves beyond the individual, beyond the family, by infecting many, it becomes a community problem. It can then become an epidemic, or at its very worse, a pandemic—moving across borders and continents. Then it becomes an international problem. Illness has the potential to disrupt every carefully established pattern of our lives, whether individual or communal. The consequences can be devastating. To prevent epidemics, the CDC must anticipate them. It must spot new pathogens as they develop and emerge. How a pathogen is transmitted can reveal how it can be stopped. But the CDC must also consider how a pathogen can affect a society. Its investigations are frequently followed by articles in the *Morbidity and Mortality Weekly Report*, which the scientific press reads avidly, and in medical journals, where its experiences can build on what is known about a particular pathogen layer upon layer, the papers piling up like a *mille feuille* pastry, telling the story of disease.

The microbiologist Dr. Mohamed Karmali likes to tell his students at the University of Toronto that Western civilization was founded because of an attack of diarrhea. An army of Greeks was able to defeat the massing troops of Xerxes because his men were suddenly attacked by dysentery and became easy picking. "That's how Greeks became ascendant to ancient Persia and went on to establish Western civilization—or at least that's my interpretation," he says.[15]

War is devastating, but it is disease that has always had the real power. Hans Zinsser, in *Rats, Lice, and History*, noted that "swords and lances, arrows, machine guns, and even explosives, have had far less power over the fates of the nations than the typhus louse, the plague flea and the yellow fever mosquito. . . . Armies have crumbled into rabbles under the onslaught of cholera, spirilla or of dysentery and typhoid bacilli."[16]

William McNeill also thinks the impact of disease on history has been badly underestimated. In his remarkable book *Plagues and People*, he proposes, for instance, that the Spanish could scarcely have conquered the Aztec nation so quickly if the invaders had not brought with them European diseases to which the Indians had no resistance. Not only were these indigenous peoples physically weakened by such homely childhood illnesses as measles (to which they had no immunity), but they believed that the invaders who could resist the illness that was felling their people must surely be gods. In their weakness and awe, the Indians were easy prey.[17]

McNeill looks through all of history for such examples of the impact of disease. Clearly disease has shaped our patterns of social organization, our cultural habits and traditions, our choices about where to live, and the sustainability of our cultural patterns. Each time we make a significant change in how we live, we disturb the balance that has evolved over time; disease is the predictable result. Says McNeill,

> Clearly any change of habitat, such as that involved in coming down from the trees to walk and run in open grasslands, implies a substantial alteration in the sort of infections one is likely to encounter. As the rain forest types of infestation and infection thinned out, however, new parasites, and fresh diseases, especially those contracted from associations with the herds of the savanna, must have begun to affect the bodies of burgeoning humankind.[18]

Periodically as populations of humans have reached a certain concentration wherever they chose to live, disease or war has intervened, not willfully, but because human activity puts a strain on some aspect of the environment, either biological or political. Good health depends on good food, good water, and shelter—and on not being exposed to disease. The more closely together people live, the more apt they are to share infections. "Great cities," McNeill points out, "were, until recently, always unhealthy."[19]

When the human population and a disease live together for a long time, a balance develops that allows both host and pathogen to exist. The pattern is quite simple and predictable. A human population is confronted with a new disease, which, in its first, most virulent form, can infect and kill many. But from the perspective of the pathogen, this aggressive behavior is a short-term and fruitless strategy. By killing its hosts too quickly, it destroys the means by which it can be spread and thus survive to reproduce. Generally it modifies its behavior—remember, it can change quickly—into one better adapted to long-term survival. Less virulent strains of the pathogen, because they allow the host to live and thus to spread the disease more easily and for a longer period, tend to survive and reproduce. Eventually an equilibrium establishes itself in which both pathogen and host survive. The number killed by disease is not so great that the community in which the pathogen has established itself will collapse. It is this strategy that ensures the pathogen's survival.

Before vaccinations made them little more than a memory, the childhood diseases of measles, mumps, and chicken pox followed such a course. The evidence suggests that they too were far more virulent until they established the pattern of infecting the young in a population early with an infection that allowed most to survive; thus was a community established with a population sufficiently dense to support the disease by providing a continually fresh supply of unexposed children. While vaccinations have made a tremendous difference, the incidence of widespread outbreaks of some common childhood diseases was already dropping by the time they were in wide use—evidence that a more benign relationship was in the process of establishing itself. It is when something—often something quite minor or seemingly innocuous—disturbs this precariously balanced equilibrium that new pathogens and new diseases can find an ecological niche or an entry point into a human population, with devastating results.[20]

It was once assumed that small changes had small consequences.

We now know differently. It is now well understood that small changes can have huge unforeseen consequences. The chaos theory, as it is called, is part of the "new science," which has abandoned the old linear, reductionist way of thinking in an attempt to understand the whole and to appreciate the complex synergistic relationships between the parts of the whole. But people understood the concept long before it was fully developed and given a name. The poet George Herbert expressed it memorably in 1651. "For want of a nail the shoe is lost, for want of a shoe the horse is lost, for want of a horse the rider is lost"—and possibly the battle, the war, and the empire. The chaos theory applies perfectly to epidemics, and especially to foodborne disease.

As comfortable as the periodic states of equilibrium can be, they are doomed to be disrupted. Nothing remains stable, not in the microbial world and certainly not in the human sphere. Each step forward by the human population, it seems, has been rewarded by a new and different exposure to pathogens.

Early on in the development of agriculture, notes McNeill, "food production permitted a vast and rapid increase in the number of people, and soon sustained the rise of cities and civilizations," with all the risks of disease that implied. New means of food production also exposed us to an assortment of new pathogens. Domesticating animals or maintaining them in herds not only exposed us to the microbes they carried but created a better situation for the microbes themselves.

Disease-producing parasites and other pathogens, says McNeill, were quick to take advantage "of new opportunities for occupying novel ecological niches that opened up as a result of human actions that distorted natural patterns of plant and animal distribution." When humans succeeded, agriculture translated into larger numbers of fewer kinds of plants and animals. Parasites and pathogens could flourish.[21]

A perfect illustration of how a niche is created is the story of vampire bat rabies in a remote area in northeastern Peru. It demonstrates also how humans, animals, and microbes interact with the elegance of a macabre ritual dance. Pig raising was introduced into the area in the hopes that it would provide an economic stimulus to a poor agrarian population. After a few years, when it was found that the pigs threatened the traditional crops the people raised, pig raising was abandoned. But the pigs had been an unexpected blessing for the vampire bats in the area. They had provided an abundant supply of fresh blood, and as a result, the bat population had grown fat, happy, and numerous. When the farmers abandoned pig raising, a large and hungry bat population had to look for another source of nourishment. They found it—in the human population. Since doors and windows were commonly left open in this remote area, the bats could easily enter the homes at night. Suddenly there was a large increase in the number of human illnesses from rabies carried by vampire bats, and because the remote area was far from medical care, the death rate was high. In a comparatively short time, 4.5 percent of the population was infected with fatal results.[22]

It would be stunningly helpful if people could look at change with an eye to how it might create niches for emerging diseases. The reality is that this almost never happens. Who considered beforehand that the introduction of air cooling systems might create pleasant environments for *Legionella pneumophila*, the microbe responsible for Legionnaires' disease? Or that the introduction of superabsorbent tampons would encourage *Staphylococcus aureus* and result in toxic shock syndrome? Change in a society is often expedient or accidental; it may be economically or culturally driven and seemingly unrelated to health. The effects reveal themselves later—and the consequence may be an outbreak of disease.

The Niche

Timbuktu, that synonym for the exotic and remote, is actually a city in the African country of Mali. In 1984 a prolonged and intense drought was plaguing the area and the Sahara Desert was moving relentlessly southward, transforming what had been grazing and farming lands, even wooded areas, into barren scenes of devastation. When traumatized people moved in search of food, cholera broke out in the camps and overcrowded villages; 25 percent of the people were dying. It was once again human misery at its worst. It was also one of Dr.

Robert Tauxe's first international outbreak experiences, and it would confirm for him the role of even minor shifts in facilitating the entry of foodborne disease into a population.

There is a simple solution, literally, for treating cholera. The World Health Organization has developed and distributes small packages of oral rehydrating salts (ORS). This inexpensive mixture—it costs just pennies—of salt, sucrose, and potassium can be mixed with clean water and the solution administered, spoonful by spoonful, to a dehydrated patient. It works. Miraculously, it saves hundreds of thousands of lives at a cost even the Third World can afford. There were boxes of the ORS packets in Mali, but they weren't getting to the refugees. Less than 1 percent of the patients were being rehydrated. One of the CDC's first tasks was just to get things moving, to coordinate among the various agencies, impressing upon them that the ORS packets were more important than vaccines. Soon the supplies were getting where they were most needed. But the good epidemiologist—the one, says Tauxe, "who looks beyond the potato salad"—always wants to know more. Why were these people getting sick? If the CDC could discover why, perhaps it could help stop the epidemic.

The CDC team hooked up with the French and Belgian group Médicines Sans Frontiers, which was already working in the area, and in dusty Toyota land cruisers they took to the bumpy roads in search of cholera outbreaks in the more remote villages. The linkup was helpful: The CDC team could share MSF's base camps and communications network. Each night at a specified time the radio would hum as these doctors reported to one another on conditions in their outposts. In all that human misery, Tauxe remembers the reassurance of the beautiful, clear nights as they slept on the roofs of the dwellings and heard the haunting sound of the call to prayer that began each day.

The chief food of the refugees, Tauxe discovered, was millet, cooked into a kind of mush. Because firewood was so scarce, it was prepared in large pots only every two to three days. It was left out in the open—refrigeration did not exist—and scooped out, sometimes with bare hands. Clearly the bacteria could be transferred by dirty hands, especially when water for washing was in such short supply. This pattern of cooking and keeping millet, however, was not new; never before had it caused the spread of disease. The question Tauxe says he always asks himself in situations such as this is, "What has changed?"

"I asked them how they prepared the dish, and they said, 'Well, we would usually add goat yogurt or tamarind [a native fruit],'" remembers Tauxe. But there was no tamarind because the drought had killed the trees, and no yogurt because the goats' milk had dried up. Tauxe wondered whether this small change in the recipe might make the difference. Back in Atlanta he put his theory to the test. He hunted until he found a Nubian goat herd, took milk back to the labs, and prepared the dish. Then the food was inoculated with the *V. cholerae* bacillus in the labs. His hunch proved correct. The fermentation process produced an acid strong enough to prevent the growth of the bacteria. A seemingly simple and inconsequential change, the omission of one ingredient, had led to the horrifying epidemic of disease.[23]

One reason the CDC went to Mali, Peru, and elsewhere in the world, looking at the relentless march of cholera and other diseases, is that the developed and the developing worlds now overlap, says Tauxe. They do this in several ways, but in particular, more and more food now comes from the developing world. "That means," says Tauxe

> that the people who pick our green onions and strawberries and pack them into boxes and ship them to us so that we can eat them, are food handlers—our food handlers. Even if we could wave a wand and make our food sterile and completely safe, as we import more and more the problems of the developing world are going to be our problems. Those are the hands that feed you, and it might actually matter whether they are washed or not or whether they have a latrine. If we are interested in the safety of our food, then we have to be interested in the living conditions of the people who handle it.

What brought that point home was Tauxe's trip to Guatemala a few years later as the epidemic spread. He was standing in the middle of a town where the number of cholera cases was growing daily, and he looked up to see a truck with Vermont license tags. On the side it said, "Vermont strawberries." It was there, apparently, to pick up Guatemala's crop and to bring it home to New England so that we might have strawberries in winter.

3/The Changing Nature of Foodborne Disease

What is food to one, is to others bitter poison.

LUCRETIUS (99–55 B.C.),
De Rerum Natura

Sue Mobley's two encounters with foodborne disease occurred more than forty years ago, but the memory of each is as vivid as if it happened yesterday. Mobley lives in a large southern city, one that was slower than some to relinquish its past, and both of her experiences involved the gentle world of ladies' luncheons. The first happened in 1955 when she was home from college after graduation; her grandmother seized the opportunity to include her in a gathering of her friends. Together they went to a ladies' church meeting to which everyone contributed something to eat. One woman brought a chicken loaf, a dish typical of the 1940s and 1950s that no one has made since. It was delicious, but within a few hours everyone who had eaten it was seriously and debilitatingly sick.

No one ever found out precisely why or how the chicken loaf became contaminated, but what makes the outbreak typical of what we were once accustomed to seeing is the speed of the attack and the clear link between the event, the food, and the illness.

One bout like that would have been enough for a lifetime, but

Mobley wasn't to be spared. About three weeks before her wedding in 1958 she attended a bridal luncheon in her honor at a private club. In the dining room next to her group was another gathering of older women. Both groups were served chicken Mornay, which she remembers as delicious. Her mother and mother-in-law-to-be left the luncheon together to go to a dress shop for a final fitting of the dresses they planned to wear in the wedding, and Mobley went home. Within a few hours she was feeling strange, and before long she was acutely ill. In serious distress, she contacted the doctor and then discovered that her mother and her fiancé's mother had become so suddenly and violently ill that they had been taken directly to the hospital from the dress shop. The story was repeated that afternoon in shops, beauty parlors, food markets, country clubs—wherever the luncheon guests had gone. The elderly ladies from the next dining room had become ill, too, and there was special concern for them.

"There was no room for me in the hospital," she remembers, because it was filled up with elderly women, so the doctor came to her house and administered an IV. The illness lasted at least a week. "We felt okay by then, but we still weren't eating very much—which actually helped me fit into my wedding dress, so looking back, it wasn't all bad."[1]

The husband of one of her bridesmaids was a physician, and he led an investigation of the club's kitchen. The source of the infection was determined to be an open and infected wound on the cook's hand. The food had been contaminated by *Staphylococcus aureus*. When the human intestinal tract is confronted by these toxins, the effect is usually "swift and violent"—precisely the words Mobley used to describe her experience.

The picture in our minds of what an attack of foodborne illness looks like probably resembles what Mobley experienced—a sudden onset of dramatic symptoms that occur shortly after eating food infected by unclean human hands or soil, kept at conditions favorable to bacterial growth, and, finally, reheated insufficiently to kill the bacteria. Until recently our public health measures and food safety education were based on this model.

But the typical outbreak of foodborne disease today may look quite different. Outbreaks such as those Sue Mobley experienced still occur, but the picture is changing and has been changing for some time. The cases may be sporadic and isolated rather than striking an entire group, as hers did. The cause may be trickier to identify. The pathogen responsible may be one that was unheard of forty years ago, or it may appear in a new and unexpected food vehicle, revealing a new dimension to its disease profile. It may produce a different clinical picture or symptoms, and it may act differently in the environment.

Recently we have experienced an impressive number of outbreaks from emerging pathogens—many of which Drs. Lederberg and Shope identify in *Emerging Infections*—microbes that have only in the last two decades begun to cause serious problems but are likely to cause more trouble in the future. Among the most troubling causes of foodborne diseases today are a string of tongue-numbing names: *Campylobacter jejuni*, *Escherichia coli* O157:H7, *Listeria monocytogenes*, *Salmonella enteritidis*, *Vibrio cholerae*, *Vibrio vulnificus*, *Yersinia enterocolitica*, Norwalk and Norwalk-like virus, *Giardia lamblia*, *Cyclospora*, *Helicobacter pylori*, and *Cryptosporidium*. The diseases they cause have gastrointestinal symptoms that range from mild to severe vomiting, diarrhea, intense cramping, fever, fatigue, lethargy, or debilitating neurological changes. A quick look back into the literature of foodborne pathogens reveals that even fifteen years ago none of the pathogens causing trouble today, with the exception of *Salmonella* and *V. cholerae* was mentioned. And even so we are confronting new strains: a new, more virulent *Salmonella*, for instance, more pervasive in the food supply than the older version. Bovine spongiform encephalopathy (BSE) in cattle, familiarly known as "mad cow disease," has been linked to a variant of the human and fatal Creutzfeldt-Jakob disease (CJD), casting suspicion not only on beef but on myriad products containing meat by-products, such as gelatin (used in candy, yogurt, capsules for pharmaceuticals, and cosmetics). Stunningly, it appears to be caused not by any of the familiar forms of infectious agents but by a mysterious and previously unknown protein particle called a prion.

Comparing lists of the microbes responsible for foodborne disease, one sees a distinct shift in the past fifteen years. The old names are still there, still responsible for disease, but they are way down the list in terms of how many outbreaks they cause; the newcomers have virtually taken over. A study of outbreaks in England and Wales between

1970 and 1979 showed that *Clostridium perfingens*, *Staphylococcus aureus*, and *Bacillis cereus* played the leading role.[2] Now *Campylobacter* and *Salmonella enteritidis* top the list. In the United States in the 1960s, *Staphylococcus aureus*, *Clostridium perfingens*, and *Salmonella* (non-*enteritidis*) were most prevalent, in that order. By 1979 *Salmonella* had edged up until it was responsible for more than one-third of outbreaks, and *Campylobacter* and *Listeria* weren't players. Only recently have *Listeria*, *Campylobacter*, and *E. coli* O157:H7 even been recognized as foodborne pathogens. Now *Campylobacter*, largely from contaminated poultry, is the leading cause of foodborne infections in the United States. Close behind are *Salmonella*, *E. coli* O157:H7, and *Shigella*. *Clostridium perfingens* still plays a role, but the old familiar *Staphylococcus aureus* of my friend's experience has dropped way back.

Moreover, the common manifestations of foodborne disease have changed. Foodborne illness was previously recognized by cramps, vomiting, and diarrhea and often occurred two to six hours after eating food left too long at room temperature. These were the typical signs of "staph" poisoning. Now some pathogens, such as *E. coli* O157:H7, may not cause illness until three to seven days after eating the infected food, making it difficult to connect that food with the pathogen. The primary cause of hemolytic uremic syndrome has been recognized only in the last dozen years or so as infection with the verotoxin-producing *E. coli* and *Shigella*. Infection with *Listeria* can cause miscarriage and fetal damage or death, events that may not be linked back to a foodborne pathogen. Some urinary tract infections in women are the result of foodborne pathogens. It is now well understood that Guillian-Barré syndrome can result from *Campylobacter* infection. Reactive arthritis, Reiter's syndrome, Miller-Fisher syndrome, and Chinese paralytic syndrome may follow infection with *Campylobacter* as well. Ulcers are newly understood to be the result not of stress but of bacterial infection. Some suspect that Crohn's disease is related to microbes; such conditions as irritable bowel syndrome may sometimes be related to a microbial (perhaps viral) infection.

Even among familiar "bugs," new strains, or serotypes, can increase in importance because something in the environment has changed. In the past two decades, more and more *Salmonella enteritidis* (*SE*) has been isolated in Europe, the Americas, and possibly Africa and Asia.[3] In 1980 it accounted for 6 percent of reportable *Salmonella* infections in the United States; by 1994 *SE* was the leading serotype, accounting for 26 percent of all *Salmonella* infections.[4] It is an increase in which contaminated and poultry play an important role. *E. coli* O157:H7 was unheard of before 1982. Found primarily in ground beef (although also found on fruits and vegetables and in water contaminated by cattle waste or infected humans), it is now a significant and serious cause of illness in many countries, including the United States, Canada, and Great Britain. In some areas today *E. coli* O157:H7 is the second or third most common pathogen found in the cultures of patients with diarrhea. And its presence is being exacerbated by changes in the way we produce and distribute food.

The pattern of foodborne outbreaks has changed as well. Most foodborne outbreaks in the 1950s, 1960s, and early 1970s were caused by people, whether in their homes or in institutions, preparing food in advance, heating or cooking foods insufficiently, and storing them at improper temperatures. And these are still important causes of foodborne disease today. But more and more frequently contaminated food and equipment and food from unsafe sources are responsible for foodborne disease.[5] The new pathogens are coming not so much from the humans handling the foods, although that is still a factor, as from the foods themselves.[6]

When we talk about "emerging foodborne pathogens," the key criterion is that a given infectious agent is causing more disease than before. Some are distinctly new pathogens. Some have been around a while and are simply causing more trouble now for a variety of reasons. Some are pathogens that apparently have been present but are only now being identified. Some have been well known but only recently recognized as causing disease. Why are they appearing? Environmental changes, in the broadest meaning of the term, very likely account for most emerging diseases.[7]

The Source of Polluted Food

We have greatly altered the world's ecology through deforestation, urbanization, industrialization, and changes in how we produce foods, thereby creating new opportunities for the spread of microbes. Some of these changes we confront directly as we face the pollution of the environments from which our foods are harvested.

Polluted waters can easily transmit disease by contaminating shellfish. Just how easily was demonstrated in November 1993, when cases of foodborne disease began popping up in Louisiana. Eventually fifteen separate groups of individuals in Mississippi, Maryland, and North Carolina would be made ill—127 people all told. The outbreaks were associated with eating raw oysters. Norwalk virus, an emerging pathogen, was identified as the agent in stool analysis. While there is no such thing as "stomach flu," influenza being a respiratory disease accompanied by fever and body aches, there are foodborne diseases caused by viruses—most often transmitted by food and water, just like bacterial pathogens. Viral causes are often missed in diagnosing foodborne disease because most laboratories don't routinely test for viruses.

Oyster beds are incredibly sensitive to what happens around them. In the 1993 case the implicated oysters were traced to two adjacent beds. The first thought was that they had been contaminated with raw sewage, but the nearest sewage outlet was sixteen kilometers away and fecal counts at the site were not particularly high. The investigators then turned to the local fishermen. They interviewed workers on forty boats at three docks and on twenty-six boats fishing in the area. On twenty-two of the boats they discovered that either the toilets flushed directly into the sea or buckets were used as toilets and the contents tossed overboard. Two of the fishermen had been ill with gastroenteritis, and both tested positive for antibody titers to Norwalk virus. Each reported defecating and vomiting into a bucket that was then dumped overboard.

The investigators would later estimate that in a case like this where the infected material was dumped directly into the sea, one infected person could contaminate an area one hundred meters wide and one kilometer long. That was about half the size of the area being harvested. The oysters consumed the virus along with plankton, but it accumulated and remained in their bodies for some time. Tests in the laboratory indicated that the virus could be present in the oysters for as long as twenty-five days and recent unpublished studies seem to indicate that the virus can survive cooking—very bad news indeed.

In this case the all-too-common practice of tossing human waste overboard was responsible for a profound economic loss in oyster sales. Between 15,000 and 20,000 bushels that had been shipped to Canada and seventeen other states had to be destroyed, not to mention the cost of halting local harvesting for a week and the medical and personal costs of those who became ill. Louisiana took a new look at its regulations. While state law required holding tanks for human waste on fishing vessels, there were no pumping facilities at the dock and inspections were few and far between.

Increased pollution of the marine environment coupled with improved recognition is probably responsible for an emerging pathogen called *Vibrio vulnificus*. This free-living bacterium is now ubiquitous in marine environments. It may not be new, but it has been only recently recognized as a cause of gastroenteritis, wound infection, and septicemia. Individuals who are immune-compromised, have liver disease, or have diseases that involve a chronic overload of iron are at special danger from *V. vulnificus*. Eighty-one percent of patients with primary septicemia caused by *V. vulnificus* and 88 percent of those patients who eventually died of this infection reported eating raw oysters before getting sick.[8] The bacterium is also dangerous to swimmers who swallow water or have open wounds that can become badly infected by the pathogen. Perhaps it has always been so abundant, or perhaps it has not. There are many aspects of the increases in certain diseases that we simply don't know. There are, however, more and more people living near our ocean waters, more and more people using the waters, and more and more factories and farms whose waste waters eventually reach the estuaries.

Now, as new pathogens emerge, whether they have been released as we have cut down the rain forest, or whether they have been created through mutation, conjugation, or perhaps in the future by genetic engineering, many of the ways we previously approached food safety no longer apply. Food-preservation techniques that have worked for centuries, such as the dry curing of sausages, are no longer completely safe because somewhere something we have done has created or made available a niche for a new and dangerous microbe.

The new microbes are putting our public health systems to the test. Water departments are learning to cope with pathogens they have only recently heard of. Water cannot be separated from food when looking at foodborne disease. It is, after all, not just what we drink but what we provide to animals and plants and what we wash and prepare our food in.

The medical world, the scientific community, the food industry, the regulators and inspectors, the media, and, most of all, the consumer have all been caught off guard by the changing nature of foodborne disease. We are right now in the midst of a frantic rush to catch up. Applying the old models of foodborne disease to the present situation, something that individuals and even medical professionals sometimes do, has led to many outbreaks being either missed or attributed to something else. Using the old models also allows foodborne pathogens to catch us off guard by popping up in unexpected foods. Neither do they account for the possibility that some individuals are more disposed to foodborne disease than others because of physical vulnerabilities or a particular way they interact with food. For many people, however, only bewilderment and consternation have replaced complacent assumptions when it comes to matching wits with these new pathogens.

New Niches for Old Pathogens

The pattern of outbreaks of one emerging pathogen, *Yersinia enterocolitica,* reveals how a population may be exposed to a pathogen because of food customs—what they eat and how they prepare certain foods. *Yersinia enterocolitica* was first recognized as a cause of foodborne disease in the 1960s; it was subsequently isolated ever more frequently in animals and then, quite naturally, in the humans who ate those animals. In 1976 the World Health Organization called the spread "dramatic" and "impressive," both in animals and in humans, and by 1995 the organization had selected it as one of the important emerging foodborne pathogens.[9]

While *Yersinia* generally causes only self-resolving diarrhea, it can leave patients with long-term sequelae that include reactive arthritis—which can be severely incapacitating—as well as septicemia—which can be fatal. The pig is the animal that regularly harbors the serotype O:3, a frequent cause of *Yersinia* illnesses in Europe, and O:9. Outbreaks of O:3 have increased in the United States as well. The climb in *Yersinia* infections has precisely paralleled the trend in developing countries to raise swine intensively—in large numbers and in close quarters, conditions that favor the spread of any bacterial disease. In articles in both the *Scandinavian Journal of Infectious Disease* and *The Lancet*, researchers have associated the emergence of an antibiotic-resistant strain of *Yersinia enterocolitica* to antibiotics given to pigs.[10] And in the United States, a recent study found that "at some time during the grower/finisher period, nearly 93 percent of all grower/finisher pigs received antibiotics in their diets."[11]

The reasons for the emergence of *Y. enterocolitica* as a human disease are now familiar. The microbe can be transmitted from contaminated drinking water, streams, and wells; from vegetables that have been washed in contaminated water; and from the animals that harbor it. The bacteria are often found in the head and tonsils of the pig, whose use in finished, processed products such as luncheon meats presents a danger. The environment the microbe prefers makes it a natural for some processed foods. *Yersinia* is tricky because it can grow under refrigeration; in fact, it grows at minus one degree Celsius. It also grows well in vacuum-packed products, particularly in the absence of competing microorganisms.

The pattern of the outbreaks, however, has revealed a link not just to the increase in swine rearing but also to cultural food traditions. Habits unique to groups of people or geographical areas can determine whether a particular food or pathogen makes them sick. What seems to be important with *Yersinia* is how much pork is eaten, how it is eaten, and how the pork is raised.

Yersinia has traditionally been more prevalent in Europe than in the United States. In Belgium the large number of cases of infection with *Yersinia* was found to be related to the habit of eating raw pork and of nibbling raw ground pork while preparing pork dishes.[12] Even though the same animals were exported to France, the French do not customarily eat raw ground pork and thus escaped the high rate of infection. The extent of infection can also indicate the popularity of certain foods. Cases of infection with *Yersinia enterocolitica* in Norway weren't related to consumption of raw pork but simply to the handling, preparation, and consumption of a lot of pork since the meat is very popular in Norway.[13] Because fewer pork products are eaten in the United States (although pork producers have been trying to change that with their "other white meat" campaigns), and because most Americans know to cook their pork well, the problem is not as acute. But it does happen.

An outbreak of *Yersinia* from U.S.-raised pork occurred in Atlanta and was connected by CDC investigators to the African-American custom of preparing chitterlings, a dish made from pork intestines and generally served in the winter.[14] Those preparing the intestines, a time-consuming task, were apparently spreading the infection by touching their young children with contaminated hands—a not unnatural occurrence in a busy household. To combat this route of infection, an education campaign was focused on the communities where this dish was traditionally prepared. Warnings were given of the danger of becoming contaminated during preparation and the need to wash hands before touching other family members. Whether these lessons were absorbed or whether people just abandoned their traditional fare is not certain, but the illnesses decreased.

Outbreaks caused by *Yersinia* are difficult to spot precisely because the illness generally is relatively mild and occurs sporadically. Occasionally, though, one is recognized in the United States. In October 1995 the Vermont State Health Department had a call from the Dartmouth-Hitchcock Medical Center in Lebanon, New Hampshire. The hospital had cultures from three residents of the Upper Valley in central Vermont that had yielded *Yersinia enterocolitica*. By October 30 two more cases had been reported, one in a Vermont resident and the other in a New Hampshire resident. Eventually ten cases would be identified, their ages ranging from six months to forty-one years. The symptoms were abdominal cramping and fever, and over half reported diarrhea and nausea as well. The disease was not mild; three patients were hospitalized. One underwent—unnecessarily—an appendectomy. At that point the Vermont commissioner of health, Dr. Jan Carney, and Dr. Geoffrey Smith, the New Hampshire state epidemiologist, invited the CDC to help with the investigation. EIS officer Dr. Marta Ackers packed her bags.

The case-control study the epidemiologists carried out pointed to milk from a particular dairy—only one of the ten affected people had not drunk milk from there. The milk was pasteurized, and an evaluation of the process revealed no defects. The milk samples they obtained were also clean. But pigs were also kept at the dairy, and the employee who cared for the pigs also handled both empty and filled milk bottles. The investigating team speculated that the contamination happened after the milk had been pasteurized, perhaps by contaminated bottles.

Then *Yersinia enterocolitica* serotype O:8 was also found in one of the pigs. That would have been pretty conclusive except that further tests showed that the biotype of the nine people who had drunk the milk was alike but different from that of the tenth person, who had not drunk the milk. And it was also different from that of the pig. Of course, the pig had been tested sometime after the outbreak, and the microbial culprit might already have been shed.

"If you don't find it, does it negate the epidemiology?" Vermont epidemiologist, Susan Schoenfeld, asks rhetorically.[15] She doesn't think so. She remains confident of the validity of the team's findings. But it does leave the door open for the conclusions to be challenged—which the dairy owner did.

Yersinia enterocolitica has shown further versatility as an emerging pathogen by finding a new route to cause infection. In 1989 the CDC investigated seven cases of *Yersinia enterocolitica* sepsis.[16] The blood systems of these patients had become infected with the pathogen. These cases, which occurred randomly all over the United States, were traced back to donors in whom blood tests showed evidence of a recent infection with the microbe. It was not necessary for them to have been actively ill to pass it on. Some donors reported having had gastrointestinal illnesses as long as four weeks before donating blood, and one had become ill the very day he donated blood. Because *Yersinia* thrives at refrigerator temperatures, the stored blood was an effective medium for its growth. Of the seven infected blood recipients, five died.

Cases of bacterial infection from transfused blood are rare. In the decade between 1979 and 1989 there were only twenty-six cases, but one was linked with *Campylobacter*, another foodborne pathogen. As foodborne infections increase, however, the transmission of the bacteria by the blood route is likely to increase as well. Because the infectious agent apparently can be in the blood both before actual symptoms occur and long after they have passed, screening for such infections is a potential nightmare.

As more pork is raised intensively in the United States, as more pork is eaten, and as people grow increasingly careless about food preparation, we are likely to see more of *Yersinia enterocolitica*. Preventing an increase in outbreaks of disease will require a multipronged attack that seeks out the root causes in intensive hog rearing while at the same time researching methods to keep the pathogen out of the food supply. A focused consumer education campaign on the dangers of undercooked pork or cross-contamination in the kitchen is a stopgap measure, but a necessary one.

Old Foes in New Places

The epidemiologists in the Enteric Diseases Branch of the Centers for Disease Control and Prevention find that they can always be surprised by a foodborne pathogen. Just as soon as they think they know what foods and conditions a pathogen favors, it can turn up somewhere else. *Clostridium botulinum*, which produces the serious neurological and potentially fatal disease commonly known as botulism, is a good example of an old pathogen causing new outbreaks because it has discovered a new niche.

Botulism is an extremely serious disease. It begins with difficulty swallowing and double vision that may or may not be accompanied by vomiting and muscle weakness; it then develops into a descending paralysis. The fatality rate is usually about 7.5 to 10 percent. Often misdiagnosed, botulism is treated with an antitoxin that can be obtained only from the CDC. Since that agency carefully tracks those requests, it is one disease for which there are pretty good numbers—except for those cases that are never diagnosed.

When home canning was more widely practiced, botulism was a real danger. The spores of *Clostridium botulinum* are found in the ground, and since plants are grown in soil, the spores are frequently found on vegetables. In the normal course of food preparation, vegetables are eaten before the spores develop and produce toxin. Unfortunately, the conditions preferred by *C. botulinum*, low oxygen and low acid, are perfectly met in home-canned low-acid foods that may not have been processed long enough at high enough temperatures to destroy the spores. (The toxin can be destroyed by boiling the canned food for ten minutes, and yet illnesses still occur from insufficiently processed home-canned foods because cooks bypass this final cooking step.)

Ironically, it is quite often consumer demand that sets the stage for foodborne illness. People like sweet foods, and growers have obliged by selectively breeding for sweeter and lower-acid tomatoes. Tomatoes used to be the staple of home canners precisely because they were acidic enough to resist the growth of C. *botulinum*. While it might have been dicey canning beans, a non-acid food, one could be certain that the acidity in tomatoes would prevent the unseen, odorless toxin production of C. *botulinum*. (The inherent guarantee of acidic foods also explains the many recipes for pickled foods in old cookbooks. Okra and green beans were safer when pickled. Chutneys with vinegars and spices were a safe medium for a mixture of vegetables. Cauliflower, a non-acid vegetable that might have been susceptible, was often added to mustard pickles.) That guarantee for home canners has now disappeared. Several outbreaks of botulism have been caused by salsa made with the new varieties of tomatoes that are less acidic and therefore less able to retard the growth of the pathogen.[17] The innocent breeding of a sweeter tomato has had consequences no one thought of.

Sometimes a new or different cooking technique can set the stage for an emerging pathogen. I grew up wrapping potatoes in aluminum foil to bake. I can't think why. Looking back, I think this must have been some wildly successful marketing strategy on the part of the foil producers, because potatoes baked au naturel are much better. Nevertheless, many Americans continue to think that wrapping potatoes in foil is the only way to bake them. Although the use of the microwave oven has eliminated the technique in many homes, many restaurants continue the practice.

One weekend in the spring of 1994 two people were hospitalized in El Paso, Texas, with what doctors suspected was botulism. Dr. Laurance Nickey, director of the El Paso City-County Health and Environmental District, had cultivated a good relationship with the local medical community, and Dr. Azalia Martinez of the Sierra Medical Center contacted him at home with the news. The department acted quickly and on Sunday the restaurant where the two cases had reported eating was voluntarily closed; Dr. Nickey's team collected samples of food literally from garbage cans. The Texas Health Department would join the investigation on Monday and at their invitation, the CDC quickly followed. When Nickey had notified the other emergency rooms in the city on Sunday to keep an eye out for patients, more were discovered. All had eaten at the restaurant on the same two days. Eventually twenty-three cases would be identified.

To find out what food was implicated, the epidemiologists did a case-control study. They interviewed 198 of the people who had eaten at the restaurant, comparing what the well diners had eaten to what the ill diners had eaten. Seventy-three percent of the ills had eaten skordalia, a cold potato and garlic dip, from the hors d'oeuvres tray, while only 4 percent of the well diners had. Twenty-three percent of the ills had also eaten an eggplant dip, which had been next to the skordalia on the tray, while only 2 percent of the wells had. Botulism toxin type A was found in both the skordalia and the eggplant dip. When the investigators inquired, they found that the potatoes for the skordalia had been baked, left in the foil, and held at room temperature in aluminum foil for eighteen hours after baking. The skordalia had apparently contaminated the eggplant dip in the container next to it because the serving spoon had been used in both.

This was the largest reported outbreak of botulism in the United States since 1978. But, in fact, it wasn't the first involving potatoes baked in foil. There had been six other outbreaks since 1978. In every case the baked potatoes had been held at room temperature. The oxygen-free environment that C. *botulinum* prefers had been provided courtesy of the foil wrap. Holding cooked potatoes that had been cooked in aluminum foil at room temperature was now a confirmed hazard.

While the foil provides the suitable environment, other factors contributed to the increase in these outbreaks. As with accidents from all causes, not just one but a series of things have to go wrong. When a large restaurant in Denver was implicated in one of the outbreaks of botulism, all seven of the patients remembered eating potato salad. The investigation was complicated, but it eventually revealed that potato salad was prepared only every five to seven days. The potatoes hadn't been washed—just wrapped in aluminum foil and baked at 260 degrees Celsius for forty-five minutes. Each evening any baked potatoes left over were put, still wrapped in aluminum foil, in an unrefrigerated closet in the kitchen. Before finally being peeled and sliced and made into potato salad, these potatoes had been stored for up to five days at room temperature.

It is satisfying in an epidemiological investigation to find the smoking gun—to have a sample of the suspected food vehicle and discover in it the offending pathogen. But such discoveries are fairly rare. The food may have been discarded as in this case, or the pathogens might be in pockets of a dish rather than throughout. Nevertheless, epidemiology can establish the likely food culprit through elaborate and careful interviews comparing what the ill and the well ate. The effectiveness of the technique is confirmed on those rare occasions when the pathogen is found.

The CDC investigative team in this Denver outbreak didn't find any of the bacteria on any of the other foods they sampled, such as the onions and the peppers, but they did find it on two raw potatoes. To see whether the outbreak strain could produce toxin in potatoes and potato salad, spores were prepared and cultured from the patients' stool samples. Something the investigators needed to know was whether the spores of C. *botulinum* could survive the baking process the potatoes had gone through. Raw potatoes were inoculated with the infectious material, allowed to dry, then held unwrapped at room temperature for three days. These contaminated potatoes, and some uninoculated baked potatoes as controls, were then wrapped in foil and baked for forty-five minutes at 260 degrees Celsius, just as in the restaurant. When the inoculated baked potatoes were tested, the investigators were able to find C. *botulinum* type A organisms in all of them.

The bacteria should have been killed when subjected to a temperature of 120 degrees Celsius for twenty minutes. Why did the spores survive the baking? It seemed likely, the investigators concluded, that water in the potato produced a layer of steam between the potato surface and the aluminum foil during baking that could have kept the temperature of any part of the potato from rising above 100 degrees Celsius. Nevertheless, the jolt of heat was actually helpful to the growth of the microbes because it gave the spores the "heat shock" that helps germination. But could the pathogen survive in the potato salad? It seemed unlikely, but clearly it had. While the acid (vinegar) content of the mayonnaise and mustard should have retarded the growth of the pathogen, the investigators speculated that in this huge amount of salad there were "pockets" that weren't properly mixed, allowing the bacteria to grow and produce toxin. Leaving the salad out at room temperature, which happened several times, would have helped things along.[18]

Potatoes are exposed to C. *botulinum* as part of their natural habitat, growing in the soil. Years ago they arrived in the stores and in restaurant kitchens still dirty—a reflection of how they grow. The cook had to give them a thorough scrubbing and cut out the eyes. Now they arrive looking clean. Some food preparers don't bother washing them again.

In another fairly recent botulism outbreak in which the vehicle wasn't the traditional home-canned food (70 percent of outbreaks) or the rare commercially canned product (3 percent of outbreaks), twenty-eight people were hospitalized. Twelve patients were ill enough to need support from ventilators, and one patient died after six months in the hospital. Twenty people needed to be treated with antitoxin.

This outbreak began on October 15, 1983, in Peoria, Illinois, when three people who had eaten the same food from a local restaurant were admitted to the hospital with symptoms. The suspect meal included a hamburger topped with sautéed onions between two slices of melted American cheese on rye bread, along with french fries, tomatoes, lettuce, and a pickle. Since the pickle was the only canned food in this meal, health officials immediately went to the restaurant and confiscated the container for testing. When the number of patients began increasing the next day, the restaurant was closed. Eventually the case-control study would implicate not the pickles but the sautéed onions. The sliced onions were sautéed in margarine for about ten minutes. Then garlic salt, paprika, and powdered chicken base were added. The mixture was turned into another pan, which was put on the back of the grill for use during the day. The onions were not reheated before being used on the hamburger. The margarine, which coated and excluded oxygen, the warmth of the grill, and the presence of C. *botulinum* type A on the onions to begin with (it was found on raw stored onions from the restaurant) created the perfect situation in which the spores could generate and form toxin.[19]

C. *botulinum* has been popping up in some strange places recently. As Ewen Todd from the Bureau of Microbial Hazards of Health

Canada points out, in the typically restrained fashion of scientists, "In a sophisticated world with increasing demand for varieties of food preserved and sold in different ways, there appear to be more types of foods at risk for growth of C. *botulinum*."[20] What the investigations of outbreaks often show is that some vulnerable foods lack sufficient or appropriate preservatives, that they have been kept at room temperature, and that warnings to refrigerate after opening are not clear or large enough for food preparers or consumers to notice.

It could also be that people today are more casual and complacent about potential threats from what they eat. They assume, incorrectly, that the days of serious infectious diseases that can't be cured by antibiotics are behind us, and that technology has solved the problems of bacteria in food when, in fact, it has created some new ones. They assume that vegetables that look clean *are* clean. They assume that the magic wand of modern food production has been waved over whatever they buy, making it uniformly safe. The food industry participates in this fantasy, fighting USDA attempts to put warning labels on meat and printing "Refrigerate after opening" advice in tiny letters in inconspicuous areas, as if the very mention that a product is perishable might put the customer off.

My son is a cook, too—and a good one. When I compliment him on his skills, he brushes them aside with, "Anybody who can read can cook, Mom." One thing these new cooks seldom read about are food safety tips. They are certainly not present in the current crop of stylish cookbooks that blend a variety of imported cuisines with the changing tastes for lighter, fresher foods. Cooking has turned into a kind of hobby activity for many. Providing the daily fare necessary for survival has been handed over cheerfully to food processors and manufacturers who do the peeling, chopping, mixing, stirring, baking, and cleaning up, leaving the consumer to do only the warming up. Cooking can now be a pleasure, not a drudge, a creative expression of individuality. The books that supply this market create a festive, appealing fantasy in which the bad news about foodborne pathogens would be unwelcome. Corporate denial breeds consumer denial. It is an inadvertent collusion that is setting the stage for foodborne disease.

Not long ago I watched as my son prepared a meal for his fiancée, his sister and me, and the family of his bride-to-be. He had a cookbook propped on the kitchen counter as he prepared a pasta and shrimp dish, and I watched him use a product I myself had never used—chopped garlic in oil. I'm of the school that says that fresh must be better, and garlic is easy enough to squeeze through a press. But the younger generation is experimenting with new and different products, so I said nothing. The dinner was delicious. Not long afterward I heard about chopped garlic in oil as a vehicle for C. *botulinum*.

It was detected when two members of a Canadian family were properly diagnosed with botulism in Montreal and their illnesses traced back to food they had consumed earlier in Vancouver, British Columbia. Those eventually affected came from four Canadian provinces, three states in the United States, and the Netherlands—a reminder of how difficult it can be to identify an outbreak when the cases are widely dispersed as they may be with modern travel, being what it is. A case-control study revealed an association between the ill patients and eating either a steak or beef-dip sandwich. These turned out to be the only two items in the restaurant that used garlic butter, which had been made from bottled chopped garlic in soybean oil. Although the garlic was labeled with instructions to refrigerate, unopened bottles had apparently been stored in the restaurant unrefrigerated for eight months. The researchers found that one bottle of garlic was thought to be spoiled, but it had nevertheless been used off and on during both outbreak periods. "It was used for days, then left unused on the rear of a refrigerator shelf, [and] later recycled into use," the investigator wrote in their article in the *Annals of Internal Medicine*.[21]

The outbreak was serious but fortunately not fatal, as it might have been. Twenty-four patients were hospitalized; of those, seven required ventilation, but none died.

There was now a real pattern in the outbreaks of botulism. Chopped garlic in oil was a new vehicle, but it looked a lot like the other outbreaks investigators had been seeing. In every instance they had found that a vegetable tuber or root (which comes in contact with soil and thus with the spores) had been cooked, then coated with oil or otherwise sealed against air. C. *botulinum* spores that had survived the low-temperature cooking were happily producing toxin in this anaerobic environment. The bottled garlic needed to be refrigerated or to have acid added. Fortunately, the FDA responded: Chopped garlic now contains acid, and my son is in no danger, even if he leaves the chopped garlic out of the refrigerator too long (which he assures me he does not).

But other vegetables in oil have begun to be a problem. In Italy not long ago eggplant prepared in oil and sold in jars was found to be the source of a number of cases of botulism. Indeed, oil flavored with garlic cloves has caused outbreaks of botulism, and this product is no longer allowed to be sold, but recently I had dinner with friends at a restaurant where oil with garlic marinating in it was poured with a flourish onto a plate for us to dip our bread into. I inquired as to how long it had been kept, then ate it, despite my misgivings. Denial and social pressures have an impact even on those who know better.

One huge difficulty with ensuring a safe food supply is the erratic nature of food safety education. An Italian restaurant at the Mall of America outside of Minneapolis had been serving garlic-flavored olive oil. Someone pointed out to me that it had recently discontinued this practice. I phoned the manager to ask why, and he told me that the county health department was now forbidding the practice. There is enormous variability in what health departments do. The FDA puts out a food code that states can adopt or not, but for the sake of uniformity most states refer to it. The warning about garlic in oil had come from that agency. In Maine the health division of the Human Services Department has only seven inspectors to maintain standards in the thousands of restaurants in the state as well as the cafeterias of schools and institutions. Obviously they do not reach them all (a reminder that the cuts in state budgets can have a direct and important effect on our daily lives). When they do inspect, they inform restaurants of the dangers of garlic in oil, says the supervisor, and they explain why it is dangerous. They also set standards for refrigeration. But he feels sure, nevertheless, that much garlic-flavored oil remains unrefrigerated. Making certain that it has been may now be the job of the restaurant patron, but that requires knowing of the danger, taking it seriously, then overcoming a number of social hurdles. Most of us don't want to make a fuss, don't want to be thought rude, don't like the idea of questioning those who are serving us. There is a psychological hurdle as well.

There is strong mental resistance to the idea that our food might be contaminated. It is like the ambivalence we have over sexually transmitted diseases. Just as it is difficult to accept the idea that the act that creates life can possibly bring illness or death, so it is difficult to accept that what is meant to nourish us could harm or kill us. It creates a contradiction in the mind that we too often reject or ignore. In fact, the resistance to the idea of bad food may be even stronger. Sexuality has always been wrapped in mystery and prohibition. Food is basic and essential and wrapped not in mystery but in comfort, familiarity, and trust. Contaminated food is the ultimate betrayal.

But we cannot direct all our anger at the source. How we think about food, how we prepare it, where and how we consume it—all these factors play a role in how safe it is. And yet our entire relationship to food has changed and is continuing to change at a dizzying pace that cannot help but encourage the emergence of new pathogens.

Part II/Living and Eating in a Material World

4/The Paradox of Change

Society never advances. It recedes as fast on one side as it gains on the other. . . For every thing that is given something is taken. Society acquires new arts and loses old instincts.

RALPH WALDO EMERSON,
Self-Reliance, Essays, First Series, 1841

The New York State restaurant catered to the sophisticated diner who appreciated unusual foods served in season. It was spring, and the restaurant had a special offering: local hand-harvested fiddlehead ferns. The fiddle-shaped head of the ostrich fern, just as it emerges from the ground, is a tender delicacy and a traditional food in the northeastern United States and the coastal provinces of Canada. A few specialty stores stock them in season, but in Maine and the Maritime Provinces gatherers pick the ferns from places they alone know and will not reveal. They sell them beside the road from rusty cars with crudely printed signs in their windows. Buying them is a yearly ritual, a true marking of the passing of winter when other signs—such as the weather—may not be convincing. The season is brief, a mere two weeks, and the limitation enhances the pleasure. The experience must be snatched at—like the dandelion greens that will turn bitter in a few weeks, or the rhubarb, closely monitored as its sturdy leaves unfurl, at its sweetest and best when the stalks are just barely ready for picking. Spring is flooded with such transitory delights, each more pleasurable for the anticipation—and for the deprivation that preceded it.

Thus far, no one has tampered with the fiddleheads; as far as I know, no one has attempted to cultivate them, and no one has extended the season. They are canned by a small company in Maine, but it's hard to imagine who buys them. To eat them canned would be to miss the whole point.

Old-timers boiled them good and hard and served them with a bit of butter. But stylish dining has undergone a seismic shift in the last decade or so. The trend is to prefer lightly cooked vegetables. Aiming to please, the New York restaurant sautéed the fiddleheads for a scant two minutes in garlic and butter. Of the forty-five people who ate them, thirty became ill, experiencing diarrhea and/or vomiting within twelve hours. Some were hospitalized. Those who were sickest had eaten the most. The local health department was called in, then the state, and finally, the CDC.[1]

The ferns were tested and found to be negative for *Staphylococcus aureus* and *Bacillus cereus*. The investigators talked to the gatherer and went back to where they had been picked. The ferns were tested further, but the lab found no pesticides and no evidence of chemical contamination. The only conclusion they could draw was that fiddleheads might actually be toxic after all. But since people had been eating them virtually forever and would have discovered that by now, it was likely that cooking them thoroughly destroyed the toxin. This was confirmed by the experience of a nearby restaurant that had served fiddleheads from the same source but had boiled them for ten minutes before sautéing them. None of its diners had become ill. What was new here was that people had begun preparing fiddleheads in nontraditional ways without a thought as to whether there was something protective in the traditional way of preparing them.

We consider lightly cooking vegetables a good thing because it preserves vitamin content. The paradox of change is that whenever we do something differently, even when it seems an improvement, we open a door into the unknown.

Brave New Food World

We have changed what we eat, how we eat, where we eat, how we buy food, how we prepare food, and how we consume food. We eat out more. We spend less time on preparation when we eat at home. We shop less frequently and in a different way. We eat faster and on the run. Technology has brought changes as well. We eat foods processed in ways that would astonish us if the information were available, foods with ingredients that would shock us if we knew what they were. Who would guess, for instance, that some cookies and many candies contain beef in the form of gelatin? Or that seaweed is an ingredient in some ice creams? Many of the ingredients used to make processed foods look and act the way manufacturers want them to look and act are not even recognizable as edible substances.

How we relate to food, which is a part of our culture, has been profoundly altered over the past few years. We use techniques and appliances that became common only a generation ago, and new technologies have shaped our expectations of foods in ways we could not have imagined. By allowing quick reheating, microwave ovens have erased the old impetus to be at the table, on time, and to sit down together. Random snacking is the norm in many families. "Mealtime" has lost the ritualistic connotations it once had as an event at which family and cultural bonds were cemented, a sacrosanct gathering where interruptions and distractions were not tolerated. Except for celebrations, eating has been downgraded for many people to its utilitarian function, a refueling process.

As to what we consume, we've accepted new diets and revised standards for what's appropriate, interesting, or healthy. With our new understanding of the relationship between cardiovascular disease and fat, many have abandoned the traditional meat-and-potatoes menu for a diet that includes more grains, fruits, and vegetables. Between 1970 and 1990 whole-milk consumption declined by 59 percent, and red-meat consumption by 15 percent, while the consumption of low-fat milk, cheese, poultry, fresh vegetables, and fresh fruits all increased. Chicken, once a symbol of prosperity and reserved for Sunday dinner, became everyday fare; the consumption of poultry grew by 87 percent. Moreover, Americans now favor the most luxurious part of all, the breast, boned, skinned, and packaged like a huge pink oyster, until it is easy to forget that chickens have other parts.

Our shifting tastes, however, have not been accompanied by expanded knowledge. The example of the fiddleheads is a reminder that preparation traditions can be easily lost—and that losing traditions can be dangerous. At the same time we have become separated from the source of what we eat. "Food production," the new, technical term for what we once called farming, is no longer universally familiar and close at hand but rather is most often a remote and often secretive activity that resembles factory manufacturing more than farming. Its connection with what we actually eat now seems tenuous and indirect. The link between the live bird and the chicken nugget has been lost somewhere in the processing and distribution. Food distribution seldom depends on a local farmer but rather on an international network of suppliers and processors, a complex tangle of transportation hookups crossing oceans and borders more freely than one might imagine.

A relationship with what we eat that had remained relatively constant for literally thousands of years has been utterly transformed in the last fifty. These shifts in culture and lifestyle and our relationship to food have had unexpected consequences and implications for health that go well beyond a nostalgic longing for family dinners. Experts tell us that there is a connection between these changes and the dramatic rise of foodborne illnesses.

The Infinite Consequences of Change

Whenever there is the slightest resistance by the public to some proposed technological innovation, when some ask hesitantly, "Do we really need this?" or, "Will it actually be good for our lives?" you can depend on one thing. Whether the product is a home appliance, a new computerized gadget, a new trade agreement, a shift in the consumer price index, or a food technology, the hesitance will be brushed aside with the dismissive observation, "People always fear change." Thus answered, with a condescending smile and a rueful shake of the head, the unfortunate resister is relegated to the dustbin of history with the incurable romantics, the Luddites, the terminally nostalgic, and all those otherwise "unable to cope with modern life." Any voice of caution is thrown on the defensive, then marginalized, then finally silenced, as if no questions can be tolerated in the land where technology is king and progress is dogma. The 1971 edition of *The Joy of Cooking* quotes Clifton Fadiman as saying that processed cheese (cheese mixed with a starch base to extend it) represents "the triumph of technology over conscience," a remark so true and so honest that it never failed to make me laugh.[2] But today technology has triumphed over reason itself. To question is to be suspect. Fadiman's quote has been removed from the present edition of this cooking classic. The food industry is intolerant of cheek.

The irony is that there is very little evidence that people do fear or resist change. In fact, our collective dedication to the principle that "new is better" is almost religious in its fervor. Far from being reluctant to change as a culture, we seem to be passionately dedicated to innovation, excited and stimulated by the new. The Western world—indeed, the entire world, since every nation accepts elements of Western culture to a greater or lesser degree—has wholeheartedly, and with only token resistance, embraced change. The evidence is everywhere. We do almost nothing in the way we did it one hundred years ago. We happily fax and E-mail, use automatic teller machines, make purchases with plastic cards instead of money, tote our cellular phones, have elaborate conversations with machines and mechanical voices, and do hundreds of other things our ancestors would have considered bizarre.

Our food traditions are as transformed as the rest of our lives. A tribesman in the remotest Amazonian rain forest drinks a Coca-Cola, a couple in Minneapolis stops at a Japanese restaurant for a meal of raw fish, in Moscow a family goes out for pizza. The French are eating less bread and drinking less wine; the Japanese are eating more beef, more eggs, and more fat; in Mexico the Big Mac is making inroads on tortilla sales; and everywhere in the world the butcher shops, the greengrocers, the small cafés and pubs where people gathered to eat and drink and share their lives are being replaced by supermarkets and fast-food clones, breeding and spreading their standardized versions of edibility, shaped to the taste preferences of the lowest common denominator, in a vast, worldwide marketplace.

Interestingly, it takes a zoologist to note the potential impact of this shift. Stephen Jay Gould in his book *Full House* observes,

> In a society driven, often unconsciously, to impose a uniform mediocrity upon a former richness of excellences—where McDonald's drives out the local diner, and the mega-Stop & Shop eliminates the corner Mom and Pop—an understanding and defense of full ranges as natural reality might help to stem the tide and preserve the rich raw material of any evolving system: variation itself.[3]

Gould looks at food as an evolving system, and that is absolutely appropriate. The considerations that apply to the ecology of other environments apply equally to food. He understands that the loss of diversity is profoundly important, that it is not simply a loss of cultural depth and choice. Whenever there is a lack of diversity, when a standardized food product is mass-produced, disease can enter the picture. The global homogenization of food has an impact not only on taste and experience and the richness of our lives but on foodborne disease worldwide. How we think about food, how we experience it, relates directly to its safety.

Change, as the silencers will declare loudly—and correctly—is inevitable. Change is the essence of life. All life is movement—a seething cauldron of perpetual transformation at every level from the microbial to the social, from birth to death to new birth, constantly renewing and recycling itself, constantly adapting. Yet we seldom give the slightest thought to the effect of any one change on our traditions, culture, long-term emotional and physical health, and well-being. And certainly we give little thought to the potential effect of changes on that delicate microbial equilibrium on which our very lives depend: The exquisite balance of that miniature world of which we tend to be almost entirely unaware until it makes its presence all too forcefully known in an outbreak of disease.

Microbes reproduce with incredible speed and with their remarkable adaptability, they can respond quickly to new situations. Every bit as consistent as adaptation is opportunism. Or to put it another way, everything from *Homo sapiens* to bacteria, viruses, and fungi is looking for a break, looking for some shift in the prevailing equilibrium that will allow it to expand its territory or exist more comfortably with a steady supply of nutrients and fewer enemies. Finding a niche is what we all, from microbe to mankind, are programmed for. Just as a family looks for its niche on a quiet shaded street in a peaceful neighborhood near a good school, just as the fast-food franchise looks for that perfect site on a busy corner with no competition, so the pathogen looks for just the right spot to settle in and grow.

Shifts not only in the way we eat but in the pattern of our lives have unexpected effects on our exposure to pathogens. Foodborne infections, commonly transmitted by the fecal-oral route, are acquired more and more often these days in nursing homes, where many of our elderly live, and in day-care centers, where an increasing number of children now spend their days. Pathogens are lurking in our new diets and waiting to take advantage of our new eating habits. Whatever the human activity, the pathogens remain just offstage, waiting for the scene to shift to their advantage.

The microbes that find new havens as the result of shifts in behavior or environment have caught us off guard, undermined our assumptions, and shaken our confidence. It's hard to be sure about anything anymore when it comes to food safety. The general level of food sanitation has increased in many ways in the last hundred years. Worldwide there are more uniform standards for processed foods. Food preservation techniques are highly developed. Refrigeration is virtually universal except in the most remote areas. And yet even some of the improvements in food safety have opened the door to new microbes. Ironically, our new healthier diets, new food preparation techniques, and economic factors can increase our exposure. Moreover, food traditions and new pathogens can combine to produce foodborne disease. Sometimes it may be the failure to change traditions in the face of a changed product that can cause the trouble, but at other times it may be the abandoning of a tradition that predisposes a population to disease.

Forgotten Wisdom, Ignored Traditions

Despite the fact that food traditions develop over many years from the experience of those who remember what works and what does not, they are considered "unscientific." And yet science is now confirming the validity of some traditional practices that apparently developed when our powers of observation were more carefully honed, precisely because our lives depended on them. Today we have to wait for the confirmation of science, for instance, to be assured that the tradition of eating beans and rice together creates a more complete protein despite the fact that many cultures have been consuming the combination for centuries. Before North America was settled by Europeans, Native Americans soaked corn in a wood ash–lye solution, a process that released the niacin and prevented pellagra, a disease marked by skin eruptions, digestive and nervous disorders, and eventual mental disintegration. They did this long before there was a scientific explanation for the practice. Southern colonists learned this tradition from the indigenous population the hard way, after experiencing the unfortunate results of a pure corn diet. They invented grits—ground, lye-soaked corn. Southerners still eat grits today even though the original reason may have been long forgotten.

The veneration of progress has left us science-dependent. Traditional wisdom seldom gets the respect it deserves, and this disrespect filters down until we begin to forget why we did things in a certain way or to doubt even our own observations, experience, and common sense. Some items of traditional wisdom have been shown to be mistaken, but we should not dismiss all traditional wisdom without first determining what elements have validity. Sometimes we discover too late that tradition has played an unsuspected protective role in ensuring safe food.

Native Americans in Alaska today have one of the highest rates of botulism in the world. As the communities became more Westernized and education programs on food safety were conducted, it was thought that the outbreaks of this potentially fatal foodborne disease would decrease. Instead, the cases have remained about the same, and there have been outbreaks in previously unaffected areas. A number of different foods, all of them native, have been implicated, including fermented fish heads ("stinkheads"), fish eggs ("stink eggs"), fermented whale ("muktuk"), and fermented beaver tail ("stinky tail").

The outbreaks stopped for a while, then between 1984 and 1985 four occurred in Bristol Bay, an area of about four thousand Aleuts,

Yupik Eskimos, and Athabascan Indians who live in villages with populations ranging from thirty to five hundred people. At the CSC's Enteric Diseases Branch, Drs. Nathan Shaffer and Robert Tauxe decided to investigate to see what food preparation or consumption pattern was creating the risk factor. It was a serious problem. About one out of every ten people who contracted the infection died, despite good medical care. The CDC had the help of Dr. Robert Wainwright of the Arctic Investigations Laboratory of the CDC and Dr. John Middaugh of the Alaska Division of Public Health. What turned out to be the culprit, not unexpectedly, was change.

"With the introduction of Western conveniences and consumer goods, Alaska Native lifestyles have changed dramatically over the past 25 years," wrote the authors in the article they subsequently published.[4] Still, the patterns of village life, the traditional activities of hunting and fishing and gathering, continued to be determined by the season. In the summer, as they had for hundreds of thousands of years, most of the villagers moved to the fishing camps, where the catch was cleaned and preserved by drying, salting, or freezing. The salmon fish heads that remained were traditionally fermented and eaten.

The epidemiological investigative team reviewed the original Bristol Bay outbreak investigation reports, then interviewed those who had actually prepared the food implicated by the study. What they found was that fewer and fewer women were preparing these traditional dishes, and that indeed, the habit of eating these fermented foods seemed to be declining. Some of the women prepared the dishes so infrequently that they had a hard time remembering how they had done it the last time. It also appeared that the fermentation of beaver tails, implicated in three of five outbreaks, was something fairly new: It apparently was introduced into the Yupik villages in the 1960s by an elder who claimed to have "invented" the food after finding a beaver carcass that had fermented naturally during the spring thaw.

Most important, the team found that within the living memory of Yupiks the preparation methods had changed. Older Yupiks remembered that the fish heads "were traditionally fermented in clay pits dug in the ground; families used the same pit year after year."[5] About fifty years ago they began using newly available wooden barrels to hold the fish heads in the pits. Then the families stopped using the pits and began storing the barrels aboveground. The introduction of plastic bags and buckets changed things further. Some of the preparers began using them to hold the fish heads, and occasionally the bags and buckets were kept indoors to speed fermentation.

One woman had wrapped fresh beaver tails in a plastic bag and placed them near a warm stove for more than two weeks. Another woman, who had placed her barrel of fish heads covered with moss out in the open aboveground for more than two weeks, remembered too late—eight women became seriously ill—being "warned as a child that the sun's rays had a 'death meaning' and that the fermented foods needed to be kept away from the 'killing rays of the sun.'"[6] This was traditional lore, the turning of important information about food preparation and safety into a story that was more likely to be remembered.

Other Alaskan tribes had different methods of preparing fermented foods. The investigators learned that the Aleuts ate less fermented food, but that when they did prepare it, they generally placed a wooden barrel filled with fish heads into a pit in the ground in September at the end of the fishing season. One told the investigators that they did so because it was too hot in July to make fermented food. The Athabascans prepared the dish by floating fish heads on a string in the river for one or two weeks. What was discovered by the investigation was that all these traditions were protective. All of the botulism outbreaks were in the Yupiks, where the preparation traditions had been forgotten or changed.

Botulism spores are commonly found in the Alaskan environment, but the changes in the methods of fermenting foods appeared to favor the *Clostridium botulinum* growth, with its production of dangerous toxin, because the temperatures were warmer. In the lab other researchers had demonstrated that type-B spores could produce toxin in eighty-five days at 3.3 degrees Celsius, but in seventeen days at 5.6 degrees Celsius. When the food was prepared in the clay pits dug into the cool permafrost, they were naturally insulated—like the icy waters used by the Athabascans. The process of toxin production was slowed down so that the food fermented first, before the toxin was created. The new methods had sped up toxin production, with catastrophic consequences.

But the Alaskan health officials knew that, although there had been a decline in the eating of fermented foods, the practice would continue. The thing to do was not to advise against eating something with such a long tradition, but, wisely, to warn against the use of plastic bags, to encourage the cold-temperature fermentation techniques of the past, and to discourage the fermentation of nontraditional foods, such as beaver tail. In other words, the native peoples needed to remember what they had known for eons, knowledge that had been drowned out by the din of Western enculturation, which denigrates such traditions.

Conversely, when cultures trade cuisine traditions without a full understanding of the food safety techniques that should go with them, the results can be equally dangerous.

East Meets West

Bean sprouts are frequently used in Asian cuisine, but it has been only for twenty years or less that they have been regularly used in the West and made available fresh in produce counters. Generally the seeds are imported and sprouted in the country in which they are to be sold so that the sprouts can be supplied fresh. They are usually added to cooked dishes, but in the West they are often eaten raw in salads and other uncooked dishes, for their health benefits.

In Oxford, England, in March 1988, laboratories noticed a cluster of cultures positive for *Salmonella saint-paul*. All of them were distinguished by the presence of an O5 antigen. It was a serotype not recently seen. Because all the illnesses were from this unusual serotype, it looked as if they might have a common source. When the patients were interviewed, there was an association between the illness and the eating of sprouted mung beans. A case-control study confirmed the link. By June, 143 cases had been identified. When a public advisory was issued to cook bean sprouts for fifteen seconds before eating, the illnesses came virtually to a halt. Only three people became ill after the warning.

The investigators isolated the same *Salmonella* strain from both seeds and mung bean sprouts in stores in a number of different cities in the United Kingdom. But that was not the end of the story. Sweden was having an outbreak of the same strain—as well as other strains. One infected person had eaten the sprouts, not raw, but in a stir-fried dish.

It seemed that the seeds, which were imported from Australia and Thailand, were contaminated with low numbers of the bacteria, which found the conditions suitable for sprouting—dark, warm, and wet—the ideal situation in which to multiply. Because the beans were an agricultural product, the bean-packing houses did not have to meet food safety standards. Our widespread international food trade was bringing us a bonus—a supply of an unusual *Salmonella* strain. But it was also, ironically, the unusual strain that enabled epidemiologists to recognize the outbreak.

Infections from bean sprouts continued to be a problem in Sweden and Finland despite new regulations. Sprouting seeds from Australia and Asia have caused outbreaks in other countries in Europe, while in the United States an outbreak of foodborne disease was linked to sprouted alfalfa seeds. During the long-running *E. coli* O157:H7 outbreak in Japan in 1996, the finger was pointed at radish sprouts, at least in one locality, as a possible source of infection. And it now seems that stir-frying isn't enough to guarantee that the pathogens are destroyed.

Stir-fry. The technique of cooking vegetables lightly with a bit of oil over high heat, often used in Asian cooking, wasn't described in the standard cookbooks twenty-five years ago. We have absorbed new cooking styles and new eating styles, along with new foods, from other cultures. And often we have very little experience with them. Twenty-five years ago *sushi* wasn't a household word either. The eating of raw fish by the Japanese was considered one of those eccentric "foreign" practices. Now eating in any major city is a cosmopolitan adventure and sushi-eating is a rite of passage for young sophisticates. I would eat sushi in Japan; I am less enthusiastic about doing so outside of Japan, where there is less experience with preparation.

All fish are prone to infestation with parasitic worms that can infect humans. In Japan, where there is a centuries-long tradition of sushi preparation, the preparers must be trained and certified. In the United States no certification is required. You don't know whether the preparer is well trained or simply looks Japanese. In Japan the fish are inspected carefully for these parasites. Are the preparers on the lookout for them here? The way to kill most of them is to cook the fish—obviously unacceptable for sushi—or freeze them, and this is often done. Fish sold to be consumed raw in the Netherlands must first, by law, be frozen. But the public perception is that freezing lowers quality, so I always wonder, when I ask a server in a Japanese restaurant whether the fish was frozen or fresh, if I am getting the right answer or the answer he thinks I want. In fact, while some worm infections come from professionally prepared restaurant sushi, the great majority do not.

When the twenty-four-year-old young man came to the hospital in New York City, he'd been experiencing severe abdominal pain for about ten hours. It grew worse in the hospital, and eventually the extreme tenderness and other signs in his abdomen made doctors suspect appendicitis. They operated and found the appendix normal. But before closing the incision, "a pinkish-red sinuous worm that was 4.2 cm long was noticed moving onto the surgical drapes."[7] It was identified as a member of the Dioctophymatoidea family. After surgery the patient recovered nicely. He reported to the doctors that although he normally ate sushi in restaurants, he had consumed fish most recently at a friend's house. She had purchased fresh, unfrozen fish from a market earlier in the day.

Just as eating sushi grows ever more popular, the list of parasitic diseases acquired by humans after ingesting infected fish served raw or insufficiently cooked is growing, researchers say. Anisakiasis is the disease from the Anisakis larva found in raw fish that most people have heard of. Considered an emerging pathogen, it is only one of numerous parasites of different sorts in fish. Despite the warnings, eating raw fish is now so common that even those supposedly knowledgeable about these things still take risks. After a party at which sushi was served on the West Coast, for instance, four doctors, who should have known better, became infected with fish tapeworm.

Is this a major problem? Not really. The number of cases is still quite small, although when surgery is required, "the consequences of infection can be much worse than the unpleasant sensation of feeling the fish move as one bites into it," notes Dr. Peter Schantz of the CDC.[8] Japan is not immune to these infections either, but in both countries the real problem seems to be preparing the dish at home. I often overhear people at my local market buying fish to eat in raw dishes.

Experience may be the missing ingredient in preparing foods outside our tradition, but even eating foods in which we have a longer tradition may be presenting new problems because a variety of factors have come together to create a situation that sets the stage for foodborne disease.

We Are Not Who We Were

Forty years ago Yale students dressed in jackets and ties at dinner and were served traditional meat-potato-vegetable meals on china plates at three precise times a day. Now the university is considering replacing its dining halls with a food court run by the familiar fast-food giants. They are one of the last schools to change. Colleges across the country have had to come to grips with a new kind of eating. "Students come to campus used to eating what they want when they want it and where they want it and we have to adjust to that," Emily Lloyd, executive vice president for administration at Columbia University, told the *New York Times*. "The sort of 'Leave It to Beaver' household where everyone, at the same time, eats meat loaf, string beans and mashed potatoes followed by apple crisp doesn't exist anymore."[9] Students, studies show, seem to be eating four to six small meals a day rather than two or three, adapting to their class schedules and their lifestyles, which are casual and unpredictable.

Most of us have modified our eating habits, but the young especially are truly different in what they eat and how they expect to eat it. Even as the choice of foods in supermarkets grows, the average diet of young people becomes more and more limited. Previous generations grew up on school cafeteria lunches that, as ghastly as they might have been, at least exposed young people to some variety. Sometimes they actually ate what was offered. Now those menus have been replaced by a steady diet of teen foods: hot dogs, hamburgers, and pizza, a diet as limited as that of an eighteenth-century Irish crofter. My stepson admitted to eating pizza and pizza only for lunch for three solid years. In the age of two working parents, young people may or may not have a broader choice at home, where food preparation may well be left to them. This broad unfamiliarity with food variety among the young brings its own hazards. When they branch out, they have almost no experience with how things should taste, how they should look, certainly not how they are prepared—experience that might protect them.

The generation growing up now has not had the same preparation that my generation had. I and my friends remember picking up tips on food safety in the kitchen as we watched our families preparing food. I remember being told to be sure to refrigerate the potato salad and the cream pies, warned to cook pork thoroughly to avoid trichinosis, warned not to use cracked eggs because the shell might be contaminated with *Salmonella*, and warned to avoid dented or bulging cans because they might indicate contamination, perhaps with C. *botulinum*. I mentioned this last warning not long ago, and a friend of mine wondered idly whether her two grown sons knew about cans. She hadn't told them, she realized. Now that women are no longer the chief food preparers, many young men, like her sons, are cooks themselves, but they may lack that important information. I see dented cans on grocery shelves that wouldn't have been there a generation ago.

The old routes by which young people—or more accurately, girls—might have been educated in food preparation and safety have disappeared as well. Home economics classes have been dropped almost everywhere—perhaps because they seemed sexist, perhaps because they seemed unnecessary with all the prepared and instant foods available, perhaps because they just didn't seem as important as other subjects. In many schools they have been replaced with health courses, but just as often home economics has given way to multimedia courses. High school students are much more likely to be able to design a home page than design a home meal.

"Education concerning food sanitation is nowhere near where it used to be. Kids are not getting this information in school," says Michael Pariza, director of the Food Research Institute at the University of Wisconsin.[10] Hand-washing is still the single most important factor in preventing the transfer of foodborne pathogens, and once it was drummed into the heads of children by parents and teachers. Now teachers and parents alike are worried about such pressing social problems that hand-washing seems like one of the niceties of a gentler past—it somehow pales beside the urgency of drug prevention, monitoring weapons in the classroom, and safe-sex education. A *Journal of the American Medical Association* article recently warned of a "new generation of food handlers [that] probably doesn't wash its hands nearly as often as it should."[11] That is undoubtedly an understatement.

The combination of these transformed food habits, lack of familiarity with foods, and declining education in health and food safety basics has had a predictable result in young food consumers and preparers. In the early 1980s the CDC learned that 41 percent of the students who came to the health center at the University of Georgia complaining of gastroenteritis (vomiting and diarrhea) and were subsequently cultured, were found to be infected with the bacterium *Campylobacter jejuni*. The rate for the population was unusually high. This was a serious problem because it translated into costly hospitalizations and missed classes, not to mention personal misery. *Campylobacter* infection can also leave those infected with unfortunate sequelae such as arthritis or Guillian-Barré syndrome. Two other studies in different parts of the country had previously found a high incidence of illness caused by *Campylobacter* among college students as well. Looking at all these findings, the Enteric Diseases Branch at the CDC felt that the phenomenon called for a study. What was unique about college students? Where were they being exposed?

The investigators organized a case-control study, matching the ills carefully with well students. All the participants were interviewed in person about what they had eaten during a specific time period. They were asked about drinking raw milk or untreated water; traveling out of the country; having contact with pets or intimate contact with someone who had diarrhea; and how their food was prepared and eaten. They were asked about delays between cooking or purchasing and eating and whether the chicken they ate was raw or undercooked. The study revealed a high degree of association with the eating of chicken, whether prepared at home, eaten in the schools' food service facilities, or eaten fully cooked or undercooked. Given the fact that up to 83 percent of chicken has been demonstrated to be contaminated with *Campylobacter*, this finding was less than totally surprising. But what truly shocked the investigative team was how ignorant the students were about the most basic food preparation techniques. They didn't know how to cook, but they didn't seem to know that they didn't know. They operated on assumptions that weren't valid—that food was safe in its uncooked state. Many of them thought nothing of eating rare chicken, for example. Of the forty cases that reported eating chicken, nine—more than 20 percent—reported eating it raw or undercooked. And even when they cooked chicken, they didn't know how to clean up or how to avoid contaminating other cooked foods.[12]

The conclusion, that unsafe food handling had led to both contamination and cross-contamination, was supported by an earlier study that had found that people with *Campylobacter* infection were less likely than controls to have cleaned the cutting surfaces they used or to have used separate cutting boards for raw and prepared foods. Ignorance is not bliss when it involves the safe preparation of food. Unfortunately, it's not just their own food young people are mishandling. These young people—and some even younger—are probably preparing your food, whether you know it or not.

Leaving It to Others

The parent corporations of hamburger and other fast-food franchises know well the problem with foodborne pathogens. Yet outbreaks still happen. A glaring weakness in the fast-food defense system is its reliance on young, inexperienced workers who often lack the most basic information and understanding of food-handling safety. They are given rudimentary training, but it falls into an intellectual vacuum of experience. Rapid turnover in employment makes the problem even worse, as does the lack of sick leave for hourly employees who may continue to work when they are ill. In June 1996 a hidden-camera report on ABC's *Prime Time* revealed the appalling conditions and food-handling practices in some of the best-known chains. Pizza dough was cradled in bare arms against clothing, loose flowing hair hung over mixing bowls, and dropped utensils were picked up off the floor and reused without washing. The revelations were frightening. Almost without exception the workers were teens.

While giant fast-food chains are guilty of undermining the diversity and thus the overall health of food, they do have an advantage in being able to institutionalize some food safety techniques. Some major chains attempt to compensate for rapid turnover, lack of sick leave, and inexperienced workers with fail-safe technology that doesn't allow for mistakes. Some have carefully calculated computerized grilling times for burgers, for instance, that cannot be overridden by a rushed or pressured employee. But managers cannot stand in the bathroom like "Mom" and watch to be sure their workers wash their hands thoroughly. Instead, McDonald's requires, for instance, that employees sanitize their hands in a solution every thirty minutes. Still, widespread outbreaks from fast food have been linked to the hands that applied the tomatoes and lettuce—even the hands that picked up the onion rings and the french fries. McDonald's has found a way around this problem by insisting that workers wear gloves to prep vegetables and dress burgers and by creating a device to load fries into containers so that hands never touch the food. But while an overlay of technology may help produce a safer product, the underlying problem—young, undereducated, and underpaid workers—has not been addressed. The time has come to accept the irony that our lowest-paid workers are preparing the most important product we buy—the foods we consume.

Adding to this problem is the growing tendency of many restaurants and food service institutions to purchase packaged prepared vegetables in an effort to cut the costs of prep help in the kitchen. Shredded lettuce contaminated with *Shigella sonnei* from a single distributor caused 347 cases of gastroenteritis at a number of restaurants. Lettuce served in the salad bars at two university campuses in Texas caused widespread simultaneous outbreaks of *S. sonnei* gastroenteritis. Because they occurred at the same time, the cause was obviously the packaged prepared food, not the later handling in the different cafeteria locations. These outbreaks demonstrate the potential for mistakes in food handling to be multiplied when restaurants and food service facilities rely on outside suppliers. The contamination might just as easily have occurred on-site; if it had, the outbreak would have been limited.

Surprising as it is, we who prepare meals at home have been subcontracting out food preparation as well. More and more frequently we buy foods that have been partially or totally prepared by others without giving much thought to who is doing the preparing. In fact, the packaging and marketing of food today—the aura of the imagery that surrounds food products—leaves us with an impression of safety. It may or may not always be justified. The common assumption is that foods from large producers are safer. Certainly the giant food corporations have every incentive to maintain their carefully developed reputations by distributing safe foods. They have the resources to maintain up-to-date information and to hire food scientists. But mistakes can be made, and when they are made by large distributors, they become large problems.

Food-Tech Nightmares

Technology has undoubtedly made some foods safer. Tuberculosis and other diseases such as typhoid and diphtheria were widespread before the pasteurization of milk. Refrigeration has helped to keep foods from spoiling. And many of today's processed foods meet high safety standards more easily because of complex and highly sophisticated production technologies. And yet each technological change has the potential to create new niches for opportunistic microbes. Competing microbes are eliminated by pasteurization, which thus creates a fertile ground for re-contamination: A pathogen that intrudes after the process can grow virtually without opposition. There have been serious and widespread outbreaks of foodborne disease when milk has been contaminated after pasteurization, so additional safeguards have to be built into the technology to protect the product.

Irradiation, in which food is subjected to blasts of gamma radiation, electron radiation, or X rays, all of which reduce but do not eliminate bacteria, is touted as a magic bullet for contaminated foods, yet it creates a similar situation. The microbes in the product not killed by the process multiply if exposed to the right conditions. If the irradiated product is exposed to a new source of contamination, the pathogens can multiply rapidly in an environment artificially free of competitive microbes. And it is a cruel irony that refrigeration, while preventing many pathogens from multiplying, has actually encouraged a host of new ones. *Aeromonas*, *Listeria*, *Yersinia*, and certain strains of *C. botulinum* and *E. coli* can grow and sometimes thrive in the chilly environment.

"For every action there is an equal and opposite reaction," we learned in basic science class. It would seem that Newton's laws work for microbiology and food technology as well. One can predict with some degree of certainty that for every technological improvement in food safety, there is a microbe that will thrive in the environment thus produced. For instance, certain packaging techniques that replace oxygen with other gases to retard spoilage, while controlling some pathogens, have created hospitable environments for others.

Those who question change are thus not simply being nostalgic or romantic. There may be sound reasons for their caution. They are acting on instincts honed over millions of years, instincts carved from memories that say, "Wait. Let's think what this might do. Perhaps there was a reason we put goat's milk in the millet or cooked the fiddleheads for ten minutes." These were instincts that kept us alive and well over eons when science, as we know it today, didn't exist.

Sometimes it is a new way of preparing food that creates the problem. Microwaves, for example, don't cook evenly. Frozen hamburgers, fish, warmed-up dishes—all may have cool areas in them that could promote the growth of pathogens or leave unkilled microbes that grew in a dish left unrefrigerated. In fact, the meat industry no longer recommends cooking meats in microwave ovens. Epidemiologists suspect that microwave cooking may be responsible for many cases of unexplained sporadic illnesses when too few people are ill to warrant an investigation. Not long ago microwave ovens were clearly implicated in a larger outbreak.

In Alaska in 1992, at a large community picnic, roast pigs were on the menu. They were roasted in Seattle and shipped up to the event by air. It was one of those situations where a series of things have to go wrong for foodborne disease to occur. After thawing at room temperature for several hours, the pigs were broiled over a gas flame. An inexperienced cook tested them for doneness by probing with a knife—an inexact method at best. Then the pigs were wrapped in plastic, boxed, and left at room temperature until they were shipped, unrefrigerated, by plane, to Alaska. One pig arrived and was served to guests. The other pig was held up in transit and arrived seventeen to twenty hours after cooking. It was this pig that caused the trouble. Most of the guests at the picnic had already left by the time the second pig arrived, so it was divided between the thirty or so left at the event. Of the forty-three people who eventually consumed the meat, none of those who reheated the meat in a frying pan or oven became ill, but twenty-one others did. Of those, eleven had eaten the meat un-reheated. Among those who heated the meat in a microwave oven, the score was ten out of ten. Everyone who used the microwave became ill, even though they considered the meat hot when it emerged. The problem? Cold spots remain in microwaved foods, and there is no residual heating effect once food is removed from the oven. Studies conducted in labs have confirmed this finding.

Changes in our eating habits and food preparation techniques are driven by many factors. Our love affair with innovation may be instinctive—most of us actually like novelty—but it is also cultivated by the economic interests it benefits. Creative industry, as we all know, can provide for needs we didn't know we had, and profit by doing so. And we, as consumers, participate in the process willingly and enthusiastically. To be successful a new food must either manufacture consumer interest or respond to it. The amazing number of products the food shopper is confronted with attests to the success of innovation and marketing. Who would have imagined fifty years ago that we would want cheese that squirts from a can, mock crabmeat, or no-fat potato chips? Who would have guessed that entire aisles in huge food markets would be filled with nothing but fizzy flavored water or one hundred choices of pet food? Our great-grandmothers would have looked with bewilderment on this stupendous array of choice and change presented in its Technicolor marketing glory, searching, probably in vain, for something that looked like what they understood as food.

It would be wrong to imply that change is automatically bad. But it is important to remember that change is never neutral and does not occur in a vacuum. Any change precipitates other changes that may not have been anticipated. The balance is tilted, the equilibrium shifts. Synergy—the total impact when elements are combined being greater than the effect of any one element—is very difficult to study and often neglected. Some call it the "black hole" of scientific research, but interest in the field is increasing. While scientists once studied the impact of a single chemical, some now study combinations. AIDS drugs that have an impact when used alone, researchers say, may have a hundred times the effect used together.[13] Something in the synergy amplifies the effects, and the implications are startling. Similar effects have been found with other chemicals, and it is a possibility that must be taken into account when considering the potential impact of any change on human health.

Although it is easy to forget, food is made up of as many complicated components as are pharmaceuticals. Each year there are thousands of changes in the ingredients in processed and prepared foods. Most of them have no obvious ill effects. Occasionally there is a mistake. Responding to consumer demand for low-calorie products, a British manufacturer of hazelnut yogurt decided to use a sugar substitute in the product. The result was an outbreak of C. *botulinum* that made twenty-seven seriously ill and killed one. What happened? The simple and apparently innocent shift from sugar to a sugar substitute had altered the water activity of the product so as to allow spores of C. *botulinum* to germinate and produce toxin; with sugar, their growth had been inhibited.[14] Because of an apparently insignificant change in the manufacturing process, a previously safe product became a very dangerous one. The mistake was in seeing the yogurt as simply a collection of ingredients, each acting independently. The synergistic effect of the sugar, a seemingly unimportant ingredient, on the whole product hadn't been considered.

We are beginning to see the unexpected consequences of genetic engineering as well. Small changes in the molecular nature of food products—or even the bacteria involved in the production of these products—can have unforeseen consequences. In 1989 public health officials in Minnesota spotted an outbreak of eosinophilia-myalgia syndrome, which can be fatal, and connected it to a product containing tryptophan, which some patrons of health-food stores had been buying as an apparently safe relaxant.[15] After one death, the FDA recalled the product from the market. The investigation pointed to one Japanese company as the source of the problem. The company used a fermentation process involving *Bacillus amyloliquefaciens* to produce tryptophan, but two factors had changed in the production: The company had reduced the amount of powdered carbon used in purification, and it had introduced a new genetically altered bacterial strain of *Bacillus amyloliquefaciens* that differed only in its enhanced ability to synthesize two chemicals. This change apparently caused more of the agent responsible for the outbreak to remain, thereby creating a dangerous product that resulted in illness and death half a world away.[16]

The frightening potential of genetic engineering to create dangerous foods was demonstrated again when researchers at the University of Nebraska at Lincoln modified soybeans with genes from Brazil nuts to produce a soybean with more protein. Unfortunately, the new soybeans were found to be capable of setting off a serious—potentially fatal—allergic response in people sensitive to Brazil nuts. Such genetic modifications could present a serious and unacceptable risk to people who are accustomed to avoiding nuts but wouldn't expect to experience the same reaction from soybeans, which are now used in a great many products and are almost impossible to avoid. The company sponsoring the research wanted to use the protein-enhanced soybeans for animal feed but decided not to market the product because it would have been difficult to keep the soybeans from inadvertently entering the human food supply, where they could provoke these potentially fatal allergies. How many companies are that responsible?

Ever since the commercial potential for genetic engineering was understood, there have been worries that risks like this could be in our future. Now it has been confirmed. Critics of moving genes to food plants from other plants say that this study, the first to demonstrate that allergens can be transferred from one food to another through genetic engineering, provides an argument for tighter regulation on the part of the FDA to protect the public. Proponents say that the company's decision not to market the product demonstrates that voluntary monitoring and reporting is sufficient to protect the public. But should the public rely on the honesty and goodwill of global industries to prevent other potential calamities?

For many years science has been concerned with breaking down everything into smaller and smaller pieces, looking further and further into the building blocks of life. Now the new science based on chaos theory has begun the process of putting the pieces together again, looking at how the pieces interrelate, how one part may affect another. No matter how well we understand one part of the system, until we understand how the parts relate to each other, we will never totally comprehend the system. And without comprehending the whole, tinkering with this or that part is nothing more than educated guesswork. New drugs and new food techniques are created on the basis of theories of how they might work, reinforced by experiments and testing; then they undergo trials to see what the effect on the whole might be. Often these experiments, tests, and even trials can only guess at the long-term effects. We, the public, become the ultimate testing ground.

The Surprises of a Healthy Diet

The science of nutrition is relatively new, but it has fostered a vast and expanding industry that caters to the growing public awareness about the role of various foods in developing healthy bodies. Nevertheless, even good advice can have unexpected repercussions—especially when it is coupled with an error in processing. A campaign was begun in England a few years ago to encourage parents to replace sweet snacks with less sweet and more healthful snacks. At about the same time an Italian company began running television ads for its salami sticks, really an adult product. A glitch in the processing had allowed the sticks to be contaminated, and 101 cases of *Salmonella* infection were the result. The two campaigns—one commercial advertising and one public health—had converged. Most of the ill were children—the median age was six—and most lived in southeastern England, where the ads were running. Nineteen children were hospitalized. One child had an appendectomy, and a sixteen-year-old boy developed acute ulcerative colitis. A quick investigation and identification of the source allowed the product to be recalled and the public to be alerted, preventing more illnesses. The nutritional science that tells us what is good for us can have surprising consequences when it combines with other changes to expose us to new pathogens.

Growing up in a salad-eating family, I can't remember a single meal without it. I remember in the 1970s longing for salads when I had to eat a quick meal out; the ubiquitous salad bar had yet to be invented. A few years later the fast-food scene had been transformed, and by 1988, 71 percent of fast-food and family restaurant chains offered salads or salad bars.[17] I was grateful and gave no thought to what the availability of salad in these venues might mean for foodborne pathogens.

Ironically, the growing demand for fresh vegetables has not been as healthy a trend as it appears. Our year-round desire for fresh vegetables, our growing habit of eating them uncooked, our youthful, uneducated food preparers, our forgotten hand-washing skill, the popularity of salad bars and buffets—all play their special role in the increasing number of outbreaks of foodborne disease.

The most important pathogens where vegetables are concerned include *Shigella*, *Salmonella*, *E. coli*, *Campylobacter*, *Yersinia*, *Aero-*

monas, *Listeria monocytogenes*, *Staphylococcus aureus*, spore-forming bacteria such as *Clostridium botulinum* and *Clostridium perfingens*, and viruses, usually hepatitis A and Norwalk, as well as parasites such as *Giardia*. The lists of vegetables from which pathogens have been isolated is impressive. Outbreaks of *Salmonella* have been caused by tomatoes, mustard cress, bean sprouts, cantaloupe, and watermelon. *Shigella flexneri* gastroenteritis has been linked to onions. Listeriosis has been linked to fresh cabbage and lettuce. *Bacillus cereus* (as well as *Salmonella*) on seed sprouts has caused gastrointestinal illness. Infection with hepatitis A has followed consumption of both green onions and strawberries imported from Mexico. The list goes on. While advances in growing, processing, preserving, distributing, and marketing produce has allowed the industry to supply fresh vegetables in new ways to those who want and can pay for them, Larry Beuchat, a researcher at the Center for Food Safety and Quality Enhancement at the Department of Food Science and Technology at the University of Georgia, notes that "some of these same technologies have also brought an increased risk of human illness associated with a wide range of pathogenic microorganisms." Beuchat thinks that part of the reason for the increase in foodborne disease in the summer months has to do with the fact that we consume more fresh vegetables at that time of the year, a truly depressing thought.[18]

Young adults tend to graze rather than dine, as college caterers have discovered, and they do some of their grazing at salad bars. Health promotion campaigns and body consciousness among the young emphasize the need for fruits and vegetables. Salad bars eliminate the tedious preparation by leaving it to others, and that is a big factor in their popularity.

Cross-contamination can occur more easily in the kitchen—whether home or institutional—than can be imagined. One outbreak of *E. coli* O157:H7 in a major restaurant chain was thought to have begun when the meat cutter stopped what he was doing to lend a hand to a petite waitress who was having trouble lifting a massive container of ranch dressing. Subsequently, as each of the employees put their hands in the same place on the container to tip out the dressing, their hands became contaminated. Whatever food they touched next, such as the lettuce, had the potential to become contaminated. The later outbreak was traced, once again, to the salad bar.

Some food service managers believe that plastic hoods over the islands offer some protection against foodborne disease, and perhaps they do. But the food may be contaminated before it reaches the bar. Then the prepared food selections—mixed salads or cut vegetables or fruits—may sit at room or near-room temperature; utensils may be moved from one food to another, spreading contamination if it is there; and with hot dishes, the warming trays may actually contribute to the growth of bacteria. In my local supermarket the salad bar has been removed—"too unsafe," I was told by the manager.

Illnesses from pathogens on fresh vegetables can be very serious. Recently in an outbreak of *E. coli* O157:H7 the culprit was not the hamburger that went into the spaghetti sauce at a Boy Scout outing, as seemed most likely, but the iceberg lettuce on sandwiches the Scouts ate. How had it become contaminated with a pathogen usually found in cows? The *E. coli* O157:H7 might have been in the irrigation water, or it may even have originated with infected workers who defecated in the fields. It might have been shipped in a dirty truck, or it might have been on the hands of the supermarket employee who placed it on display. Have you ever seen the pre-packaged chicken drip liquid onto the conveyor belt at a supermarket checkout? Have you ever seen someone place vegetables on that same conveyor belt? Cross-contamination has the potential to occur at points in the long journey from field to table to which one has scarcely given a thought.

One of the fastest developing trends in the fresh produce industry is the packaged salad, sometimes containing a variety of attractive fresh greens and often even a packet of dressing and croutons. These packages are marked "pre-washed" or "triple-washed," labeling that encourages use without further washing. With their attractive packaging, these salad packages carry the reassuring veneer of technology. Not long ago I watched with growing unease as a friend of mine prepared a salad in my home from greens she had brought with her. She took the hydroponic lettuce from the open-topped plastic bag and tore it into the bowl without even a cursory washing. We all survived.

This same trusting friend of mine was watching not long ago as I filled a pan with water and emptied into it a "triple-washed" package of mixed exotic greens. This time my caution paid off visibly. We were both startled to see a truly significant amount of black soil at the bottom of the pan when I fished the greens out. When I called the distributor to ask whether the salad was, in fact, ready to eat, he replied that *he* would wash the salad before serving.

The trouble with the popular new salad mixes is that they fall, as bean sprouts do, into the gap between agricultural produce and processed foods, and as a new product, there are as yet few standards. In many places—actually most places—no one is checking to see whether the greens are properly washed in clean water in sanitary conditions. The owner of an organic farm that supplies a popular farm stand in my area never sells salad mixes because he himself got a severe foodborne disease from salad mix when vacationing in Hawaii. Some states are looking at regulations, and many growers would prefer that the standards be clearly defined. But the truth is, most health departments aren't keeping close track, and most people have no idea where their salad mix comes from—or the origin of much of anything else they eat for that matter—and recent outbreaks of disease have been linked to these salad mixes, one of them from the virulent *E. coli* O157:H7.

Although my friend is a reformed salad washer, many consumers remain careless and complacent in their preparation of fruits and vegetables. Consumer infatuation with the new, especially if it's easy and attractive, is so intense that seldom is the safety of a new product questioned, even when a product makes no claims to being safe. Oddly, no one I know seems to be at all curious as to why precut or prewashed vegetables seem to last so long in the bag these days. It's almost as if they have forgotten that vegetables are prone to rot, even under moist refrigerated conditions. Actually, there is an important difference between "spoilage" bacteria and pathogenic microorganisms. The difference can be important because it may be spoilage that is keeping us well.

The success of bagged, ready-to-eat vegetables is one of those true symbiotic achievements between the food industry and the consumer. These new products meet both the consumer's desire for hassle-free veggies and the producer's longing for extended shelf life and added value. (Salad mixes, for instance, sell for much more per pound than the ordinary head of lettuce.) Some greens are sold in micro-perforated bags that keep them in top condition longer, but other greens or vegetables remain unspoiled for an even longer time in the bag because of something called modified atmospheric packaging (MAP):

The oxygen is removed and replaced with nitrogen or another gas. Some researchers have been concerned that the absence of oxygen creates a very pleasant environment for our old friend C. *botulinum*, as well as *Listeria* and *Aeromonas*, any of which might be present on the salad leaves.

The researchers have noted that signs of rot in choppped or sliced vegetables, which hopefully would put the consumer off eating, appear before the botulism has had a chance to develop toxin in bagged, prepared vegetables. But now some tests have shown that with MAP pathogens can grow in great numbers before vegetables show signs of spoiling. All we can do is hope that pathogens are not present—or avoid buying salad greens this way. It is ironic that in a world of increasing choice, safer food may mean limiting choice.

5/Trade: A Passport for Pathogens

> Caravans passed overland across the oases and deserts of central Asia by regular stages, while ships traveled freely across the Indian Ocean and its adjacent waters. Regular movement to and fro across such distances implied exchange of infections as well as goods.
>
> WILLIAM HARDY MCNEILL,
> *Plagues and Peoples*, 1976

On July 20, 1994, the CDC got word of twelve confirmed cases of *Shigella flexneri* in northern Illinois. The lab had identified an unusual serotype of the virulent pathogen that is much more prevalent in underdeveloped countries. The Enteric Diseases Branch had a new crop of epidemiologists-in-training, and as often happens, they pulled two of them out of the course and plunged them into the investigation. One of them, Dr. Kim Cook, already had outbreak exposure—of a sort. He had been a student at the University of Georgia in the 1980s when the CDC did its study on *Campylobacter*. He was one of the deathly ill students who had probably eaten undercooked poultry, an experience that played a role in his career choice. After medical school and further training in public health, he had joined the Epidemic Intelligence Service (EIS).

Dr. Cook and his colleague Thomas Boyce knew even before they left Atlanta that this outbreak was not going to be tidy. While two of the very sick cases had eaten at the same restaurant, the other cases appeared to be sporadic. The Illinois Department of Public Health couldn't find a link and needed help. Boyce and Cook took a flight out the next day, checked in with the public health officials, and then set out to travel the state, looking for the common factor that the unusual serotype guaranteed had to be there.

They found that of the apparently sporadic cases, two of the ills had eaten at a church potluck supper. Among the rest of the sporadic cases there seemed to be no connection. Indiana was coming up with cases as well. They were linked to a wedding reception, and the investigators were able to implicate a single food item. It gradually became clear that among all the food items the patients had eaten, there was a common ingredient: green onions. The onions were traced to a single farming area in Mexico. The epidemiologists never made it to the area, but the *Shigella* could have come from the soil, the ice used to transport them, the water used to clean them, the truck used to ship them—from almost any point in the long trek from the Mexican countryside to church suppers and wedding receptions in the American Midwest.

"The trend among recent outbreak investigations is fruits and vegetables," says Cook. "We are seeing more than we ever did in the past. The developing world is just one plane ride away, one truck trip."[1]

Trade has always had an impact on disease. In *Plagues and Peoples*, William McNeill describes the caravan trade between the eastern Mediterranean, India, and China in the two centuries following the beginning of the Christian era. It surpassed all previous exchanges of goods. But with the trade came new danger. "Chances of an unfamiliar infection spreading among susceptible populations certainly multiplied."[2] The only difference today is that modern transportation, refrigeration, and food technology, by prolonging the shelf life of food, has allowed the trade in a previously perishable commodity to increase exponentially. Imported food is now the norm virtually everywhere.

Even trade agreements have played a role in creating a favorable environment for foodborne pathogens. If the global food market is a passport for pathogens, the GATT and NAFTA trade agreements are visas. They do this by lowering the barriers to import and export and by making regulations difficult to maintain. Safety regulations that impose a more stringent standard, such as the European Union's (EU) ban on hormone-fed U.S. beef, have been challenged as unfair barriers to trade. The criterion for defending these standards is science, yet science is not equipped to tell us with certainty what changes to our food, or what some subtle addition to or subtraction from it, are likely to produce in the human body thirty years—or even ten years—down the road. The link between bovine spongiform encephalopathy (BSE), or "mad cow" disease, and the human version, Creutzfeldt-Jakob disease (CJD), has been suspected by some scientists for a number of years, and yet even as evidence grew and caused concern in scientific circles, British beef continued to be exported to EU countries. It took human tragedy that could no longer be ignored for the accumulating scientific evidence to garner enough weight to overcome the powerful economic forces that govern import and export.

With the GATT and NAFTA, food safety standards will tend to level off at the lowest common denominator and microbes will get an even freer ride. Foodborne disease was something that few in Congress had probably given a second thought to when that body hurriedly passed the trade agreements. Nobody stopped to ask the epidemiologists what they thought, and Congress probably wouldn't have wanted to hear the answer in any case. The GATT was "fast-track" legislation, requiring an up or down vote, and the agreements themselves were available to lawmakers and the press for only a brief period before passage was rushed through. With bipartisan support, including that of President Clinton, voices of concern and opposition were drowned out, ignored, or actively suppressed.[3]

The Expanding Trade in Food Pathogens

Although we consider the possibility of pathogenic bacteria on fresh vegetables when we are in developing countries, how many of us stop to think about bacteria on vegetables from the many other developing countries from which we import produce? Writing in the journal *Clinical Infectious Diseases* in 1994 on the changing epidemiology of foodborne disease, doctors from the Minnesota Department of Health wrote:

> Meeting the increased demand for fresh fruits and vegetables in the United States has required the seasonal importation of produce from

> Mexico, Central America and other tropical areas. Seasonally, more than 75 percent of fresh fruits and vegetables are harvested outside the United States and delivered within days to grocery stores and restaurants. During the winter months from 1989 through 1992, 33 percent to 70 percent of cantaloupes, 57 percent to 72 percent of green onions, 69 percent to 79 percent of cucumbers and 20 percent to 64 percent of tomatoes purchased by consumers in the United States were harvested in Mexico. With the pending formation of a free trade zone between Mexico, the United States, and Canada, it is likely that produce imports from Mexico into the United States will increase substantially in the future.[4]

A quick check of a grocery store in a small town in Maine reveals fruits and vegetables from around the world: Costa Rica, Honduras, Colombia, the Dominican Republic, Chile, New Zealand, Argentina, Taiwan, Jamaica, Spain, Morocco, Brazil, Indonesia, Holland, Guatemala, and Canada. It shouldn't take a research scientist to figure out that importing fresh vegetables from abroad is an effective way of importing the microbes in the environment in which they are grown.

In fact, food safety is a good argument for helping less developed countries with the basics of a sanitation infrastructure. "We can make great strides in doing what we can to make our own food supply produced in this country safer and safer and safer," says the CDC's Robert Tauxe, "but as we import more and more from other countries, it's going to be the efforts to make things better and safer in [exporting] countries that is going to pay off. What is done there is bound to influence what happens in this country."[5]

Dr. Craig Hedberg and his associates in the Minnesota Department of Health, the researchers who looked at the impact of various social and economic changes on foodborne disease in their state, point out that competition between U.S. and foreign producers may also result in further cost-cutting measures in areas of the United States that rely on low-paid migrant workers. The chances of the product becoming contaminated in the field or during packing or distribution then increase.[6] Even where there are sanitation standards in the United States, they may not be maintained. On one outbreak investigation Hedberg found portable toilets in the fields, as the law required, but most did not have toilet paper or running water, making them virtually useless for controlling the spread of disease.[7]

The number of different pathogens that can be found in salad and breakfast bars is impressive, whether from the global food trade or homegrown inept handling. Finding out which one—and even what subtype or serotype is responsible—can be vital in identifying an outbreak and discovering its cause. The importance of laboratory testing and epidemiology was amply demonstrated in the summer of 1991 when forty-nine *Salmonella* cases in Illinois and Michigan, twenty in Minnesota, and seventeen in New Jersey were identified as infected by an unusual serotype called *S. poona*. When analysis of the serotypes, using a pulse-field gel electrophoresis technique, revealed an identical pattern, investigators knew that they probably came from a common source. State health departments often act independently, but Illinois and Michigan did their case study together and discovered that a high percentage of people who became ill remembered eating cantaloupe. The distribution of the cases followed the distribution pattern of melons shipped from the lower Rio Grande Valley region of Texas.

In Minnesota the laboratories found the same connection. In the case-control study carried out by the health department's epidemiologists, 62 percent of the first thirteen cases reported consuming cantaloupe from a salad bar or in a fruit salad. Grocery stores, restaurants, and distributors again identified the cantaloupes as having come from the lower Rio Grande Valley. Again, the distribution of the disease followed the distribution of the melons. In New Jersey the disease was associated with eating fruit salad at a party. The caterer said that it had received cantaloupes from Arizona, California, and Texas. Eventually twenty-eight states would report cases and the number of those afflicted would reach more than four hundred laboratory-confirmed cases, with many more uncultured and cultured but unreported cases suspected.

Salmonella isn't that common on fruits and vegetables—it is much more likely to be found in animal products—but it does occur. In Germany in 1991, 600 people were made ill from a cold fruit soup. The year before 245 Americans in thirty states were sickened by *Salmonella chester*, and another 174 suffered gastroenteritis after eating tomatoes contaminated with *Salmonella javiana*. As always, many more cases were likely to have escaped detection.

The rough outsides of melons can be contaminated on the ground. While the melons in the 1991 outbreak were traced to Texas, all melons look alike when they are unpacked from their crates. The CDC suspected that the melons actually came from Mexico and had been re-crated and labeled by Texas distributors. Although large produce companies wash and dip melons in a chlorine rinse, that is seldom done when the melons are field-packed. Cutting a melon through with a knife can transfer bacteria from the rind to the surfaces of the fruit. If the melon slices are left out at room temperature—in a salad bar or fruit salad, for instance—the bacteria can grow on the cut, contaminated surfaces. The CDC recommends thoroughly cleaning melons before cutting them and using clean and sanitized utensils and surfaces. In fact, the agency says that it is prudent to wash all fruits and vegetables before they are handled and consumed. But how many people are really scrubbing their cantaloupes?

The food scientist Larry Beuchat points out that produce can be contaminated before it is harvested by feces, soil, irrigation water, green or inadequately composted manure, wild and domestic animals, dust, or human handling.[8] After it is harvested it can be contaminated by feces, human handling, harvesting equipment, transport containers, dust, wash and rinse water, sorting, packing, cutting, ice, transport vehicles, improper storage, improper packaging, cross-contamination from other foods in storage or display areas, improper display temperature, or improper handling after wholesale or retail purchase. The dangers of transportation are seldom considered by consumers, and even regulatory agencies are only now beginning to take it into account. A friend of mine who owned an orchard told me of standing beside her harvested apples as a truck came to pick them up only to find that the trucker's previous job involved transporting chickens. The interior of the truck contained chicken excrement, filth, and dirty feathers. She refused to allow her apples to be shipped by that trucker, but how many more shipments have gone out in filthy trucks when the grower wasn't present? How many producers and truckers and distributors don't care? How many consumers are really careful about washing their apples?

The points of danger for contamination are at many places in the long drawn out process of getting the product from the field, wherever in the world that field might be, to your table. It only makes sense that if the distance between producer and consumer is less, if fewer hands touch it, and if those hands are clean and healthy, the consumer is more likely to get clean food. One obvious advantage of buying produce directly from the producer at a farmers' market is that you are often looking directly into the truck that transported it.

But it is not only fresh vegetables but canned and frozen foods as well that come from great distances or even from across the globe. We are importing processed products from around the world today, and any thought that federal agencies are inspecting them all is an illusion.

Frozen Trouble

Marcella Ruland is a high school social studies teacher who lives with her husband Tim in Columbia, Maryland, a planned community about halfway between Washington, D.C., and Baltimore. On a beautiful weekend in August 1991 a Thai friend called to invite them to her house for an impromptu afternoon party. She told the Rulands that she would buy steamed crabs and if everyone brought something they could create a feast. Marcella prepared a fruit salad and made some Thai ice tea. Another friend prepared "a very beautiful Thai specialty—a rice pudding covered with coconut milk," Marcella remembers.[9] The group sat around the backyard enjoying a perfect day and wonderful food. A lot of crabs were eaten before the guests went home late in the afternoon.

By Sunday, Marcella had developed cramps, some vomiting, and unrelenting diarrhea. It came in waves every ten minutes or so. It was also debilitating. She felt extremely weak.

The Rulands belong to an HMO, and by Monday morning Marcella felt ill enough to go to the urgent care center. The doctor there scribbled something about a "gastrointestinal disturbance" on her chart and sent her home without taking a stool culture. Feeling better for simply having been seen, she went back to bed, somewhat reassured. But she continued to feel light-headed, and her diarrhea continued. By Thursday she was no better. Because her husband was at work, she took a taxi back to the urgent care center. There they discovered that Marcella was seriously dehydrated, and she was transferred to the hospital, where she was put on IVs and encouraged to eat. She managed to get something down, and with the IVs restoring her fluids, she felt well enough to get out of the hospital the next day—or at least she said she felt better. Marcella is a weaver, and she was supposed to assist with the judging of the weaving at the state fair. She was determined to be there.

She struggled through the judging but felt extremely weak and dizzy and knew she wasn't thinking clearly. She could sense that she was acting what she would later describe as "stupid and crazy." These were signs of dehydration and an imbalance in fluids and salts. At home the diarrhea was constant. She seemed to be up every two minutes or so and sometimes couldn't even make it back to bed. She remembers having as many as thirty episodes of diarrhea a day, bouts that left her weak and confused. She knew she should drink something, but alone in the house, she found she didn't even care. She had no energy, no inclination to do anything. Lethargy overcame her. All the time precious fluids were leaving her system. The next morning, on Saturday, she struggled into the shower only to faint. Her husband, very alarmed at this point, took her back to urgent care.

This time they were more aggressive. Finally they ran blood tests and stool cultures. Her potassium, they found, was at critical levels. She was readmitted to the hospital. The medical staff wanted to give her an oral solution of potassium and send her home, but since she couldn't keep the solution down, she was put on a potassium IV and kept in the hospital. Realizing now that she had some kind of infection, although they didn't know what, the doctors bombarded her with broad-spectrum antibiotics.

The stool cultures took a couple of days. She had a new doctor now, and this time she stayed in the hospital on the IVs, but it would be Tuesday before the results came in. The nurses had been acting strangely that morning, she noticed, obviously avoiding her. They knew what the tests had revealed, she found out later, but wanted to wait for the doctor to tell her. When the doctor finally arrived and told her what the lab had found, she became hysterical. She had something that people in the United States "didn't get," something you heard about in Third World countries, something she knew could kill her. The commercial laboratory had isolated *Vibrio cholerae* O1. Marcella was moved into an isolation ward where her treatment was changed to ciprofloxaci. With the appropriate antibiotic and steady IVs, she began to improve. After five days in the hospital, she recovered quickly. But another part of the story was just beginning. The lab had not only reported the results to the doctor but notified the Maryland Department of Health and Mental Hygiene. The implications for public health were considerable if the bacteria causing Marcella's illness proved to be the epidemic strain of cholera.

The yellow piece of paper saying that a lab was reporting *Vibrio cholerae* O1 landed on the desk of Jean Taylor, a young epidemiologist with the Division of Outbreak Investigation at the Maryland department of health. "Check it out," someone had written across it. She wasn't overly alarmed. The lab results had yet to be confirmed. Nevertheless, she notified the CDC at once.

In Atlanta the immediate concern was that the new South American strain of cholera might have reached Maryland waters. But first the diagnosis needed to be confirmed by the agency's own labs. If it was *Vibrio cholerae*, they needed to know whether it was the strain that produces cholera. It was possible that Marcella Ruland was infected with the non-toxin-producing *Vibrio cholerae* bacteria that causes illness but is not usually associated with the frightening large epidemics. Even if it was confirmed, there was probably a logical explanation. The patient might have recently traveled abroad or eaten undercooked shellfish. It was apt to be a sporadic case.

Still, Taylor wasn't willing to wait for the laboratory results. "Even if it wasn't cholera, it could still be significant if they found that other people were involved," she says.[10] The duty of the health department is both prevention and control of human disease. A case investigation is carried out whenever there is any reportable illness to prevent it from expanding into an outbreak.

A number of things needed to be done almost at once. Taylor began setting up the epidemiological investigation to discover how Ruland contracted the disease. But she also needed to find out the extent of the infection. Was it in the general population? There were several simple ways to find out. She notified hospitals in the area to be on the lookout for cases that could be cholera. She also arranged to meet men from the utility department to inspect the sewer system at dawn the next morning. Wearing her oldest jeans and sneakers, she rendezvoused in the morning mist with burly men carrying crowbars. They were there to pry open the heavy metal covers over the sewer lines so that she could plant Moore swabs.

The Moore swab has a specialized role, but its form is simplicity itself. It is nothing more than a roll of gauze with a string attached. Taylor tied one end of the string to the manhole cover and let the swab float and bob in the effluent. There it would provide an appealing draw for the microbes in the wastewater. When it was retrieved and cultured, it would reveal the level of cholera infection—if any—in the general population. Taylor inserted the swabs at four critical points around the city, where they would stay for twenty-four to forty-eight hours. Then they would be pulled up and put in a clean container to be cultured, and a new swab would be inserted. Over a six-day period she would plant and retrieve fourteen samples. It was a smelly job.

The epidemiologist is always working on several aspects of a case at once. Taylor had already learned from the county health department nurse who first interviewed Marcella Ruland that she had not traveled abroad recently. But the picnic, which had been attended by six other people, was already looking suspicious. It was in the right time frame. Marcella Ruland's illness had begun a day later. The incubation period for cholera can range from hours to five days, but somewhere between twenty-four and seventy-two hours is usual. Marcella and the others had eaten a good number of crabs, and although she remembered the crabs as having been thoroughly cooked, they might have been contaminated in some way.

"The crabs made my antenna go up," says Taylor, "but we were not ruling anything out. It was just that they seemed the most likely source." Cholera, after all, is usually transmitted by polluted or contaminated water.

The first thing the epidemiologist does when setting up an investigation is to establish a methodology and then a case definition. A CDC epidemiologist was now on the scene and involved in the investigation. How were they going to approach this case? Epidemiology is a science in which basic common sense plays an important role. Hard science comes close on its heels. While they couldn't be certain where Marcella Ruland was exposed, the most likely explanation was the picnic. They would start there. First they would try to find out whether any food was left from the gathering. Then they would interview all those present, asking about illness, about what they ate, and how much, and about what they did not eat. They would collect stool and blood samples from those attending the party, sick or well. They would carry out an environmental inspection of the crab vendor's facilities. The crabs had been purchased from a van on the side of the road. The vans would be inspected as well, looking for ways the crabs could have become contaminated. But first, the food.

"Leftover food is the gold in any foodborne outbreak," says Taylor. Every epidemiologist hopes to find it. To her distress, she learned that what remained from the event had been thrown out just the day before. She also learned that there had been a second gathering later that day that Marcella Ruland had not attended. Taylor would call that Party B and the first, Party A. At Party B the food served included leftover crabs from the first party, fresh fruits, vegetables, chips, and beer. There was no rice pudding left to serve at Party B. It was vital to find out whether the Party B guests were sick or well. Everyone involved cooperated, although eventually, after all the questioning and prodding and pricking, their patience would wear thin. Slowly the sequence emerged.

Several others attending the first party had become ill but only one other than Marcella Ruland had sought medical care. Since the investigation didn't begin until almost two weeks after the two parties, it was not surprising that none of the attendees were still shedding the pathogen in their stool, but the labs could look for antibodies in their blood, signs that a body has waged a war against a specific pathogen.

None of those attending Party B reported any illness. Their memories were confirmed by the blood tests. Cholera antibodies were present in the blood of all but one of those who had attended the first party, but none of those who attended the second. In an epidemiologist's terms, attending the first party was a risk factor for getting sick. Everyone at the first party had eaten the same food; the only difference was that some had eaten more of one food than another. The problem was the size of the party. "The numbers were really too small to have statistical power," said Taylor, and besides, everybody seemed to have eaten everything. But not everyone got sick. It was puzzling.

With a sanitarian to accompany her, Jean Taylor set out to investigate the crabs. She felt burdened by the enormity of the responsibility. Make a mistake in identifying a food as the cause of an outbreak, and a restaurant or a caterer or even an entire industry can be severely damaged. There could be no mistakes. Maryland is famous for its crabs.

The crabs, the investigators found, had been harvested locally, steamed by a local vendor, and transported immediately to the party. They had been consumed hot. There seemed nothing amiss in their preparation. Nor had the Moore swabs revealed any *Vibrio cholerae* of any type in the sewage system. Lab tests had shown that Marcella Ruland's cholera was indeed toxin-producing, but to the relief of the CDC, it was an Asian strain—not the South American strain they feared. There was still the possibility that it had somehow found its way into the Maryland waters, but tests from the Rappahannock River, which opens into the Chesapeake Bay just to the south, where the crabs had actually been harvested, were negative. It didn't seem to be the crabs. Taylor breathed a sigh of relief. "With that, the focus changed," she says. It seemed unlikely that the dipping sauce was the vehicle. That left the rice pudding.

The pudding had a sauce of coconut milk. When the investigators asked how the dish had been prepared, they learned that on the day of the party three packages of imported frozen coconut milk had been put in a pot over medium heat on a gas stove. The milk was stirred occasionally as it thawed, brought to the point of boiling, and immediately removed from the heat. It was stored at room temperature for five hours until it was served.

Although there was nothing left to test, there was something Taylor could do. The coconut milk had been identified by brand and source. Investigators went back to the store where it had been purchased, bought more, and turned the packages over to the FDA. Their labs isolated toxin-producing *V. cholerae* O1, the causative agent of cholera, and the same organism found in Marcella Ruland. One significant part of the puzzle had been confirmed; the vehicle had been found. The discovery demonstrated that the pathogen could survive freezing and that, in fact, the coconut milk was very dirty indeed. The lab also found *Escherichia coli*, *V. cholerae* non-O1, *Vibrio fluvialis*, *Vib-*

rio alginolyticus, *Aeromonas hydrophilia*, *Salmonella derby*, *Salmonella lexington*, and *Salmonella london*. When they tested an additional twenty-five bags of coconut milk from nine different shipments, they found no more of the pathogenic cholera bacteria, but they did find a high total aerobic bacterial count.

The epidemiologist must often be content with a numerical association that compares the percentage of ills who ate a specific food with the number of those who did not to see what the risk factors are. "In this case," Taylor says, "epidemiology alone was not sufficient to solve the puzzle because of the small numbers, both of guests and of those infected. Without finding the pathogen in the food, we might have been able to say that Party A was a risk factor, and we might have said that crabs were a less likely source and that the coconut milk was the most probable vehicle, but we would have been unable to say definitively that the coconut milk was the cause."

A lot of things went right in this investigation. It was lucky that Marcella Ruland's diagnosis had come to the attention of health officials and that they had been able to buy more of the coconut milk. And as for finding cholera in the coconut milk, Taylor called that simply "amazing." They would discover, too, a dose-response relationship in who got sick and how sick they got. Marcella Ruland had really liked the rice pudding. She'd eaten three helpings. Her husband, who didn't get sick, had eaten only a bite or two.

The implicated frozen coconut milk had been exported for several years by a trading company in Bangkok. It was distributed primarily in Maryland, but some had been sent to a distributor in Texas. When the Thai Ministry of Public Health was contacted, it found that the manufacturer was not licensed by the Thai Food and Drug Administration. The product had been shipped only to the United States.

An investigator in Bangkok was sent to look at the manufacturing facility. She discovered that the coconuts were peeled by hand at a house forty kilometers outside Bangkok, then trucked to the city. The coconut meat was not refrigerated. The baskets and the floor of the facility were washed with canal water. The meat was washed twice with tap water, and the coconut milk squeezed out in a press. The milk not immediately used for coconut ice cream was put into five-kilogram bags, frozen, and transported to another site in Bangkok. Here it was poured into stainless steel containers, taste-tested, then bagged and frozen for shipping to the United States. At various sites in the process the investigators used swabs—sterile cotton rubbed across surfaces, then sterilely bagged and tested—to look for the presence of microbes. They would culture positive for *Vibrio fluvialis* and *V. cholerae* non-O1. Tests of the workers revealed similar pathogens. While the investigators never found the specific pathogen, they demonstrated how the milk might have been contaminated.

Once the problems with the product were revealed, the Maryland distributor of the product issued a voluntary product recall classified by the FDA as a Class 1 recall. The remaining packages were destroyed. The press was notified. Anyone who thought that they had been made sick by consuming the product was urged to contact the local health department. The public was urged not to use any of the product still in their freezers. Further importation was halted.

The crisis wasn't over. The product had gone to two hundred stores, and Taylor recalls poring over lists in the distributor's offices to determine precisely where so that the stores could be notified to take the product out of their freezer cases. The investigation was a classic demonstration of why it is critical to have full cooperation among the laboratories, the local health departments, the state, the CDC, and the FDA.

The team speculated that more people hadn't become ill because coconut milk is usually used in dishes that are heated sufficiently to destroy these pathogens. But the procedure used in preparing the milk for the rice pudding hadn't heated it long enough. Canned coconut milk, used in drinks like piña coladas, is safe because it is heat-treated sufficiently to kill any pathogens present during the standard canning process. But the investigators couldn't exclude the possibility that others had been infected by the juice, become ill, and never been diagnosed. Marcell Ruland's infection had come very close to being missed. It was not only possible,[11] it was likely that others were ill as well. Of those made sick in this outbreak, only two sought medical care, and of these, only one was cultured—and then not until the third doctor's visit. The potential had clearly been present for an epidemic form of cholera to enter the environment from an imported food product. That it did not was a tribute to a decent sewage system—not to excellent medical care.

Who Is Minding the Store?

The entry of a grossly contaminated frozen product into American grocery stores has profound implications for the everyday safety of consumers. "You go into stores," Taylor says, "and you think the food you're buying is safe." It was an assumption that had proven to be wrong.

Public employees tend to "hang together." Only rarely does one agency openly criticize another. But in this outbreak the investigators would break ranks. Too much was at stake. The article reporting the outbreak, which appeared in the *Journal of Infectious Diseases*, was harsh in its evaluation of the safety measures taken by the FDA, the agency responsible for non-animal-derived imported foods. Taylor and her coauthors (one an FDA employee) minced no words about the implications of their findings. This outbreak, they said, showed yet another way that a pathogen could get into a population far away from its original source. Frozen food had not previously been suspected as a vulnerable food product. Now two things had gone wrong. Unsanitary food production methods had coupled with what the investigators called "insufficient standards guarding imported food safety in a receiving country."[12] The watchdogs weren't watching. Consumers were making unwarranted assumptions that the frozen food was safe for use without further cooking; nothing on the label had told them differently.

The outbreak, they pointed out, brought up the question of the safety of all imported foods. The United States has no authority over foreign manufacturers and no money to spend checking that manufacturing practices abroad are adequate. Nor were FDA inspections uniform, the article said. FDA policy was to inspect products with a history of violations. With no history and thus no perceived risk, the frozen coconut milk had not been considered suspect enough to test. A newcomer had slipped through. Since the FDA samples only between 2 and 10 percent of imported foods, as much as 98 percent of imported foods could be entering U.S. stores without any examination or with no knowledge of the manufacturing conditions. An enormous hole in food safety had just been exposed—but in a journal so obscure that the warnings were invisible to the food-buying public.

The investigators' report listed only some of the recent outbreaks from imported foods. There were the cantaloupes and the green onions, but staphylococcal food poisoning had also been caused by canned mushrooms from the People's Republic of China, and an *E. coli* outbreak was associated with French Brie cheese. The report made it very clear that the conditions of food production in one country could have direct health implications for consumers in another. New ways of living and of eating, new foods and new ways of preparing them, a vast new worldwide trade in food products—all contributed to the rise in outbreaks, often caused by unusual pathogens with which American physicians had little experience. It was all part of the new pattern of foodborne disease. Now one more contributing factor was revealed—the weakness of the regulatory shield.

There is at every level—among the producers, the distributors, the retail sellers, the restaurants, the food associations, right up to and sometimes including the government agencies we naively think are there simply to protect us, and among the politicians who make the rules—a collective effort, sometimes developing into a fierce determination, to maintain the illusion, against all developing evidence, that our food supply is uniformly safe. That belief is adhered to sometimes out of genuine ignorance, sometimes out of misguided but benevolent intentions, sometimes out of deep denial, and sometimes out of pure, unadulterated greed.

As late as 1992 or 1993 it was quite common for politicians to say, "We have the safest food in the world," although there was never any statistical basis for that chauvinistic observation. The many recent U.S. outbreaks caused by all sorts of pathogens have made that claim sound questionable at best, and to the truly informed it is patently absurd. The politicians have been reeducated. You seldom hear that remark anymore. Now, if they say anything at all, they announce that we have the cheapest food in the world. As we shall see, that is nothing to be particularly proud of.

6/The Movable Microbial Feast

> Our era has seen an escalation of the rate of change so drastic that all possibilities of evolutionary accommodation have been short-circuited.
>
> SVEN BIRKERTS,
> *The Gutenberg Elegies: The Fate of Reading in an Electronic Age*, 1994

There are two kinds of movement that affect our exposure to foodborne disease. Both are efficient. We bring the disease-causing microbes to ourselves through importation, as we have seen, or we travel to where they are.

Virtually all of us want at some point to be other than where we are. We may want to be warmer or cooler, we may want more stimulation or less, we may want to see something different—a way of living, an unfamiliar landscape. We may simply want to be entertained. Often what we really want is to be different, and travel allows us to do that for a while. We can become the explorer, the archaeologist, the botanist; for a brief moment in a resort hotel or on a cruise ship we can enjoy a more luxurious life than the one we normally live. We can realize our fantasies, if only temporarily.

The architect Philip Johnson has said that we travel for the architecture of a place, and there is a great deal of truth in that observation. But we also travel for the food. To see is not enough. We want to taste the strangeness of a place, to digest it. What we give little thought to, however, is that more than just the flavors are different. When we travel, we confront not just new vistas and cultures but new microbes.

Before vaccinations against infectious diseases, travel was more dangerous than it is now. Travelers, if they lived to make the round-trip, could bring home smallpox, diphtheria, typhoid, and a host of other infections often never conclusively identified. Smallpox has been vanquished. There are vaccines for other diseases. But modern medicine has tended to give us a false sense of security. The availability of vaccines has created a complacency that has perversely led us to think that infectious disease is a thing of the past. Today's frequent flyers confronting today's emerging pathogens can come home with souvenirs no one counted on. In April 1996 a Swiss tourist on a boat trip in Brazil contracted yellow fever. Seriously ill, he nevertheless returned by a regular flight to Switzerland. He was admitted to the hospital but died two days later. He had not been vaccinated. Fortunately, he had not had a layover in Puerto Rico or he might have been bitten by a mosquito and passed the disease on there to an unprepared population.[1] Pathogens are famous stowaways.

There are no vaccines to protect us from most foodborne diseases. (There is a vaccine for cholera, but it is not considered especially effective.) We are almost as vulnerable today to enteric disease pathogens as we were a hundred years ago—and with the speed and extent of modern travel, perhaps more so. Every trip exposes us to a new microbial environment to which our bodies struggle to adapt. We give complex sanitation systems more credit than they deserve, seldom questioning whether the running water is from a clean source and properly treated; whether the pipes that bring it to us are properly maintained and flushed to rid them of bacteria; whether the sewer line is corroded or cracked; whether the flush toilet flows out into the bay where our seafood has just been harvested; whether the truck that brought our vegetables or the hands that picked and prepared them were clean. As outbreaks occur from contaminated municipal water systems, processed foods, imported fruits and vegetables, and shellfish, we are beginning to appreciate the shaky foundation on which this confidence is built. When our travels take us abroad, the uncertainties about the reliability of sanitation are even more difficult to resolve.

And we tend to have a completely unjustified faith in the care with which food is prepared, wherever we are.

One steamy afternoon driving into a small French town where we planned to stop for the night, I watched a woman trudge down the street in a sleeveless dress. Under her arm were eight or ten baguettes, with no wrappings, right up against a sweaty armpit. I laughingly pointed the sight out to my husband. Several hours later, as we sat down to dinner, the same woman stepped out of the kitchen and dropped a basket of cut bread on our table. We ate it, hoping against hope that what we were getting had been the loaf in the middle.

Confronting the Unfamiliar

Travel has increased enormously since mid-century. In 1950 there were 25.3 million international arrivals a year; by 1995 there were 567 million. Fluctuation from one year to the next is related to the economy: The more money people have in their pockets, says Somerset Waters, editor of the *Travel Industry World Yearbook*, the more they tend to travel.[2] Today when the growing affluence of developed nations is coupled with the relatively low cost of transportation, tourism is more egalitarian than ever before.

Waters has long been interested in the area of medicine that treats the diseases of travelers; in fact, he invented the name—"emporiatrics"—that the growing field now uses. During the 1960s he and a group of researchers looking at why people who traveled to Mexico so often became seriously ill identified the *E. coli* serotype responsible for what is called "traveler's diarrhea." Even today, with improved conditions, the CDC advises travelers against eating street foods, raw fruits and vegetables, and unboiled water in developing countries. The catchphrase is "boil it, cook it, peel it, or forget it."

Local populations may have built up resistance to some pathogens—at least in low doses—but the same pathogens can be especially virulent in people who have not been previously exposed. I lived in Europe in the 1970s and remember coming home to visit my family in the United States and getting mildly sick each and every time. The water probably wasn't any worse; it was just different. And most people don't realize that even acceptably clean water comes prepackaged with a host of resident creatures. My body had accommodated itself to a certain bacterial community. It had to shift gears to adjust to a new population.

Actually, you don't have to travel very far to change your microbial surroundings completely, thus putting yourself at risk. In outbreaks associated with raw milk in the United States it is almost invariably the visitor to the farm—the school group, the grandchild or other relative—who is made ill while the farm family that drinks the milk every day usually remains well. *E. coli* O157:H7 is harbored in the guts of healthy beef and dairy cattle, so you would logically assume that dairy families and workers might often be ill. In fact, this almost never happens. One such case was an infant who was allowed to play on the barn floor while her mother was milking. Not only was her immune system not as well developed, but it might have been her first confrontation with the virulent bacterium, and playing in the dirty straw and then putting her fingers in her mouth would possibly have exposed her to a heavy dose.

A shift in employment can involve both traveling and a new exposure. In June 1980 the staff of a poultry slaughter facility in southern Sweden experienced an explosive outbreak of *Campylobacter*. The school holidays had just begun, and the regular staff had been replaced by inexperienced teenagers just out of school. Seventy-one percent of the cases were in the holiday workers, while only 29 percent were in the regular staff. When both groups, well and ill, were cultured, five asymptomatic carriers were found among the regular staff. These five workers, who had undoubtedly been exposed to the pathogen before, apparently were able to tolerate its presence without getting sick.

The same pattern can be seen in outbreaks from seafood. It is more often the visitors who are made ill from the catch of the day than the local residents. We seem to develop a give-and-take relationship with the microbes in our immediate environment, providing the dose remains low and our immune systems remain healthy. It's an arrangement that serves both humans and pathogens equally well.

The potential for foodborne disease in unfamiliar territory is well known. Less often considered are the serious challenges to our digestive health we may face in the very process of getting where we are going. Cruise ships, airplanes, and trains are all closed environments with passengers whose food choices are, practically speaking, limited to what is offered. When what is offered is contaminated, the result can be devastating. These settings also present unique possibilities for the distribution of foodborne pathogens. Outbreaks can be widespread, difficult to detect, and challenging to investigate.

Cruise Fever

It was August 1994 when the CDC was told about an outbreak of gastrointestinal illness on a cruise ship off the California coast. The numbers were already impressive. Dr. Kim Cook, epidemiologist-in-training, was next in line for an assignment and sent to investigate.

Cruise ships provide unique conditions for disease: an isolated environment where many people are exposed to the same risk factors. And there are plenty of those. CDC doctors have described cruise ships in their manuals as "large floating salad bars," with all the inherent danger that term implies. They are staffed by low-paid employees who may not be granted sick leave. And passengers may be doing things they don't usually do—such as lolling around in a hot tub, a wonderful breeding ground for microbes, or sampling virtually throughout the day a variety of exotic foods someone else has prepared. The cruise ship is a closed system. If something pathogenic is in the water or the air conditioners or the kitchen, it infects what is essentially a trapped population.

The problem of cruise ship outbreaks isn't new. In 1973 the CDC had been alarmed by the increase in such diseases as typhoid fever and shigellosis on luxury voyages. What they found was that most ships reported very little trouble. On the other hand, some reported a lot. Clearly, where foodborne disease was concerned, some were being run better than others. In fact, interviews with the passengers on the troublesome ships revealed much more disease than was ever reported in the ship's log. The investigation showed how important basic health measures are in such closed environments. Those ships that had the most trouble had not passed routine sanitation inspection. Given how ideal the cruise ship setting is, it is surprising not that outbreaks occur but that there are so few—testament to the fact that most ships are clean and well run.

Detecting an outbreak may be difficult if the infection point is near the time of return and all the travelers disembark and go their various ways, but if the contact with the offending pathogen, whether it produces a foodborne disease or Legionnaires' disease, occurs during the voyage, it won't go unnoticed. When a large percentage of a boatload of vacationers become ill at the same time, it's hard to miss. The boats are required to contact the CDC when disease breaks out. The outbreak to which Cook was assigned had taken place on a five-day round-trip from California to Mexico. More than five hundred passengers had reported having diarrhea. Cook's job was to figure out what was causing it.

The CDC, as part of the U.S. Public Health Service, is given responsibility for outbreaks on cruise ships under something called the Vessel Sanitation Program (VSP). The Enteric Diseases Branch has outlined a special procedure, the "EIS Cruise Ship Primer," which young EIS officers can use to guide their investigations on these facilities. It takes them step by step through the process, warning them even that the ship's officers might not be as forthcoming and helpful as one might wish. It's a common problem in every investigation. No commercial establishment likes finding out that it has an outbreak on its hands, and some may go to great lengths to pretend it isn't happening.

Sometimes the investigating team meets the ship as it arrives in port. Then there are only a few hectic hours in which to collect samples and distribute questionnaires. Less commonly, the team meets the ship in mid-cruise and boards it. The approach is the same, but there is more time to do the job—but not much more. "During either investigation," the manual advises, "you will not have much time to eat or sleep. Bring some snacks!" CDC staffers, dealing as they do with the darker side of life, nearly always make a stab at humor, however weak. "Remember," they warn the neophyte investigators, "the airline will take any opportunity to lose your luggage on the way down and your specimens on the way back."[3]

Many of the preparations the team goes through before it sets out are the same, whether the investigation is of *E. coli* in Oregon, cholera in Peru, or shigellosis on a cruise ship. Cook reviewed the manual and got to work. As soon as the investigation was firm, he notified the enterics lab of the outbreak. The lab put together the supplies the team would need to bring back properly prepared samples.

He needed to think at once about the most obvious things—such as arranging for plane tickets and ground transportation, and coordinating travel plans and hotel reservations with the sanitarians, who would check out the ship's water and sewage systems.

Cook would have to find out what people had actually eaten, and memories can be faulty. It might be possible, even before leaving Atlanta, to have the ship fax a menu so that the team could begin preparing a questionnaire for the passengers to help jog their recollection. They might simply reduce the menu and use it as the questionnaire—passengers could then just circle what they had eaten. One thing epidemiologists have to consider if the outbreak peaks in the first three days of a voyage is what the passengers ate and drank before boarding the ship. And, of course, it complicates matters drastically when passengers disembark in ports midvoyage and dine ashore. In one recent outbreak some of the sick were found to have eaten contaminated food on board the ship and others to have been infected by what they ate in the port town where they docked.

The team must take with them a host of humble yet important items that cannot be purchased at sea. In fact, they are taken on most outbreak investigations to prevent having to search for them during the investigation. Cook looked carefully over the list of reminders. A personal computer is helpful but not essential; he would definitely need, however, a calculator, graph paper, big clips and rubber bands, yellow stick-on paper, rubber fingers for counting patient questionnaires, a stapler, and an extra roll of strapping tape for securing coolers; CDC stationery for writing an official report to the captain; copies of old preliminary reports on ships' outbreaks to use as models; articles on foodborne illness, especially aboard cruise ships; special ship outbreak tally paper for epi-curve[4] and symptom checklists; poster or shelf paper, heavy magic markers, and tape to make signs; lab equipment lists and instructions for the collection of specimens ("Read this on the plane—don't wait until you are busy on the ship," the CDC sternly advises); and, of course, the supplies the lab had packed, including one large or two small refrigerator chests for the specimens.

An investigation is a lot like an expedition. Loaded with gear, Cook and his outbreak partner took a station-wagon taxi to the airport. On the plane to California they reviewed the "EIS Cruise Ship Primer" carefully, as instructed. What would be required was a blend of expertise, authority, and diplomacy. Cook had already contacted the area sanitarians, who would meet him at the ship. He would learn from them the known details about the vessel and its sanitation record. Cook and his partner worked out how they would contact each other during the outbreak—they might well be at the far ends of the vessel, and walkie-talkies were a possibility. The sanitarians would check out the water supply, and Cook hoped that, since they would be in closer contact with the crew, the sanitarians would hand out the questionnaire to them (on a different color paper so than those of the passengers, since the crew usually eats different food prepared in a different kitchen). On board he would have to meet with the ship's captain quickly because passengers would be getting ready to disembark. And all the time he would have to assume an air of authority he didn't yet feel. The ship's officers, the CDC primer warned, might want to make it appear that nothing was out of the ordinary and "might consider it in their interest to delay" the distribution of the questionnaires. Cook had to be prepared to deal with this sort of obstruction politely but firmly. In addition, he would need to familiarize himself with the ship, its phone system, its chief officers, and how this city-on-water was run. Once the questionnaires were distributed by the ship's crew members and pickup points were established and passengers sternly advised to fill them out before disembarking—he would have to instruct crew members to begin handing out stool sample cups and instructions, then begin taking blood samples from the ills. There was an enormous task ahead.

The relationship between the CDC and the media is not a simple one. There are times when it is perfectly symbiotic, when the press's need for a good story neatly coincides with the CDC's need to get important information before the public. At the beginning of an outbreak, when nothing is really known and speculation could be counterproductive at best and damaging at worst, the press is something to avoid. In this case, word of the outbreak had already broken. Cruise ship outbreaks, of which there have been many recently, are favorite topics for newspapers. The specter of humbly distressed individuals in a luxury setting has about it that delicious irony that the perverse nature of a journalist finds impossible to resist. Cook battled the gathered press of reporters around the dock. At this point he knew virtually nothing. He threaded his way through them, resolutely silent, and then, in the relative quiet of the ship's dispensary, went to work.

It would prove eventually to be the largest cruise ship outbreak in two decades. Of the 1,503 passengers on board, 535 reported being ill, and 19 crew members were affected. The culprit was *Shigella flexneri* 2a, which produces a frighteningly fierce diarrhea more common in underdeveloped countries. It was isolated from 48 percent of the stool specimens cultured, and it had probably been present in all, but the fragile pathogen survives only 24–36 hours and Cook felt they were lucky to isolate what they did. Follow-up questionnaires sent to passengers after they left the boat revealed that 60 percent who attended the welcoming buffet became ill, while only 29 percent of those who did not attend became ill. The food most strongly associated with the outbreak was the German potato salad, but other items seemed to be implicated as well. The cause was most likely an infected food handler. It is no coincidence that employees on these ships are not particularly well paid and that they are not encouraged to take time off when they are ill. Robert Tauxe of the Enteric Diseases Branch wonders how many outbreaks could be prevented if only the management told workers that they would not be put off at the next port if they reported their illness and avoided contact with food. This CDC recommendation might well have made the difference. Several passengers on this cruise were hospitalized, and one died. We seldom consider that how food service employees are treated on the job can have a direct and dramatic impact on the safety of our food.

In some ways the sanitation situation on cruise ships has improved over recent years. When the VSP program began, none of the cruise ships passed periodic sanitation inspections; since 1978 more than 50 percent of ships have met the standard each year. But while the number of outbreaks has decreased, the proportion due to bacterial pathogens (36 percent) has not. It's worth noting that in eight out of ten documented foodborne outbreaks, seafood cocktail was implicated. And that in a study of forty-five outbreaks, five were associated with drinking water. Still, there are good and bad boats out there. While the CDC found that on only 2 percent of cruise ships was the incidence of reported gastrointestinal illness greater than 5 percent, "the actual incidence of gastrointestinal illness determined by a questionnaire survey of passengers sailing on nine cruises was found to be at least four times as high as that recorded in the medical logs."[5] When they are bad, they are very bad. When you next make a cruise reservation, it's worth checking out the ship's sanitation record, information the CDC is happy to provide by phone, mail, or WWW access.

Fresh-Squeezed *Salmonella*

Less than a year later, on July 27, 1995, Kim Cook was taking his turn on phone duty in the Enteric Diseases Branch at the CDC. Things were slow, and he was pleased when he heard it ring. It was the New Jersey Department of Health on the line. Officials there had spotted a cluster of *Salmonella* cases with an unusual serotype and thought the CDC would want to know. One patient had a duel infection with *Salmonella hartford* as well as another unusual serotype, *Salmonella Gaminara*.[6] The seven patients, none of them related and all of them children, were brought to physicians because of intense diarrhea, cramps, and fever. Most of them also had nausea, headaches, vomiting, and bloody stools. They had one other thing in common. They had just returned from a trip to Orlando, Florida, where they had visited what investigators would later refer to in their official reports as Theme Park A. The CDC makes a courteous attempt to avoid identifying the commercial establishment at which a foodborne outbreak occurs so as not to invoke unnecessary bad publicity, but in this case there was no missing the obvious. The children had been with their families to Disney World.

Cook sat up a bit straighter and felt himself grip the phone a bit more tightly. They had his full attention now. The park had one hundred thousand people on the property each day. If something there had made people sick, it could be an anomaly that no longer posed a danger; on the other hand, there might well be a massive outbreak in the offing. It was imperative to get on it quickly. The only problem was the Fourth of July, coming up fast—right in the middle of the following week. Offices in Florida would be closed; staff at the health departments and the labs would be taking their vacations. Working around the holidays was always difficult. And there was another hitch. To do an investigation, the CDC had to be invited in by the state, which sometimes doesn't happen.

But there were things to be done beforehand in any case. First Cook needed to find out how widespread the outbreak was. There were a couple of places he could check right away. Sure enough, the Orange County Health Unit, which serves Orlando, had noticed the same thing as the New Jersey Health Department—visitors to Disney World who became ill were culturing *Salmonella*. Cook felt that now-familiar surge of excitement, though he also felt guilty about it. Most of these cases were children—suffering children. Some were seriously ill. But their misfortune was also a mystery that might, if solved, take medicine one step further in understanding enteric disease.

Disney is a company that prides itself on its clean, safe, family-oriented image. It works hard to make the image a reality. But foodborne illness has a way of mocking the best of intentions. Despite a clean, well-run food operation, the food entering the safest of environments may already be contaminated. Nor is high cost a guarantee of safe food. The food implicated in the 1996 *Cyclospora* outbreak was a very expensive variety of imported raspberries served at luxury eating establishments. It doesn't have to be a homely potluck supper to make you sick—it can be something you ate at one of America's best-known and best-run commercial operations.

On Friday, June 30, the state epidemiologist from the Florida Department of Health and Rehabilitative Services issued the formal invitation. Cook and Thomas Dobbs, a medical student, waited over the holiday weekend, and on Tuesday, July 4, they began the familiar routine of packing the supplies they would need for the investigation.

The first job was to find out who had been ill. Since most visitors leave the Orlando area after their Disney World visit, that wasn't going to be easy. But once again, the unusual strain would help. Cook and Dobbs had already asked the CDC's Public Health Laboratory Information System (PHLIS) to let them know of reports of *S. hartford*. The lab, which tracks *Salmonella* reports nationwide, went back to states that had reported the serotype in the recent past to see whether any of the patients had visited Orlando. The CDC team had already launched letters by fax and E-mail to all the state and territorial epidemiologists and laboratory directors telling them of the outbreak and asking about cases of *S. hartford* since May 1995. They would also need the cooperation of the theme park.

At first the Disney executives were wary, says Cook. But he explained carefully why it was important to conduct the investigation—the possibility of finding what was causing the problem and correcting it—and how they would go about their work. They would continue investigating with or without the cooperation of the theme park, he told them, but a joint effort would make the process infinitely easier. And they would try to keep the story quiet as they did their work. (That they managed. It was July 13 before the *Orlando Sentinel* picked up the story.) With some uneasiness, the corporation agreed; Disney gave Cook and Dobbs carte blanche.

The investigative team also contacted local labs and health departments as well as the urgent care centers and the hospital adjacent to the theme park. Disney itself was urged to culture all the employees or visitors who came to the medical clinics complaining of diarrhea and vomiting. This nationwide alert eventually yielded sixty-two people from twenty-one states. One of the first things that struck Cook was that none of the people affected was actually from the surrounding area. Orlando residents visited Disney World too. Why weren't they getting sick?

He realized that one thing local residents did not do was stay in the hotels. There was something about the hotels that seemed to be putting people at risk. The epidemiologists then surveyed those who had been ill who agreed to participate and who had stayed in one of the theme park's resort hotels. A good many of the ills mentioned eating what they called "character breakfasts," which Cook and Dobbs described in their report as "special breakfasts unique to Theme Park A in which costumed cartoon characters are present."[7] After this preliminary questioning the team felt confident enough to create a case definition, which always sounds elementary but is critically important. If it is wrong, it can spoil the entire investigation. It would be: "Vomiting or diarrhea in a Theme Park A hotel visitor from May 1 to August 4, 1994, with *Salmonella* serogroup C or *S. hartford* infection." Then they set up a control group of wells, people who could be matched as to the hotel they stayed in, their check-in dates, the time they spent in the park, and their age group. Since the visitors had returned home to states across the country, the interviews were conducted over the phone. Both groups were asked by trained interviewers about what attractions they visited, what they had eaten and drunk, and whether they had been exposed to animals—other than Mickey Mouse. Although the team was beginning to suspect the character breakfasts, it was too early to eliminate anything. When the information they gathered was put into their special, epidemic computer program, an association emerged. Character breakfasts were eaten by 91 percent of the ills, as compared with 58 percent of the wells. And within the character breakfast group, 97 percent of the ills had drunk orange juice as compared to only 54 percent of the wells.

"When someone casually mentioned that the orange juice was not pasteurized, it was like a light bulb going off in my head," says Cook.

The labs at Disney were first to culture *Salmonella* from the orange juice, and when the CDC tested the samples, it found *E. coli* bacteria as well as *Salmonella Gaminara* in 83 percent of the juice containers they tested and from all of the four lots. The University of Florida Citrus Research and Education Center did its own testing and reported finding *Salmonella* in an unopened container. It was a different serotype, but indicative of contamination. In fact, it was *S. Rubislaw*, and in August this serotype would be found in a one-year-old North Carolina girl who had consumed orange juice produced by the company when her family was on vacation in South Carolina. The orange juice Disney used had been supplied by a single company, and company records confirmed that it had shipped fresh, unpasteurized orange juice to South Carolina during that period, as well.

Orange juice is the American elixir. Considered a prime source of vitamin C, it is identified with health and well-being. Americans drank 743,308 million gallons of it in 1995, down slightly from its peak in 1993. Much of it is served in the morning. At hotels it has become, like the thermal carafe of coffee, an expected part of breakfast service. Generally it is pasteurized, but more and more often now it is fresh. In fact, consumers have been buying less frozen orange juice. But "fresh" orange juice is not fresh in the way it is if you squeeze it at home—halving individual oranges and pressing the halves into a squeezer. The great bulk of "fresh" orange juice is mass-processed at "squeeze" facilities that press thousands of oranges at a time, then bottle the juice. Since it has not been pasteurized, the juice is perishable and usually contains a sell-by date. It is also very sweet, as a result of growers having selected over the years for the sweeter varieties to please mass-market tastes. And it is more expensive. Disney, looking for quality, had been supplying this fresh-squeezed product in its character breakfasts.

The oranges that poisoned the travelers had been travelers themselves—passing through an assortment of potential contamination points.

The investigators discovered that the theme park received shipments of juice about every other day and sold all the juice received in two or three days. It arrived in single-serving containers or large containers that were poured into pitchers or glasses, or even poured into self-serve juice dispensers. There seemed nothing amiss in Disney's handling.

The oranges that produced the juice were hand-picked in Florida citrus groves and shipped to the juice company usually within twenty-four hours. They were typically juiced within forty-eight hours. Packaged and shipped, the juice, which should be refrigerated, has a twelve- to seventeen-day shelf life. The majority of juice produced by the company was sold to Disney World. Inspecting the company, the investigators found that all processing equipment was sanitized twice daily, but was it done properly? They couldn't be certain. They also discovered that the oranges had passed through only one cleaning and sanitizing step. And the Florida state health department officials noticed small openings in the building through which rainwater, insects, and rodents could enter. It was possible that contamination had reached the oranges through that route.

But when the investigators saw how the oranges were harvested, the possible routes of contamination became clearer. They were hand-picked, but from where? Often, they discovered, from off the ground. Juice oranges were shipped, without washing, to the processor. Orange growers are using chicken manure as a fertilizer more frequently. Although neither grower who had sold oranges to the processor during the period when the juice had become contaminated had used chicken manure on his fields, it was used on fields nearby. Moreover, while all the oranges sampled from one grower tested negative for *Salmonella*, the bacteria was found when swabs were obtained from the soil around the orange trees.[8]

Studies have shown that *Salmonella* can survive up to twenty-seven days in orange juice with a pH of 3.0–3.1 held at 5 degrees Celsius. When samples of the juice company's orange juice were tested by the Florida Department of Agriculture, they were found to be sweeter than that. The pH was 4.0.

The fact that one serotype of *Salmonella* was found in both May and June from the company's juice suggested to Kim Cook and his colleagues that the contamination was ongoing, perhaps as a result of inadequate sanitation of the processing equipment. But the fact that "substantial numbers" of *E. coli* and different *Salmonella* serotypes were found in the orange juice "also suggests there was periodic introduction of pathogens into the plant, possibly from the exterior of the oranges." Critically important to producing clean juice, the team said, was the cleaning of the oranges before squeezing. Responding to the problem, the company introduced a new sanitation method. They began spraying the exterior of the oranges with hypochlorite solution, but without pasteurization, Cook's report said, juice from "Company A may continue to be contaminated and should not be consumed. Production of juice should be suspended. Juice should be recalled and tested for bacterial pathogens."[9]

Only sixty-three individuals, mostly children, were culture-confirmed as part of what Cook would call the "fresh-squeezed *Salmonella* outbreak." None of them died, although the aftereffects of a bout of *Salmonella* can be serious. So why was the CDC interested enough to conduct an investigation? Studies done during previous *Salmonella* outbreaks had revealed that only between 1 and 10 percent of actual cases are reported. "We estimate," Cook wrote in his final report, "that between 630 and 6,300 cases of salmonellosis resulted from this outbreak."[10] Having discovered what had caused the outbreak presented the means to ensure that it didn't happen again—at least not from that route. Disney World decided that it preferred to serve pasteurized juice after all.

"It is ironic," says Cook. "People were drinking something they thought was a gourmet product, naturally produced. But if it's not pasteurized, you have to be real careful."[11]

The outbreak made a number of points about modern foodborne disease. Many travelers were made ill by a product distributed broadly in a relatively closed environment, an outbreak that was discovered only because of an unusual serotype of the bacteria—leading one to suspect that many similar outbreaks from more common serotypes are missed. The vehicle—fresh, unpasteurized juice—had been mass-processed; whatever flaws had allowed it to become contaminated affected thousands. And the product itself was different. Unpasteurized fresh bottled orange juice was a relatively new offering that now seemed inherently dangerous, both because the oranges had been exposed to pathogens either on the ground or during processing and because the oranges had been bred to be sweeter, making the juice more susceptible to bacterial growth. And finally, contaminated chicken manure was being spread on nearby fields and could have been the original source of contamination.[12]

Poisons in Flight

One thing about food poisoning remains the same. It is only when a large number of people become ill that the source is usually identified—and even then it may take a special set of circumstances. In 1988 a newspaper in Minnesota reported that many members of the Minnesota Vikings football team were suffering from diarrhea. This alerted the state's epidemiologists. They were quickly able to connect the illness to sandwiches served on an airplane flight the team had taken a day or so before. The sandwiches had probably been contaminated with *Shigella sonnei* by one of the food handlers in the airline's flight kitchen—then left unrefrigerated for perhaps as much as two days. The epidemiologists investigating the case would later estimate that nineteen hundred cases of shigellosis in twenty-four states and four countries occurred among airline passengers during that outbreak.

It almost wasn't detected. Passengers quickly disperse, and they wouldn't be likely to associate diarrhea several days later with the flight. They are not likely to know that others have become sick, and they may blame their illness on travel itself. Even if the illnesses of other passengers had been correctly identified and reported, no one would have been likely to make the connection if it had not been for the football team traveling together. Or if it had occurred in another state. The attentiveness and resources of the public health sector make all the difference in identifying outbreaks. Minnesota's public health team is one of the very best in the country.

This outbreak is an excellent illustration of the changing nature of foodborne illness. More than one hundred thousand meals a week were prepared by the airline's kitchens. The plant operation appeared to be modern. It operated under a system called hazard analysis and critical control points (HACCP), which calls for potential hazards in the assembly-line preparation process to be identified and critical control points established and monitored. Despite the system, the plant had been cited for unsatisfactory sanitation, a vivid illustration that HACCP plans are no guarantee of food safety. What is critical is how well they are designed, how well they are implemented, and how well they are monitored. Among the deficiencies the investigation uncovered were lapses in all-important food-handling practices and hand-washing—simple things that had far-reaching consequences.

Even in high-volume commercial operations, with all the modern technology available, hand-washing still remains one of the most important aspects of food safety. Nothing can replace it. The rapid turnover in food handlers, a traditionally low-paying job, also played a role in the outbreak. At this facility the average length of employment was ten months. Only after employees had worked ninety days were they given sick leave. During the investigation several food handlers admitted that they had worked while suffering from diarrhea in order not to lose a day's pay.

But what was really classically new about this outbreak was the vast system of distribution that enabled the contaminated food to be eaten by thousands of airline passengers and flight personnel. Systems like this, in which foods are processed and distributed over a vast complex network, are not limited to transportation systems.

The outbreak that takes place on an airline is unusual, but it happens—perhaps more often than we realize because of the difficulty of detection. Robert Tauxe was an EIS officer in 1984 when he investigated his first airline-associated outbreak of foodborne disease. Eight people had become ill after a flight from London to the United States on British Airways. The CDC sensed that the outbreak was likely to be much more widespread, and it undertook a study of the outbreak to learn more about the potential for airline travel to disseminate foodborne disease.

Serving meals to air travelers was first introduced on London-to-Paris flights in 1919. Since then, airline food has become a global mass-catering industry.[13] As such, it is subject to the same lapses in sanitation that any other catering service makes. Because meals are prepared en masse for one carrier, other flights of the same carrier may be susceptible to whatever has made passengers sick, and thus any outbreak demands a quick response to prevent worldwide cases. Investigating and controlling the outbreak calls for international cooperation.

The day after the CDC learned of the cases on the flight from London, more cases of gastroenteritis were reported by the Communicable Disease Surveillance Center of the Public Health Laboratory Service in London from local public health authorities. Tauxe wanted to find other patients to get a broader picture of the extent of the outbreak, and he was sure they were out there. Finding them required the help of the airline in notifying travelers who'd flown on the carrier in March. They were able to reach many and gave them an emergency number to call at the CDC if they had been ill.

Tauxe was right. The eight original patients had been the tiny tip of an iceberg. The outbreak was widespread and serious. Some passengers had begun to feel ill on the plane; for others the illness had developed more slowly. There were 186 culture-confirmed cases, 37 people were hospitalized, and one died after emergency surgery for suspected bowel obstruction. Nine of the ill were crew members. But when the CDC investigated more thoroughly, sending out questionnaires to confirmed passengers, they found that many more still had actually been ill. The outbreak had probably affected as many as 2,747 people in the United States alone.

The story didn't end there. The carrier had flights going virtually around the globe. During the period March 12–17, the time of the outbreak, British Airways had made 125 overseas flights to non-European destinations from London, carrying a total of 23,576 passengers. Passengers on these flights became ill as well. By applying the 81-percent attack rate that emerged as a viable statistic from closely studying just one flight, it would seem that worldwide well over 19,000 people may have been ill. Just counting those in the United States made it one of the largest single-source outbreaks the CDC had ever investigated.[14]

So what had happened? Because overseas flights used one terminal and domestic and European flights used another, two different kitchens were involved—which explained why no one was sick on the domestic and European flights. The British public health and airline authorities investigated the overseas flight kitchen and managed to isolate *Salmonella enteritidis* from seafood cocktail. Eventually they decided, however, that the culprit was the aspic used on several of the hors d'oeuvres.

In light of the seriousness of this outbreak, the CDC took a look at airline outbreaks of the past. In the twenty-three outbreaks since 1947 it found that hors d'oeuvres, cold salads, and custard-based desserts were the foods most often responsible. The agency also found mistakes in food handling. Sometimes the problem was an infected food handler; once it was crab that had not been sufficiently cooked. Always the outbreak could have been prevented by "cooking food completely, preventing cross-contamination, practicing adequate personal hygiene, and most importantly, maintaining proper temperatures during preparation and holding."[15]

What the CDC found were many problems that could easily contribute to foodborne disease. Food was being held at improper temperatures on nearly one-quarter of all flights. That figure wasn't surprising. A previous study had revealed that 30 percent of airport catering services didn't have adequate refrigeration facilities. Perhaps things have improved, but on a recent flight my food arrived barely warm and had to be sent back twice before it was hot enough to eat safely.

While passengers may be uncomfortable or even seriously ill, the plane's safety is not affected unless the crew becomes ill. In twelve of these outbreaks they did. Many airlines have anticipated the impact of what their crews eat on flight safety. Their policies, however, are not consistent. Some airlines have the crew eat first-class meals—which seem no more likely to be safe—and others have the crew eat different meals prepared in the same kitchen. But problems can arise whenever all the members of the crew eat the same thing. The results of in-flight food poisoning of the entire crew, needless to say, could be devastating. It has happened. On one flight from Lisbon to Boston the crew of ten had eaten the same food, and all became ill. Fortunately, they were still able to operate the aircraft and the plane landed safely, but the incident alerted the airline to a frightening possibility.[16] The advantages of serving the pilot and copilot different foods prepared in different kitchens are obvious. When it is not possible to do so, one of them could be served a frozen meal. In fact, since a preflight meal has just as much potential to cause disease, Tauxe suggests that it would be prudent for the pilot and copilot to avoid eating together before flying.

Travel has always involved disease challenges. Passengers on trains in India lean out of the windows at stops and willingly take their chances with food prepared by small, independent caterers, but travelers on jumbo jets don't anticipate having to take such risks. Food is generally much safer en route than it was in the past, but the difficulties of preparing, holding, and serving food to great numbers of people who have come to expect haute cuisine or home cooking at thirty thousand feet or on the high seas should not be underestimated. And the mass-catering used in modern transportation ensures that when something goes wrong, it can go very wrong. The worldwide increase in travel means that these mistakes, when they happen, will play an important role in the further distribution of human disease.

Part III/Emerging Diseases and the Anatomy of Outbreak

7/Super *Salmonella* and the Bad Egg

> Most and probably all of the distinctive infectious diseases of civilization have been transferred to human populations from animal herds.
>
> WILLIAM H. MCNEILL,
> *Plagues and Peoples*, 1976

If there were a contest for most successful all-around emerging foodborne pathogen, *Salmonella enteritidis* (SE) would clearly be a contender. Thirty years ago it was barely a player, and yet the most recent CDC report found that between 1988 and 1992 it had become the pathogen most frequently reported as the cause of foodborne disease outbreaks.[1] The history of *Salmonella enteritidis* reveals how a microbe can slip into the food supply, how its growth and spread can be favored by any number of human-initiated changes on the long and circuitous route from the farm to the table, and how politics, government regulators, industry, public health, science, and the media all play a role in whether an emerging foodborne pathogen is stopped in its tracks or allowed to go on to become pervasive in an industry, to dominate a microbial environment, and to spoil one of our most essential food ingredients.

Of all the foods available to mankind, surely no other is as marvelous, as versatile, or as nutritious as the egg. It is indispensable to cooking as we know it. Gently beaten over warm water, an egg can raise a cake on its own, or it can be a structural framework for a leavening agent; it can smooth and thicken a sauce or tenderize a timbale. It can bind a gravy or give a foundation to mayonnaise. It is the essential ingredient in a soufflé or a proper mousse. A meringue without it would not be a meringue. It can clarify or enrich soups, provide a glaze for baked products, or produce a fine-grained ice cream. It can be fried, poached, scrambled, coddled, boiled, or made into an omelet. Hard-boiled, it can be transported as a traveler's food; pickled or preserved, it can last for a very long time indeed as a safe form of nutrition. And until recently it was regularly eaten raw in any number of products; the white of the egg was thought to have the nature of a bactericide—capable of protecting the growing animal in the yolk from invasions of microbes dangerous to the chicken. My grandmother began her day with a milk shake—a raw egg beaten in milk. My grandfather ended his day with a freshly made eggnog prescribed by his doctor—a raw egg in milk with sugar and bourbon whiskey.

If the many virtues of the egg were not enough to enthrone it as undisputed king of the realm of culinary arts, challenged only by the versatility and nutritional benefits of milk, it is also the ultimate convenience food. The egg comes to us in a handsome package that protects its contents, delivering them, if it is treated with thoughtfulness and care, in perfect condition. And the package itself is easily disposable and perfectly recyclable. It can enrich the compost and the soil it is worked into.

And yet, what has happened to this precious jewel of a human food is nothing less than a tragedy and an outrage. Its changed taste is one thing. The flavor of a truly fresh egg from a free-range chicken can startle us with its wholesome intensity. Garbage in, garbage out, as computer mavens say, and the standard supermarket egg today reflects what the hen eats—probably a mixture of ingredients, some of which are definitely unsavory. But in the past thirty years we have also come close to destroying the egg's safety; many traditional egg and egg-based dishes can now make one very sick. Some scientists suspect that the egg, being an animal food, may never have been completely safe, but if that is true, there were few signs, missing in the scientific literature before 1965 are outbreak reports from clean, uncracked fresh eggs. What has certainly happened is that its potential to be contaminated has greatly increased, as have the chances of that contamination reaching the consumer. Because *Salmonella enteritidis* managed at some point to get into the ovary of the chicken to produce eggs prepackaged with pathogens, and because changes in how we produce eggs have actually encouraged this microbe, of the hundreds of uses of the egg, only those in which it is thoroughly cooked are now considered completely safe.

For hundreds of years the dyed and decorated egg has been the symbol of life, fertility, and renewal. Today, in the kind of reversal that is becoming all too typical of our age, the egg can be deadly.

The Front Lines

One group that has seen the rise of *Salmonella enteritidis* from also-ran status to star player in the food pathogen challenges are the epidemiologists in the public health arena. They are on the front lines of a battle they know they cannot win. They can stop an outbreak if they are alert, quick, and lucky; they can identify a problem and point to its cause; but their most important role is simply to cope with the human dimensions of a problem whose origin is beyond their reach. The root of foodborne disease lies tangled in the complex world of food production. And in that world the assumption that an efficient, industrial, mechanized model could be applied to farming without negative consequences has gone virtually unchallenged since the agricultural revolution began.

Our chief guardians against foodborne diseases are the regulatory agencies. And yet they too are ineffectual in coping with the sources of microbial contamination because regulation has failed to keep pace with the changing nature of foodborne disease and a dynamic agricultural industry. Mired in the irreconcilable conflict of its role as both promoter of agricultural products and guardian of meat safety, the USDA has poked and prodded and eyeballed animal foods at a point in the process far removed from the chief source of the problem: the disease-causing microbes that cannot be seen or smelled and that originate on the farm. Even now, after the western states outbreak of *E. coli* O157:H7 in 1993 alerted the nation to the growing problem of foodborne illness, disease-causing microbes are not officially considered adulterants or contaminants on animal products by the

USDA. And even today the USDA has no authority to regulate on the farm except in the case of an animal disease that is communicable to humans, such as tuberculosis or brucellosis. The FDA is reqired to stay off the farm as well. The regulatory agencies cannot monitor or test or even observe without permission. They can only watch from afar and hand out advice, which may or may not be taken.

The USDA and the FDA have always assumed that healthy animals produce healthy foods and that there are certainly animal diseases that can pass directly to humans. But increasingly animals are harboring and passing on microbes that seem to cause them little trouble yet nevertheless are harmful to humans. One of these is *Salmonella enteritidis*. Since the USDA has been staffed with veterinarians instead of physicians (the department has just recently hired its first medical doctor), the full impact of these pathogens on consumers has only been revealed as epidemiologists have dissected the nature of human disease, picking, prodding, and probing the outbreak evidence.

The Disease Detectives

On the expansive fourth floor of a nondescript brick building in East Minneapolis, behind a standard room divider, a young woman hunches over a telephone. Her tone is at once sympathetic, soothing, and yet authoritative as she coaxes the information she wants out of the person on the other end of the line.

"Did you eat hamburger," she asks. Pause. "Did you cook it at home?" Pause. "Where did you buy it?" She pauses between each question, waiting for the answer, marking the form in front of her. "Did you eat chicken?" Slowly Julie Hogan is building a picture of what her interviewee ate in the week before she became ill and tested positive for *Salmonella enteritidis*. Hogan is part of a group of epidemiologists and part-time public health graduate students who call themselves, with a big grin, "Team Diarrhea." Part of the crack Minnesota Department of Health, they are looking for the causes of foodborne disease in their state. They have been given the resources to do it, and they are one of the best in the country.

The Minnesota epidemiologists track more than enteric diseases, of course. Toxic shock syndrome, Legionnaires' disease, tuberculosis, sexually transmitted diseases—they are all public health concerns. But foodborne diseases keep them busy. They work closely with their lab, which runs a battery of tests on every isolate they receive, an identification that can be vital. In the normal run of sporadic cases of disease in a community, a rare serotype found in several individuals might stand out, signaling an outbreak from a single source. Or the cases from a familiar serotype might suddenly, inexplicably surge. And if the epidemiologists ask the right questions, they might find out why.

Julie Hogan's questions go on. "Did you eat eggs? How did you prepare them?" Pause. "Did you do any baking that would have included raw eggs?" Pause. "Did you eat ice cream?" The woman tries to remember. Hogan patiently draws out what she might have done to help her focus her memories—a doctor's appointment, a birthday, a trip.

When she has finished the form, the information will go into the computer, where it will be tabulated. Details of what the ills ate, and where, will be compared with the diets of the controls. Then the investigators might get lucky; the implicated vehicle might pop out. It has happened many times.

Michael T. Osterholm is Minnesota's nationally known state epidemiologist. "Epidemiology is a team sport," he likes to say. That is certainly true. He, Craig Hedberg, and Kristine MacDonald make up the top tier of a team that now includes forty people. With his outgoing personality and bottomless supply of authoritative sound bites, he has become one of the sources reporters turn to when foodborne disease raises its head. In the back of his crowded office on the second floor there are old quarantine signs for diphtheria, poliomyelitis, scarlet fever, smallpox—an indication of how public health has shifted from merely contending with disease to attempting to discover the causes and sources.

If there is one thing that stands out about the members of this department it is their broad perspective on disease. The citizens of Minnesota are their primary responsibility, but as Osterholm says, "We have a certain philosophy that considers the 'global village' issue. We maintain contacts with the rest of the world."[2] With disease moving as rapidly as it does today, with new pathogens emerging on a regular basis, showing up wherever travel or trade has carried them, a health department that does not consider the broader implications of disease is at a real disadvantage. The department's investigation of the outbreak of shigellosis in the Minnesota Vikings football team brought that point home.

Vital to the department and its work is its "marriage" with the laboratory located only two floors above. It is there, amid a rather old-fashioned clutter of cotton-topped test tubes and beakers (the high-tech equipment for analyzing the genetic structure of the microbes is in another room) that the pathogens in the community are monitored. Like body heat in a human, a certain level of foodborne disease in the community is, unfortunately, considered normal, but too much above that level can signal trouble.

In May 1989 the lab noticed a real increase in the number of isolates they identified as *Salmonella javiana*. The department contacted the CDC, which reviewed its national data on isolates of *S. javiana* from humans. At the same time the Minnesota investigators set up a case-control study and went to work. One hundred and thirty-six culture-confirmed cases of *S. javiana* infection were eventually identified in Minnesota. Three other states, Wisconsin, Michigan, and New York, reported cases as well. Sometimes it just takes a few interviews to hit on a likely vehicle. Cheese was implicated.

It would be helpful if the world of foodborne pathogens were neat and tidy, if one could count on being able to associate one particular pathogen with one food or animal and to anticipate how it will behave. Once we made assumptions based on that model. Only gradually, in our lifetimes, have microbiologists learned how quickly microbes can adapt, discovered the ease with which they can pick up resistance to antibiotics and even pass that resistance on to other microbes, and measured how thoroughly they can be distributed within an environment.

For a long time we assumed, for instance, that there was a fence, if not an actual wall, between animal and human disease. We called it the species barrier. Now we see infectious agents leap nimbly over our illusions. And so the growing importance of *Salmonella* neither begins nor ends with the egg. If the bacterium has a beginning, it is in the animal world, where, like many microbes, whether disease-causing or not, it can find a hospitable host and settle in for as long as is convenient. It is when *Salmonella* moves on that trouble begins. Traveling via animal waste—or in the case of the egg now, within the food itself—it hitches a ride in water, on human hands, or on any object those hands touch. From there it can travel via equipment, food handlers, and kitchen counters, from one food to another, until humans are infected. The serotypes of *Salmonella*, each of which is named, are uniquely adapted to particular niches. *Salmonella* is especially well adapted to poultry, and now to the egg, but it is also found on beef, pork, and lamb, and because water and the hands of harvesters may be contaminated with waste, it finds its way onto vegetables and fruit as well. At every stage, from the host to the infection, human actions and inactions, influenced by custom and tradition, negligence, politics, economics, or a hundred other factors can impact the journey.

By following the serotypes, the particular changes that favor the growth of one or another can be spotted. And so to tell the story of *Salmonella* and eggs, one must put the bacterium in context.

Outbreaks of salmonellosis from cheese are rarely reported. When the Minnesota outbreak occurred in May 1989, there had been only four since 1976. One had been linked to a soft cheese made in Switzerland from raw milk, another was from mozzarella that was cross-contaminated by raw chicken in a restaurant kitchen. And two were related to faulty pasteurization in a cheddar cheese process, one in Colorado and another in eastern Canada. When cheese was implicated in this outbreak, the Minnesota and Wisconsin Departments of Agriculture, Trade, and Consumer Protection, and the Food and Drug Administration, all entered the picture.

The government agencies that regulate food have a complex relationship. The tasks of maintaining food safety are divided between them in ways that have evolved over the years. Some of those ways seemed logical at the time, some of them were politically motivated. There are obvious overlaps and competing agendas in this confusing arrangement. And there are obvious gaps: the sacrosanct nature of the farm, for instance, and the private realm of commercial enterprises.

The problem has attracted the attention of those who watch the regulators. In 1990 a Government Accounting Office (GAO) report on federal efforts to ensure food safety noted the general agreement among the agencies that microbial contamination would become critical in the near future but concluded that the "fragmented, complex federal food safety and quality regulatory system" (consisting of thirty-five different laws and twelve agencies) was hampering efforts to do anything about the problem.[3] The report had little impact. When the GAO returned to what was clearly a growing problem with foodborne disease in the spring of 1996, it found that almost nothing had changed.[4] There have been bold and sensible suggestions to combine all food safety under one roof, but that decision would be fraught with dire consequences for those who have built small empires within the agencies, for the industries and agricultural interests that have lobbied to gain the maximum advantage, and for the politicians who have strong relationships with those interest groups. The industry fears that an all-powerful agency with a narrow scope and a single agenda could harm U.S. agriculture. The suggestions have predictably hit a brick wall. The reality of unsafe food is only just filtering into the public and political consciousness. And so the conflicts and competing agendas within the regulatory agencies, the duplication of effort, and the general confusion as to who is responsible for what, make mustering a coordinated defense against microbial invaders a real challenge.

The roles of the state agencies make mounting a quick response to foodborne disease even more difficult. Cheese is an agricultural product, but it is regulated by the Food and Drug Administration, whose job it is to ensure the safety of all foods except meat, poultry, and egg products. But state agencies, protective of their food producers and their reputations, want to get into the act as well. When a product crosses state lines, the authority is even more confused. In the Minnesota cheese outbreak, every agency involved tested the cheese samples collected from the homes of the ill, and from grocery stores, a wholesale warehouse, and the Minnesota processing plant. The Wisconsin Department of Agriculture tested samples collected from a plant in that state. Most of the tests were negative, despite the implications of the epidemiological study in the outbreak. The regulators shook their heads. They wanted more; they wanted proof of the cheese link before they would ask for a recall.

What Is Proof?

Many health officials around the world consider it essential to find the offending pathogen in both the patients and the food vehicle before taking action. And yet epidemiology can make the link without that evidence. One of the goals of epidemiologists is to demonstrate that interview techniques work—that the microbial proof, while helpful, is not essential. Despite the wide acceptance of this theory in the U.S. public health field, many industry and agricultural officials still insist on microbial confirmation before the agencies ask for a (voluntary) recall of a contaminated food—a move that can be financially devastating for a small company. This time the agencies got the evidence they needed. The outbreak strain was finally found in an unopened block of mozzarella when the product was examined and tested by the Health Protection Branch of Health and Welfare Canada in Ottawa.

The outbreak revealed the classic conditions that predispose for contamination—and for an ongoing epidemic. What the investigation told epidemiologists was that sick workers (whom they identified) probably contaminated cheese at a plant that was pressured by bankruptcy into increasing its production and lowering plant sanitation to cut costs. The cheese was sent to four processing plants for shredding, and there it cross-contaminated other cheese products. Even so, the federal agencies asked for a recall of only the single day's production in which *Salmonella* organisms had actually been found. Instead of halting the outbreak, illnesses continued through September 1989 as other batches of contaminated cheese remained in refrigerators and on store shelves and only gradually were used up. If the FDA and state regulators had accepted the epidemiological evidence, the outbreak could most likely have been brought to a halt months earlier, saving the ongoing human and economic costs of episode.

The Minnesota team began looking closely at *Salmonella*. More than forty thousand infections with the pathogen are reported to the CDC each year, yet that number is a fraction of the actual number of illnesses. While the outbreaks get attention, most of the cases seemed to be sporadic. The Minnesota team suspected they were actually part of unidentified outbreaks that were geographically spread out because they were caused by low levels of contamination in mass-distributed foods.

If there is one thing that has changed dramatically about food, it is this: Production has become centralized in giant multinational corporations, and their products, instead of being distributed locally, are now worldwide travelers. The classic model of a foodborne disease outbreak, the bride's luncheon or church supper at which the cause is immediately obvious, has been replaced by a new model—seemingly unrelated sporadic cases that in fact are caused by contamination in one widely distributed commercial product. Only something odd—an unusual cluster, serotype, or food vehicle—would attract enough attention to call for an investigation. Otherwise, most sporadic *Salmonella* cases from common serotypes will never be explained. The food vehicle will never be incriminated. Foodborne disease will become simply background noise.

Cheese and other similar food items, the Minnesota epidemiologists thought, were likely to be responsible for more apparently sporadic *Salmonella* infections than anyone realized. They urged the federal and state agencies to do whatever was necessary to address the problems the cheese-borne outbreaks of salmonellosis had identified, but like so many urgings, little heed was given to this one.[5]

The Big Picture

It is one thing to be able to stay on top of disease outbreaks; it is quite another talent to be able to spot a trend and figure out what it means. The Minnesota epidemiologists were seeing a lot of foodborne outbreaks in the state. It was unlikely that there was something unusual about Minnesota; rather, they were simply paying more attention. To Hedberg, Osterholm, and MacDonald there seemed to be a pattern to what was happening. Pulling their outbreak investigations together, they revealed how changes were affecting foodborne infections in Minnesota in an article that appeared in the *Journal of Clinical Infectious Diseases*.[6] One of their investigations focused on *Salmonella enteritidis* and eggs.

No single study in medicine or science is definitive in itself, yet each can reveal a vital piece of the puzzle. This particular serotype of *Salmonella* had emerged as a growing and serious problem in 1988 when the CDC reported a nearly sevenfold increase in cases between 1976 and 1986. The *SE* outbreaks were associated with the consumption of eggs and foods containing eggs, especially in the northeastern

United States. Most investigations were of large outbreaks, many from restaurant settings where the practice of "pooling" eggs—cracking hundreds together to use in omelets and other dishes—was clearly a problem. One contaminated egg could infect the batch and turn what might have been an isolated illness into an outbreak.[7]

Not only was the epidemic rise in *SE* infections connected to grade A shell eggs, the CDC investigators said, but unlike past outbreaks of salmonellosis, which had been related to cracked or dirty eggs, the new cases raised the possibility that *Salmonella enteritidis* was probably now in the ovary of the chicken and being transmitted to the egg by that route.

The CDC report prompted Minnesota to look at its own cases of *SE* infection. When they compared two three-year periods, 1980–1983 and 1987–1990, they discovered that the cases had doubled. At first it didn't seem that the increase was related to eggs, but looking more closely at the sporadic cases, they found that individuals infected with *S. enteritidis* and *S. typhimurium*, another increasingly common serotype, were more likely to have eaten foods containing eggs or undercooked eggs in the three days before they became sick. Often the eggs were pan-fried, the study found. Alarmingly, the small amount of bacteria in the egg at the very moment it was cracked was apparently enough to cause infection. And it was perfectly possible that the mass outbreaks, which had been traced back to contamination from sick individuals, had in fact begun with a fried egg.

The Minnesota investigations were beginning to come together, each one teaching the team—and the broader medical community as they were published—something new about the nature of foodborne disease. Eggs were now clearly a vehicle for both *S. typyhimurium* and *S. enteritidis* when used individually in home kitchens. They didn't need to be pooled or kept out at room temperature. People were getting ill from simple foods they had eaten their entire lives. The implications were obvious and frightening: *Salmonella enteritidis* was on its way to becoming a more important pathogen than anyone had really understood. A natural question was, How and why had this happened? From reports of *Salmonella* conferences held by government agencies in 1964 and 1978 it is possible to unravel something of what they knew—or should have known—about an emerging pathogen.

The end of the egg over easy is an ecological tragedy, a depressing story of the human inability to take concerted, forceful action even when directly faced with the disastrous consequences of maintaining the status quo.

The Murky Past of an Emerging Pathogen

The regulatory agencies had actually been worrying about eggs and *Salmonella* (from serotypes other than *enteritidis*) for some time. The year was 1964. In March the CDC gathered a group in Atlanta to talk about the problem of salmonellosis, for which there had been a twentyfold increase since 1946. *Salmonella* is one of the most important causes of infectious diseases. Typhoid fever in humans, long a serious disease, is produced by *Salmonella typhi* and *S. paratyphi*, and that, at least, was declining. In 1946 there had been 3,268 reported cases. In 1963 there were only 525. But while there had been only 723 cases of disease reported from other *Salmonella* serotypes in 1946, in 1963 there were 18,696—more than 5.7 times as many cases as there had ever been of typhoid fever at its worst.

Salmonella are hardy and resilient bacteria, adaptable, resourceful, and persistent; they are the cockroaches of the microbial world, seemingly designed to remind the human race forcefully of the dangers of overconfidence. As the USDA pondered the increase, the assistant surgeon general, Dr. James L. Goddard, observed to the 225 attendees from several countries how paradoxical it was "that the incidence of salmonellosis in the United States continues to rise in spite of the continued long-term interest in the problem." He asked, Is it because we do not understand the ecology of the disease? Or is it because there are no effective measures of control? Or is it that the industry isn't using them? Many of those attending the conference, even in 1964, knew the answers to these questions.

The chief concern of the gathering was chickens—those that were sold for food and those that produced eggs. One of the first indications that eggs might be a source of *Salmonella* infection in humans had come in November 1961 when commercial cake mixes in Canada contaminated with *Salmonella thompson* were found to have produced gastroenteritis. Since cakes are cooked, the infections obviously came from licking raw batter from the spoon, a time-honored tradition. It was not surprising, then, that so many children were infected. At least one died. Then an outbreak in Michigan was found to be associated with commercial eggs. Canada too, had noticed a "marked increase" in reported human *Salmonella* cases other than typhoid fever since 1955. One outbreak was from meringue powder used to make a topping for custard. Their tests of cake mixes and other egg products also revealed a significant amount of *Salmonella* in processed foods. Poultry, they said, appeared to be the principal reservoir. While *S. enteritidis*, as an individual serotype, was not yet a major problem, the Canadians noted, almost offhandedly, that they were finding it with compelling regularity. In fact, it had jumped from seven isolations in 1958 to twenty-three in 1959 and then to fifty-three in 1961. In the United States cases of *S. enteritidis* infection were increasing quietly as well, but hardly anyone had noticed, and certainly no one had asked why.

The Canadians had observed something that seemed of particular interest. They told the conference that when diseased humans were cultured, the *Salmonella* types most commonly found were the same types that researchers were finding in farm animals. This news jolted the fondly held assumption that man is quite a distinct and separate creature, at some remove from the animal world. "Is this purely coincidence?" they asked. Surely it was evidence of "a direct relationship of the disease in animals to that in man."[8] The importance of the observation would be ignored.

Even in 1964 more and more foodborne outbreaks were occurring among people widely separated geographically who had eaten the same processed foods, and that made linking the cases difficult—if not impossible. "This is the price we pay for modern conveniences, the mass production and widespread distribution of common food items, and the refinements in transportation," Dr. Charles McCall of the Epidemic Intelligence Service (EIS) observed.[9]

The conference heard the facts of a large outbreak of salmonellosis in Washington State in 1962 that made the point. The eating histories from the patients pointed to meringue pie. One of those taken ill had eaten the pie at a church luncheon. One hundred and eighty people from all over the state had been at the luncheon. Investigators were able to contact sixty-five of them and found that thirty reported the symptoms of *Salmonella* infection. None of the victims knew that anyone else was ill. When all the restaurants or events to which the pies were supplied were tracked down, over three hundred people were found to have been made ill. None of the meringue was left at the bakery, so it was impossible to confirm microbially the suspected cause. Despite the clarity of the epidemiology, the bakery continued to make pies as it had for years.

The following year there was another outbreak of gastroenteritis among students at a private college. These cases and three hundred more were again linked to meringue pie from the same bakery. But this time the same serotype of *Salmonella* was found in the frozen egg whites the bakery had used to make the meringue. Finally an embargo was placed on approximately 450,000 pounds of frozen egg whites, yolks and whole eggs. (Whether it was voluntary on the part of the company, or by order of federal or state officials, is now apparently lost to history.) Of 181 samples of egg whites, 96 were positive for *Salmonella* of many different serotypes. When the egg-breaking plant was investigated by epidemiologists with the Washington State Department of Health, it was found that "cracked and dirty eggs made up a substantial proportion of the eggs purchased for breaking." The investigators tried to trace the eggs to the farms, but the plant drew from twenty states and no records had been kept. Had epidemiological evidence been sufficient for federal agencies to halt the sale of the contaminated meringue powder the year before, the second outbreak might not have happened. Attempting to convince the government agencies of the accuracy of epidemiology would take more than thirty years.

The 1964 *Salmonella* conference explored the question of how the bacteria were getting into the food chain, and looking back, there is an innocent charm to the first suggestions that the roots might, just might, have been in prevailing animal production practices. Joan Taylor of Great Britain, a widely recognized expert in *Salmonella*, seemed certain that intensive farming was probably playing a role. Animals and poultry in the United Kingdom, she said, were bred and reared in confined areas, then transported in crowded vehicles to other localities where they were mixed with other stock. Calves from Wales had been shown to have transmitted *Salmonella* into herds in England. Intensive farming methods, said Taylor, created ideal conditions for the spread of disease. That was the case in all developed countries, not just the United Kingdom. With obvious economies of scale, intensive farming was even by then the norm.

The pathways *Salmonella* took were many and varied. British researchers already had evidence, said Taylor, that some types were transmitted vertically from parent to progeny. That is, *S. thompson*, *S. menston*, and *S. typhimurium* had gotten into the ovaries and thus into the eggs. Now they were in breeding flocks that produced the United Kingdom's poultry, she revealed. This was news to some poultry producers, but vertical transmission had long been identified in ducks.

In the 1940s cases of *Salmonella* in Europe had been traced to lightly cooked duck eggs, a revelation that had led to the eventual collapse of the duck egg industry. What Taylor was suggesting was that certain strains were being selected for—that is, if they caused disease in baby chicks and then took up lodging in the ovaries of those that survived, then those eggs and succeeding generations would carry an infection that might not kill the offspring but might well get into the food chain. If flocks were made up of animal monocultures—generations bred from highly selective stock—as indeed they were, then infection would spread through all the animals because they would be almost identically susceptible. Having created a niche for a particular serotype, these highly developed species could carry it wherever they went.

Looking at the path of transmission of *Salmonella* from animal to human, Dr. Kenneth Newell suggested at the 1964 conference that the most obvious way of breaking the cycle would be to wage a selective attack on the animal host—the animal of choice, he said, being domestic poultry. The conditions under which poultry was raised presented all the conditions for infection and disease: the animals were closely confined; subjected to stress; often fed contaminated food and water; exposed to vectors (flies, mice, rats) that could carry contamination from one flock to another; bedded on filth-collecting litter; and given antibiotics (which, ironically, made them more vulnerable to disease) to encourage growth as well as ward off other infections. This environment was remarkably like that of humans in a hospital, where microbial infections were transferred with ease. The role of antibiotics in predisposing humans to infections in that setting, by killing off colonies of protective organisms, had long been suspected, and there was no reason to think it wasn't operating the same way in animals.

By 1964 importing and exporting was already spreading pathogens with efficiency. Chicks traveled between countries, carrying the infection with them, as did processed foods. An assignment of cake mix from the United States was rejected by the United Kingdom for being contaminated with *Salmonella;* Canada did the same with a number of shipments from the United States of processed foods containing eggs and egg products. But it was foolish to think that every shipment was tested or that *Salmonella*, if present, was always detected—or, if it was detected, that the product was automatically condemned.

In June 1962 the Canadian Food and Drug Directorate prohibited the sale of egg products contaminated with *Salmonella;* the number of outbreaks dropped, but the industry was given the benefit of the doubt. In one case a lot of dried egg white was tested and found to be positive for *Salmonella*. It was tested again and found to be negative. Microbial testing is a lot more difficult than most people know. The product was released because the level of contamination was assumed to be very low. But a lot of cake mix implicated in a later outbreak turned out to have been made from that very shipment of dried egg white. Much of the problem in Canada was traced to one producer—unfortunately, one of the major suppliers of dried egg products. They were shipped across the country to institutions and to food processors. Inevitably outbreaks followed.

Taylor made another observation at the 1964 conference that was ominously prescient. She could see the possibilities of changes in animal production creating niches for emerging pathogens to infect first animals, then animal products, and then humans. That had already happened with *Salmonella menston*, she pointed out. It was isolated only 28 times from humans between 1949 and 1959. After it became associated with eggs, it was isolated 103 times between 1960 and 1962. What had allowed that to happen, she wondered? What had created a niche for that particular microbe?

The Canadians had noticed the same thing with *Salmonella thompson* in the late 1950s. They had managed to trace an increase in human disease caused by the strain right back to its origins. After spreading quickly through a particular hatchery, it had become widely distributed on Alberta poultry farms. The infected eggs those hens produced then contaminated processed foods containing eggs, and within a year of the original hatchery infection *S. thompson* was causing widespread illness in the human population.

Some conference participants thought that *Salmonella* already had a strong hold; it might just be too late to do anything. But perhaps there was still a chance if they acted at once. The trouble with *Salmonella* is that, despite a direct connection with human disease, it is not a serious threat to animal well-being. Therefore, animal producers have little incentive to get rid of it. Representatives from the animal producers, the egg industry, the renderers, and the slaughterhouses were at the conference, and in the end they simply blamed external conditions or each other for the problem. The feed producers pointed to the infected material they received from the renderers; the growers pointed to the feed, the birds in the air, the rats and mice they had to contend with; the processors pointed to the obvious fact that poultry came to them infected, then were touched by human beings and equipment that transferred the contamination from one to another. What could they do except douse everything with chlorine and hope for the best?

If nothing was to be done, at least the public might be informed of the dangers, but there seemed to be more of an interest in protecting the industry. The CDC was actually criticized for advising the public, in conjunction with the FDA, to avoid using cracked and dirty eggs—clearly sensible advice given the outbreaks with which they had been associated. "A lot of damage, economic and otherwise, was done to the industry," complained Dr. Philip Levine, who was then director of the State Poultry Disease Laboratories at New York State Veterinary College.

As to the damage that was being done to humans, Drs. James Steele and Kenneth Quist, both veterinarians, with the Department of Health, Education, and Welfare (HEW) and the CDC, respectively, minced no words. "Had there been a stronger, more direct approach to the control of salmonellosis in animals, the present incidence of salmonellosis in man would not exist."[10] But little had been done, and little else would be done in the future, to control the infections in animals.

The stage was being prepared for the emergence of *Super Salmonella*.

Another Decade of Inaction

In 1978, with the problem predictably worsening, another meeting was called. This time it was sponsored by the USDA, and once again, everyone involved from the industry to the public health agencies sent a representative. The CDC reported that in the wake of climbing numbers of *Salmonella* cases, it had begun a Salmonella Surveillance Program. Pet turtles were causing outbreaks of salmonellosis, and unlike food producers, the pet turtle industry had little political clout. The interstate shipment and importation of pet turtles had been banned, and three hundred thousand cases of salmonellosis yearly had been prevented. The total number of cases, however, did not go down. "As fast as we make inroads in the control of one facet of the *Salmonella* problem, these gains are canceled by a newly emerging problem," the CDC's Dr. Eugene Gangarosa told the attendees. He called it a "fire-fighting operation."

But some fires seemed too hot to even approach. It was clear, he said, that the major source of the *Salmonella* problem in man was food of animal origin. And infected animals were taking it into homes. The infection rate in infants was huge. Apparently mothers were transferring pathogens from the kitchen to the nursery. But no one mounted a campaign to tell mothers that what they were touching as they prepared their family's meals could be making their babies sick. *Salmonella* wasn't just a pathogen on cracked and dirty eggs, it was in the animals that produced eggs. It was in poultry, beef, and pork. The CDC noted a truly significant jump in *Salmonella* infections every November. The obvious association was the Thanksgiving turkey. The agency would warn about not stuffing the turkey until just before roasting, but confronting the turkey industry on the actual source of the problem was another matter entirely. What no one told the consumer was that the turkeys and chickens they were buying were a significant source of pathogenic organisms that could make them sick or kill them. It was the truth, but it wasn't anything anyone wanted to say—or perhaps even to hear.

Once again conference presenters noted that *Salmonella* contamination at the farm and processing level was continuing unabated. Reports were given describing the factors that had contributed to the infections in poultry. They pointed out the clues in the scientific literature, starting in the 1950s, that feed could be a significant source of contamination of poultry with *Salmonella* and other pathogens. While the feed producers disputed this, numerous studies had found particular serotypes of *Salmonella* in feed and then the same serotypes in the animals that ate the feed.

The idea that we could build effective walls between animal and human was crumbling by the minute. Instead, by recycling animal byproducts and waste, we were actively constructing pathways for pathogens. Looking back, the consequences seem obvious. Even forty or fifty years ago there were indications that the rendering cycle was as efficient at distributing microbes as it was at recycling animal protein. Some of these serotypes were making their way from feed to food animal to human. One study found *Salmonella* in twenty-six of ninety-eight samples of dried dog food, which, like poultry and animal food, contains animal protein. Seventeen serotypes were isolated. Sixteen of these were later found in dogs and fifteen in humans. Researchers in Germany found unusual serotypes of *Salmonella* in fish meal fed to hogs and later found the same unusual serotypes in humans. In Ireland in 1959 researchers reported finding the same serotypes in pig meal, in pigs, and then in humans. They said that this mode of transmission was likely to continue and that the only way to break it would be to "break the animal cycle."[11] The solution was to control the presence of bacteria in feed, but these suggestions were ignored. In 1971 *S. agona* would be isolated in Peruvian fish meal used in feeds. By 1977 it would be well established in humans in the United States via the animal food connection.[12]

The problem with trying to reduce contamination was that the USDA had no authority in the industry—feed came under FDA regulations—as well as no money or personnel to devote to the problem. The FDA had defined *Salmonella* as an adulterant in finished feed in 1967 but didn't go looking for it. Unfortunately, there were no real studies to tell them how to fix the problem and none were commissioned.

The USDA made some significant discoveries, however. The amount of contamination in feed was related to whether the rendering plant processed dead animals or not. They also found that some methods were more successful than others in controlling the presence of bacteria in rendering operations. Vat-pressure systems were more successful at reducing *Salmonella* (and presumably other infectious agents), but ironically, it wouldn't be long before almost everyone in the industry shifted to the more economical continuous process system, which allowed pathogens to be distributed throughout the material rather than to be confined to a batch. The recycling of dead animals into animal feed would continue without a pause.[13]

The USDA had made a stab at solving the problem of contaminated feed. Its voluntary Salmonella Reduction Program, begun in 1968, at first showed progress, but then it stalled. The level of contamination in finished feed remained virtually the same. The agency looked for a way out. It commissioned a study, which, as one observer noted, was a famous "bureaucratic strategy for removing an agency from an unresolved problem."[14] It confirmed that *Salmonella* in meat and in bone and fish meal did indeed contribute to salmonellosis in animals and man but concluded that eliminating it would be too costly.

The potential for contamination in what animals were eating went beyond prepared feeds. Beef cattle were increasingly being fed chicken litter and other manures, both to get rid of animal waste and as a cheap source of protein. Sometimes this material was composted; sometimes it was merely left in a shed to age for eight months to a year. Veterinarians began reporting problems with *Salmonella* infections that seemed to be related to the use of these waste products as feed.[15]

But the USDA, without authority, could only offer a series of recommendations to the industry that were remarkable for their toothlessness. The agency suggested that renderers prepare a brochure for their operators on good manufacturing practices. The FDA, for its part, said that it would investigate contaminated feed only if it had a complaint. Since the *Salmonella* didn't affect poultry, and since it was nearly impossible to trace an egg or a chicken leg from an unlikely-to-be-spotted outbreak back through the complex process of distribution, slaughter, and rearing to the feed source, the chances of the FDA getting many complaints was very close to zero.

If no one wanted to go to the trouble or expense of getting *Salmonella* out of animal feed, there was an easy way to deal with part of the human *Salmonella* problem. Dried milk and egg products still contained *Salmonella*, but by 1979 they at least were no longer causing outbreaks because they were pasteurized. Meanwhile, raw meats and fresh eggs continued to bear the federal markings that were the consumer's guarantee of wholesomeness. But were they?

At the close of the 1978 *Salmonella* conference, Dr. E. T. Mallinson took note of the contradiction that USDA-approved meat had the clear potential to cause disease. "What priority will *Salmonella*, a known pathogen, have in future official guarantees or certifications to the public about the wholesomeness and cleanliness of our food?" It would cost money to control *Salmonella*, but perhaps, he suggested, the USDA should consider that cost in relation to the cost of human disease. Intensive farming was efficient, and it produced extra profits, he noted. But that efficiency just might require spending some money to clean up the process. The problem was a lack of overall coordination for food safety at the federal level.[16]

What would happen instead was that production grew ever faster and the level of contamination ever higher. Precisely as they had in 1964, each part of the industry would again look to another segment to solve the problem, and nothing was done. In fact, the real crisis was yet to come. Looking back over this thirty-year history, one USDA veterinarian who has long been involved with *Salmonella* is struck by how little has changed since those gatherings. It was as if there were roles to play, he says, and the 'script' for each character was the same."[17]

What was perfectly clear was that the agencies responsible for protecting the safety of American food, and the industries that produced it, had been presented with every indication that business as usual could and probably would create niches for emerging pathogens, and that these pathogens were very likely to have a significant and serious impact on the health and economic well-being of the public. The opportunity had been there to do something about it, but the opportunity was missed. Every condition that predisposed the spread of disease from animal to human actually worsened. Farming became more intensive, slaughtering became more mechanical and faster, products were processed in even more massive lots, and distribution became wider. *Salmonella enteritidis*, having found a foothold in the food chain, seemed to be here to stay.

The Antibiotic Connection

Predictably, the outbreaks continued. *SE* was the culprit in more and more cases. What was creating the niche for this particular pathogen? Some wondered whether the widespread agricultural use of antibiotics was contributing. Antibiotics have a reasonable and legitimate use in veterinary medicine to control disease, especially today when intensive farming practices exacerbate the spread of animal infections. But they play another role as well in food production.

In 1948 Dr. Thomas Jukes and his colleague Robert Stokstad discovered that antibiotics could make chickens grow faster. When they announced this discovery at the American Chemical Society annual meeting in Philadelphia in 1950, the business of raising food animals was transformed overnight. It would soon take only forty-seven days to turn a chick into a salable broiler. The use of antibiotics as growth enhancers became widespread, but as always, there was a price to pay.

Says Dr. Jeffrey Fisher, author of *The Plague Makers*,

> The resistant bacteria that result from this reckless practice do not stay confined to the animals in which they develop. There are no "cow bacteria" or "pig bacteria" or "chicken bacteria." In terms of the microbial world, we humans along with the rest of the animal kingdom are all part of one giant ecosystem. The same resistant bacteria that grow in the intestinal tract of a cow or pig can, and do, eventually end up in our bodies.[18]

Alarmingly, human infections that were once easily cured are now becoming resistant to an entire shelf of antimicrobials. Some pathogenic strains are resistant to nearly everything. The situation is causing widespread concern among many in the medical community who can envision with horror a return to the pre-antibiotic era when simple infections sometimes proved deadly.

Because of the obvious advantages of cheap growth enhancers, and the need for drugs to combat the animal infections that intensive farming has created, the efforts to restrict the widespread use of antimicrobials has met fierce resistance from industry. After a large outbreak of severe salmonellosis in England in 1965, caused by a multi-resistant strain, resulted in widespread illness and six deaths, British researchers made the link between the use of antibiotics in animals and the human disease. Twenty-five percent of the strains of *Salmonella* found in the humans had the same pattern of antibiotic resistance as those that had similarly caused widespread disease in calves. A commission looked into the situation and produced the Swann Report, which concluded that the subtherapeutic use of antibiotics in animals over a long period of time was likely to select for resistant strains in animals and that this could consequently have an adverse impact on humans. As a result, the use of antibiotics was strictly limited by law in the United Kingdom in 1970.[19] The Netherlands, all of Scandinavia, Germany, and Canada set up similar guidelines. The United States considered a ban, but it was never approved. American Cyanamid, a large producer of antibiotics, was worried that the report and the reaction of other countries might inspire the United States to do the same. Its tactic: question the science. The company released a publication called *The Swann Report: An Appraisal* in which it argued that one hundred billion head of livestock and poultry had been given medicated feed in the United States and Canada since 1950, but "not a single medically annotated case of disease outbreaks in humans caused by the use of antibiotics in animals has been recorded."[20]

That is no longer the case. The widespread use of antibiotics in animals—as well as excessive use in humans—is now widely recognized as playing a role in the creation of resistant strains that adversely affect human health. Anita Rampling of the Public Health Laboratory in Dorchester, England, has suggested another effect of antibiotics. She has been able to show that one in particular may actually select for the colonization and continued low-grade invasive infection of chickens with *S. enteritidis* phage type 4, the strain that became the biggest problem in England. In her research she found that 98 percent of isolates of *S. enteritidis* phage type 4 were resistant to nitrofurantoin. This is the antibiotic used in the poultry industry to treat a protozoal pathogen that causes gut infection in intensively farmed poultry. And there was cross-resistance to another antibiotic, furazolidone, often used to treat *Salmonella* infections in poultry. The widespread use of nitrofuran drugs had killed off the competitors and allowed the resistant *SE* to survive. That, she says, "could explain the prevalence of the organism in poultry and in human enteric infection in the United Kingdom."[21]

There was little doubt that the same thing was happening in the United States. A story that perfectly illustrates how antibiotics can both select a particular pathogen in animals and subsequently pass on the pathogen to humans involves *Salmonella newport* on beef, not in eggs or on chicken, but if the reader will tolerate a small diversion, the concept will be made clearer. In 1983 Michael Osterholm worked with the CDC on a study that traced an unusual strain of *S. enteritidis* subtype *newport*, resistant to ampicillin, carvenicillin, and tetracycline, from a herd of dairy cows to a beef herd, to hamburger, and then to a group of seriously ill individuals.[22] It was an intriguing story, a reminder of the intimate relationship between animals and humans and the complex ecological relationships between exposure and disease. A quick investigation of an increase in *S. newport* revealed that many of the patients had been taking antibiotics before they got sick, and the epidemiologists thought at first that the antibiotics might be contaminated. More questioning revealed, however, that the patients had taken different antibiotics. It was a serious illness. Of the ten Minnesota cases, six were hospitalized for an average of eight days, and one died from the infection, which he received from improperly sanitized diagnostic equipment in the hospital. How was this last case connected? The investigation ended up involving six states, and the tale was amazingly complex.

It began in November 1982 when a calf died during an outbreak of diarrheal disease on a farm in South Dakota. Isolates of *S. newport* with a particular pattern of resistance to several antibiotics were found in the animal. The dairy herd owner was probably infected at that time, carrying the bacteria in his body without obvious illness. In the meantime, the beef feedlot herd next door apparently picked up the resistant *Salmonella* from the dairy herd. The beef herd was given chlortetracycline throughout 1982 for growth promotion and disease prevention and that, the epidemiologists concluded, applied the selective pressure that enabled the *S. newport* to thrive. Because it was resistant to chlortetracycline, it could move into the open niche when its bacterial competition was knocked out.

Several families in the area bought beef directly from this farmer, but no one became ill until two members of the family took antibiotics for other infections. They became sick at once. The use of an antibiotic to which the *S. newport* was resistant was the chance the bug was looking for. It had been held back in their guts until then, but with the bacterial competition knocked out by the antibiotic, it reproduced aggressively. One patient was examined at the hospital using a sigmoidoscope. It was disinfected briefly in a solution that had been in use for about twenty-five days. The next patient on whom it was used had been taking several antibiotics. He developed septicemia and died. The same multi-resistant S. *newport* had been isolated from his blood, sputum, and stool.

In the related outbreak in Minnesota the investigation found that everyone reported eating hamburger before they became ill. Two patients said they might have tasted raw hamburger before cooking it. Beef from the herd in South Dakota was traced through a meat broker to six or seven supermarkets in the Minneapolis–St. Paul area. They were among those named as the source of the ground beef the Minnesota cases had purchased. And back in South Dakota there was another illness. The dairy farmer whose calf died had apparently harbored the S. *newport* in his gut without a problem until he took penicillin for another illness in February 1983, after which he immediately became ill. There were several subsequent infections, presumably from person-to-person contact between family members.

The complexity of meat production, processing, and distribution has made it difficult to prove the connection between antimicrobial use in animals and disease in humans. But the epidemiologists now felt that their evidence was conclusive "that antimicrobial-resistant bacteria of animal origin can cause serious human disease, especially in persons taking antimicrobials, and that the emergence and selection of such organisms are complications of subtherapeutic antimicrobial use in animals."[23] They advocated a more prudent use of antimicrobials. The use of human antibiotics such as tetracycline and ampicillin has decreased in animals, but from one-half to two-thirds of antibiotic production (estimates vary) still goes to agriculture, creating resistant bacteria and selecting for emerging pathogens, both of which will create frightening problems for the human population.

Looking back at the clear link between certain *Salmonella* strains in animals and their subsequent appearance in human disease, and the likelihood that antibiotic use selected for certain bacterial strains, it seems almost without question that the continuing use of antibiotics in poultry and livestock has played a significant role in the growing problem of *Salmonella enteritidis*. *Super Salmonella* had triumphed where others had fallen. The other conditions common to intensive farming—the limited diversity of the animal species, the close quarters, the contaminated feed and water, the stressful conditions—each then played its own role in paving the way for this microbe.

The Pandemic Begins

The surge in *SE* infections in the United Kingdom began slowly but rapidly accelerated. From 1982 to 1987 the number of reported and confirmed cases increased sixfold, from 1,101 to 6,858, and it was widely understood that these cases represented only a small fraction of the real number of illnesses. By July 1988 the number had reached 4,424 for that year alone. The outbreak had been growing uncontrollably since 1986. A letter to *The Lancet*, the British medical journal, pointed to poultry and eggs as the source.[24] Foods implicated as causes of disease included mayonnaise, tartar sauce, eggnog, milk shakes, mousse, and ice cream, as well as other foods containing raw or partly cooked eggs.

It can often be hard to get the attention of lawmakers on something as mundane as foodborne illness. That took care of itself on May 9, 1988, when fifty peers in the House of Lords and twenty members of their catering staff became violently ill. It was *Salmonella typhimurium* that had infected the mayonnaise, not *SE*, but the point was made. Something was amiss with British eggs. The outbreaks continued, and the public became increasingly alarmed.

In August 1988 warnings were issued not to eat raw eggs or dishes that contained them, and purchases of eggs began to drop. If Great Britain's political, business, and financial establishments were not fully convinced, two banquets in London's financial district in late October 1988 brought them around. At the final count, 151 people, including the lady mayoress of London, became ill with salmonellosis. The cause was thought to have been raw eggs used in a cheese savory. Then, on the first weekend in December, the junior health minister, Edwina Currie, already controversial for her outspokenness, said in a television interview that most of Britain's egg production was affected by *Salmonella*. While she presumably meant that virtually every flock had hens that occasionally produced eggs containing *Salmonella*, that subtlety was not appreciated by the press. Nor by the British egg industry. Farmers threatened to sue and jointly urged Prime Minister Margaret Thatcher to sack Currie if a retraction and apology were not forthcoming. Said one egg producer, "We want her head on a plate."[25]

Farmers had reason to be concerned. Their egg sales, a business of £110 million a year, had dropped by almost 50 percent that year. Their tactic was to deny it all. Only twenty-five illnesses had been linked to eggs in the past year, they said, and a representative from the feed industry estimated that there was only a 200,000,000-to-1 chance of an egg being contaminated. Backing Currie, for the moment, the minister of health reiterated his warnings against eating raw eggs. The public and the marketplace were genuinely alarmed. Signs in a Birmingham market, reported the *Times*, London, advertised *Salmonella*-free Dutch eggs, a useless promotion since the Netherlands had an *SE* problem as well.[26]

A quick investigation of a sixty-thousand-bird flock implicated in an outbreak of food poisoning found *SE* in thirteen of fifty dead birds. *SE* was isolated from the egg contents or shell membranes of eggs from six birds with either ovarian or oviduct infection. There was no doubt where the *Salmonella* was coming from. The aftermath of Currie's remarks, however, is infamous among egg producers. By December 13, 300 million eggs worth £25 million were sitting unsold. An egg advertisement campaign to buck up sales pointed out that only raw or undercooked eggs were a problem, and then mainly for the elderly, the sick, and the young, but the message didn't seem all that reassuring.

Currie had her supporters, nevertheless. The problem with eggs had been thoroughly discussed in medical publications such as *The Lancet*, for those who read them. And Oliver Gillie, writing for the *Independent*, laid the problem with contaminated eggs squarely at the feet of intensive farming, contaminated feed, and the Ministry of Agriculture's "wholly inadequate measures" to control the infections. The regulation passed in 1981 requiring protein foods for animals to be *Salmonella*-free had been voluntary (as it had been in the United States). It had been useless: One in ten samples of processed animal feed had subsequently been found to be contaminated. For imported feed, one in six tested positive for *Salmonella*. One of the most contaminated feeds was feather meal—ground-up chicken feathers fed back to hens. One in four samples was contaminated. It was little wonder that both chickens and eggs were carrying the pathogen.

Dr. Richard Lacey accused the government of a cover-up, claiming that every week someone in Britain died from eating eggs. With 30 million a day consumed, that wasn't a bad guess. The government admitted to twenty-six deaths. Lacey told the *Guardian*, "We are beginning to see the worst *Salmonella* epidemic of our time. We need to eradicate *Salmonella* from the egg industry and this means culling the flock." He didn't blame egg producers, whom he felt should be fully compensated. "They were working within the law. I blame the Ministry for allowing *Salmonella* to take hold."[27]

Not only eggs were to blame. A study conducted a year earlier and published amid the scare did nothing to calm the public. The investigators found that 60 percent of chilled and frozen carcasses were contaminated with *Salmonella*. Politics had played a role in the increase. The *Guardian* revealed that the government had been "quietly diverting money" from *Salmonella* research since 1987. Currie, nevertheless, was forced to resign, and to this day she remains an archvillain to the poultry and egg industries—and a warning to public officials and health professionals tempted to speak too freely.

Lacey was right. The regulators deserved much of the blame. But he ignored the fact that the regulators are, in turn, relentlessly pressured not to regulate by industry and by elected representatives, who themselves have been intensively lobbied. Despite conclusive evidence that voluntary restraints are inevitably doomed to fail nearly everywhere they are applied, they continue far too often to be the government's preferred course of action, especially in the present deregulatory climate. Voluntary restraints may provide short-term satisfaction to food producers, but in the long term they constitute a misguided strategy as reports of contaminated food products and outbreaks grow and consumers become more and more aware of the dangers of food.

The problem was not confined to the United Kingdom by any means, as government officials were quick to point out when eggs from abroad were touted as safer. Over the next few years there would be reports of *SE* outbreaks in Argentina, Austria, Bulgaria, Finland, France, Hungary, Italy, Spain, Sweden, and Switzerland. In 1990 the CDC's Daniel Rodrigue and Robert Tauxe, along with B. Rowe of the

Central Public Health Laboratory in London, would jointly publish a paper entitled "International Increase in *Salmonella enteritidis:* A New Pandemic?"[28]

The United States was also experiencing a large increase in *SE* infections in 1988. But while the United Kingdom had predominantly phage type 4, the United States had predominantly phage types 8 and 13a. The CDC had been keeping careful track in its Salmonella Surveillance Program. From 1976 to 1986 infections with *SE* had increased more than sixfold in the northeastern United States. In an eighteen-month period the CDC watched with a mixture of horror and fascination as sixty-five foodborne outbreaks were reported, associated with 2,119 cases and 11 deaths. Seventy-seven percent of the outbreaks, their investigations found, were caused by grade A shell eggs or foods they contaminated. In fact, eggs constituted a much greater cause of salmonellosis than anyone had previously thought. The cause of the outbreaks had shifted as well, from bulk egg products (after regulations in the 1970s required pasteurization, which halted cake mix outbreaks) to dishes prepared from individual shell eggs. They described the increase as "extraordinary."

The CDC researchers weren't sure how the pathogen was being transmitted by eggs, since they had all gone through the egg-washing process, but they guessed that it was by the transovarian route, previously demonstrated by *Salmonella pullorum*. If the bacteria penetrated the ovary in chickens, they would produce infected chicks, and the infection would persist for several generations. When the CDC tested bulk eggs from all over the country, 10 percent of those in the northeast yielded *Salmonella enteritidis*. Elsewhere in the United States they found none. But the regional differences wouldn't last long. The CDC thought that the responsibility for many of the outbreaks lay with the "limited public awareness that raw eggs, like other foods of animal origin, constitute a potential source of *Salmonella*."[29] It was difficult to see how the public was supposed to become aware without direct warnings from the regulatory agencies. Like the USDA, the CDC seemed anxious not to offend the egg industry. High-risk settings such as nursing homes and restaurants could gain a "margin of safety" by using pasteurized eggs, they suggested. The implication was that everyone else would be fine.

The recommendation of CDC investigators in 1990 was that a "rational control program would focus on detecting infected primary breeding flocks and eliminating *S. enteridis* from animal feeds." Poultry production methods should be modified as well, they suggested. In the meantime—and the meantime would unfortunately last a long time—"consumers should be advised to avoid recipes using raw eggs and to cook shell eggs and other foods of animal origin well."[30] If producers and regulators wouldn't act, consumers would have to. But the information, in the journal *Epidemiology of Infection*, was not where it would reach many of them.

Owing perhaps to the absence of an Edwina Currie and the lack of ongoing aggressive press coverage—the American press seemed more interested in theoretical long-term questions relating to food, such as the role of saturated fats in congestive heart failure—even the outbreaks in the Northeast caused much less interest in the general population than did outbreaks in England. One example makes the point: The state epidemiologist in Maine remembers that, of the many outbreaks of *SE* from eggs in her state between 1988 and 1990, none was reported in the newspaper.[31] The egg producers liked it that way.

The End of the Egg as We Know It

As the seriousness and incidence of *Salmonella* infections grew in England, and as the link to eggs became even clearer, a group of British researchers decided in 1989 to find out how *Salmonella enteritidis* and other strains reacted to cooking "under simulated domestic conditions." They began by inoculating ordinary eggs bought from retail shops with various *Salmonella* cultures. The eggs were then fried sunny-side-up or over-easy, scrambled in a saucepan or in the microwave, and boiled for four minutes.

Much to their surprise, all four test strains of *Salmonella* survived in the boiled eggs and in the sunny-side-up fried eggs. Basting with hot oil during cooking made no difference. Provided the inoculum of *SE* was low, it was possible to destroy the bacteria by quick scrambling at high heat, by boiling an egg for nine minutes or longer, or by frying an egg over-easy until all the yolk had solidified, but their tests showed that the bacteria remained in any form of cooking where all or even some of the yolk remained liquid.[32]

This was bad news indeed for egg lovers, for home cooks everywhere, and certainly for chefs—if they read obscure medical journals. How would they prepare soufflés, mousses, mayonnaise, Caesar salads, ice creams, hollandaise sauces? But few would hear about the study; the outbreaks continued. By 1990 *SE* was the most frequently isolated serotype, making up 21 percent of the total in the U.S. And it was spreading. From 1985 through 1991, 380 outbreaks of *SE* were reported by thirty-seven state health departments, Washington, D.C., Puerto Rico, and three ships at sea. The CDC anticipated that soon every state would be affected. During 1989 and 1990 the USDA investigated sixteen farms implicated in human outbreaks of *S. enteritidis* infection. *SE* was found in all sixteen. It was everywhere, spreading through dust and water and feed and poultry monocultures.

Aside from the CDC and the federal regulatory agencies, which were deeply worried but intent on investigating rather than alarming, the only people who seemed truly concerned over this extraordinary increase in egg-related foodborne disease were the individuals made ill and the families of those fifty between 1985 and 1991 for whom the infection was fatal. And those fatalities were simply the deaths the CDC heard about. A few years before, someone from the CDC had looked closely at death certificates of those dying after gastrointestinal symptoms and noticed the high number in which a culture had never been taken. Couple that information with the vast numbers known to be unreported and who knew what the morbidity and mortality rates really were?

Where Were the Regulators?

In 1990 the FDA redesignated the egg as a hazardous food under its model food codes. Eggs would have to be refrigerated, as were other foods of animal origin, even during transportation. It was such a reasonable suggestion, one that would certainly slow the growth of bacteria in contaminated eggs, that one could be forgiven for wondering why it hadn't been made earlier. But if consumers thought their eggs were safer than before, they were in for a surprise.

Five years later, in 1996, the USDA revealed that the rules had never been enforced because the egg industry felt they were too restrictive. The industry had sought legislation in 1993 and 1994 to amend the 1991 provision, protesting that the temperature of the air wasn't a guarantee of the temperature of the eggs. The effect was to delay the refrigeration of eggs and to send the USDA back to do more scientific studies. But the problem of transportation was a significant one. The USDA and the FDA would jointly address the issue, but it was going to take a very long time.

Despite bending to the industry on the enforcement of the egg temperature rule, the USDA was in fact seriously concerned about *Salmonella enteritidis* in the food supply. In the aftermath of the widespread outbreaks in the Northeast—in effect, an epidemic of salmonellosis—the agency launched a two-pronged attack in 1991. One was a study to see just how prevalent *SE* was in laying hens; the other was an emergency rule that would attempt to certify newly hatched chicks that traveled interstate as free of *SE*. When a flock was shown to be infected, the eggs, instead of being sold in the shell, would be rerouted to the breakers. These are firms that crack eggs mechanically and supply pasteurized liquid or dried egg to food processors, institutions, and restaurant kitchens. The eggs sent to breakers are always the undesirable eggs—those cracked or blemished, or, in this case, those suspected of being contaminated. Pasteurization, provided it is done properly and the material is not recontaminated, makes such eggs safe to use.

When laying hens are no longer producing the number of eggs producers want to see, they go to slaughter. These "spent" layers are usually transported at night in trucks that may contain between eight thousand and ten thousand birds. These are the trucks of hysterical birds that one sometimes sees flying down the highway scattering ruffled white feathers as sad little mementos of their short, busy lives. They arrive at the slaughterhouse in the morning and are supposed to be slaughtered within twenty-four hours. If the birds are not already contaminated with pathogens, they surely are by then. Studies have shown that frightened animals react like frightened humans—with loose bowels. They have also shown that animals deprived of food and water shed more pathogens in their feces. Jammed into pens, the excrement of one bird will get onto the others. They are a mess when they arrive.

To determine which flocks were contaminated (and thus the prevalence and distribution of the pathogen nationwide) the agency decided that it was practical to test, not the henhouses, but the slaughtering facilities that were dealing with the spent hens.[33] Not all the slaughterhouses agreed to let the investigators in, but investigators thought they still had a decent sample. They found that the number of birds infected with *SE* in a given region matched the number of human cases in the region with eerie precision. Canadians were also surveying for *Salmonella* infestation at about the same time, but they found far less contamination—only 3 percent overall. The difference, the American researchers thought, was that Canadians had far fewer hens per house—about 8,800 compared to U.S. flocks of 50,000–70,000 hens. In fact, when the USDA had looked at the forty flocks implicated in outbreaks, they had all been from the largest producers; none had come from the "small entities."[34] Once again, intensive farming practices had been implicated in the spread of disease in humans. And once again, the implications would be ignored.

The USDA's emergency rule on *Salmonella* didn't last long in its original form. It was modified after extensive "comment" from the "interested parties" before the final rule was published in January 1991.[35] The recommendations resulted in a series of changes that lowered the impact of a flock being identified as contaminated—and also lowered the effectiveness of the effort. Moreover, eggs found to be contaminated could be hard-boiled or exported. The probability that *Salmonella enteritidis* would begin to be spread to poorer, hungrier countries didn't seem to matter. The changes to the rule would also make it tougher to link an egg to a producer. The epidemiologists would have to come up with the shipping records to prove that contaminated eggs came from a particular flock. The USDA and the CDC would soon discover that this would present a challenge to outbreak investigations.

When something has gone badly wrong, a classic tactic to avoid responsibility is simply to change a definition. The USDA did not say that the wholesale contamination of eggs by giant producers was responsible for the growing number of illnesses and deaths. Instead, it blamed the illnesses on mishandling and improper cooking of eggs by the American public.[36] With a swipe of the pen, the way people had been cooking eggs for a hundred years at the very least was suddenly described as "improper." Our grandchildren would grow up thinking that the only safe egg was a hard-boiled one or a pasteurized product poured out of a package.

This was not altogether bad news for the egg industry; any food that goes through processing has "value added," meaning more profit for the producers. It may very will be that egg producers dream of a time when they no longer need to ship perishable eggs in refrigerated trucks to customers. This magnificent food, after all, is an awkward shape, fragile, and has a fairly short shelf life. Why not wean consumers off the fresh product entirely? Eggs could simply be crushed on-site, the liquid interior extracted, pasteurized, packaged attractively, and sold at an increased profit. It would be egg producers' nirvana—and the end of the egg as we know it.

The *Salmonella enteritidis* cases continued to climb through 1989, dropping in the Northeast in 1990 but picking up steam elsewhere. The number of *S. enteritidis* outbreaks reported nationally rose from forty in 1988 to seventy-five in 1989, then sixty-seven in 1990 and 1991. The West Coast was now being hit.

In January 1993 an outbreak in Los Angeles County was linked to egg consumption. Egg salad and pooled eggs were stored in conditions that allowed bacteria to grow. In February 1993, twenty-three more people became ill with *SE* infection after eating at a restaurant in San Diego County. This time hollandaise or béarnaise sauce was implicated. In March 1993, twenty-two people became ill with *SE* infection after eating in a Santa Clara sandwich shop. Sandwiches and mayonnaise were apparently the vehicles. Unrefigerated eggs from the shipment from which the mayonnaise had been made were still in the restaurant and some yielded *SE* contamination.

But if the USDA, under its *Salmonella enteritidis* Control Program, were to condemn the flock and divert the rest of production to the breakers, it would have to trace the eggs back to the farm that produced them. When the CDC looked at these three outbreaks, it found the same unusual combination of phage type and plasmid profile type to the bacteria in all of them. The eggs had probably come from one producer. But in each of these outbreaks the traceback was a failure. The eggs were purchased from a distributor who bought and mixed eggs from many different producers, making it impossible to follow the eggs back to their source. Thus, the farm could not be identified through the records, even though the plasmid profiles of the bacteria matched, and its contaminated production remained in the food supply.

In 1991 it appeared to the USDA that its traceback program alone was not bringing down the incidence of *SE* quickly enough. In 1992, in cooperation with Pennsylvania egg producers and the Pennsylvania Department of Agriculture, it began the voluntary SE Pilot Project. The goal was to reduce the number of *SE*-contaminated eggs distributed by testing the henhouse environment. They wondered whether rodents might have something to do with the problem.

Mike Opitz, a veterinarian at the University of Maine, did a study in 1990 with D. J. Henzler looking at the role of mice in the epizootiology of *Salmonella enteritidis*. They looked at ten farms and found half of them clean. The other five were positive for *SE*. Twenty-four percent of the mice on these farms cultured positive for *SE*. Opitz had a neat theory as to why mice had come to play such a role in the infection of flocks. Back in the 1970s during the energy crisis, he said, egg producers went to great lengths to further insulate their henhouses. The added insulation created mouse heavens—warm places where mice could nestle in and make homes. The henhouses also provided a steady supply of food and water, and the mice used the conveyor belts that carried in food and took away eggs as runways to dash from one henhouse to another in big operations where the belts linked the houses. Once the mice became contaminated, they would leave a trail of *SE*-contaminated droppings. Imagine a completely bored chicken the next morning as the food conveyor belt begins. There amid the sameness of the feed is something interesting, something different. It's a black mouse pellet, and it's carrying *SE*. It is snapped up and eaten, and infection is perpetuated. A system perfectly designed for continuous recontamination.

Perhaps the mice were contributing, but the problem of *SE* is complex, and there are no magic wands for controlling it. It is an ecological nightmare. Flocks are too big. Feed is too contaminated. Poultry monocultures allow infection to spread, and the stressful conditions and antibiotic treatments encourage it. Rats and mice carry the infection, and dead and diseased chickens are recycled into feed, as are their filthy feathers.

The attempts to eradicate *SE* have been resisted by the egg industry because an infected flock means a significant loss of income when contaminated eggs are diverted to the breakers. Compensating producers for the loss might make a difference. If the USDA had really wanted to clean up the mess, diseased flocks could have been eradicated and hygienic standards imposed. But the Food Safety and Inspection Service (FSIS) has no authority on the farms, and farmers wanted no regulations. FSIS head Michael Taylor assured animal producers as recently as 1996 that the service had no intention of seeking that authority.[37]

The USDA program has had its successes. In 1990, after at least sixty-seven people who ate in a chain of family restaurants became ill from salmonellosis, the restaurant eggs were traced to a farm in southwestern Pennsylvania.

Such successes have grown rarer. In Florida in September 1993 the Duval County Public Health Unit learned that five children and seven adults were ill with *Salmonella enteritidis* infection.[38] They had all attended a cookout at a psychiatric hospital in Jacksonville, and all but one had eaten the homemade ice cream. It had been prepared with fresh grade A raw eggs. The USDA's Animal and Plant Health Inspection Service (APHIS) tried to trace the eggs back to the farm but ran into difficulty. The distributor had bought them from two suppliers, and one of those suppliers had mixed up the eggs from several different sources. Under the changed regulations, the USDA could trace the eggs back only if one farm was identified. The industry had spotted the loophole and jumped in.

Today more and more eggs are being marketed in shipments containing eggs from multiple sources. Thus, the egg industry, in a cynicism that is almost beyond belief, has found a way to avoid detection and diversion of the contaminated eggs to pasteurization or destruction of the flock by confusing the issue and making traceback impossible. In 1990 the USDA traced 86 percent of outbreaks back to the source; by 1993 that number stood at a distressingly low 17 percent.

By 1994 the CDC warnings were no longer coy about the dangers from eggs. *Salmonella enteritidis* infections had increased as a proportion of all *Salmonella* infections from 8 percent in 1983 to 19 percent in 1992. During that period the CDC received 504 reports of *SE* outbreaks alone that made 18,195 people ill, caused nearly 2,000 hospitalizations, and led to 62 deaths. Of the 233 reported *SE* outbreaks in which a particular food could be implicated, 193 were associated with eggs, and of those, 7 percent were linked to ice cream.[39]

It was clear that the feeble attempts to inform consumers of the dangers of raw eggs had not been successful. Studies by the FDA in 1993 had shown that in a national sample 53 percent of respondents had eaten foods containing raw eggs. Of these, 50 percent had eaten cookie batter and 36 percent had eaten ice cream containing raw eggs.

The egg industry was not only unrepentant but brazen in its irresponsibility. Through lobbying it was able to have the U.S. House of Representatives cut the Food Safety and Inspection Service budget by the amount of the *Salmonella* program—to de-fund it, in other words. When the United Egg Producers (UEP) learned that the USDA was serious about continuing the program, even if it meant looking for the money elsewhere in the agency budget, it quickly sprang into action. It had been rumored, but seldom proven, that industry was actually writing legislation during the heady days of the short-lived, Republican-led "revolution of 1994"; indeed, the egg industry actually bragged about it. The UEP crowed in its "Washington Report" that it had added language to specify that no funds could be spent on the *SE* program, which would prevent the USDA from trying to keep the program alive despite the spending cut. The UEP expected this bill to pass, and it did. The program was dead. The telephone in the office of the former program director was no longer answered but instead played a recording saying that the number was no longer in use.

At the CDC in the spring of 1996 the epidemiologists sat around a table at their weekly meeting discussing yet another outbreak of *Salmonella enteritidis*. There was to be no traceback now on the part of the USDA. *Salmonella enteritidis* was apparently to be left to increase uncontrolled. When little attention is being paid by producers, processors, and distributors, the last remaining critical control point is the consumer. And when that fails, the health care professionals take over to mop up the inevitable human disease.

The Inevitable Disaster

In Minnesota the health department continues its regular surveillance for *Salmonella*. All isolates received by its lab are serotyped and maintained for reference—a library of microbes. In September 1994 that surveillance alerted the department to a problem, but it took a while to recognize what was happening. An increase in an unusual organism can point to an outbreak from a common source, but *Salmonella enteritidis* cases were by then so common that the department "has a rather high threshold for concern,"[40] says Craig Hedberg. So when the laboratory told him about an increase in *SE* cases that September, he wasn't particularly worried. But the increase continued, and report cards from physicians were coming in as well.

"As we saw these begin to pile up, we started to look at the distribution. A lot of them were from southeastern Minnesota. Then we got a report card in from a physician at the Mayo Clinic who was reporting a case of *Salmonella* infection, and on the bottom he wrote that it was associated with Jesse James Day in Northfield, Minnesota."

The town is about thirty miles south of the Twin Cities, and it's famous for having the last bank robbed by Jesse James. Each year there is an annual celebration of the big shootout when several members of the gang were killed. The department began a preliminary investigation, but after talking to ten to fifteen people, it discovered that none had attended the event. "In the meantime, cases continued to come in," says Hedberg, "and we were continuing to get inquiries from Larry Edmundson and the Mayo Clinic [telling us] that this problem was not going away and we needed to do something." The department decided to attack the problem as a regional outbreak of *Salmonella enteritidis* in southeastern Minnesota and began a case-control study focusing on the most recent cases in that area.

"That was on October 5—a Wednesday—[when] we actually started calling people. And we used the off-the-shelf case-control study form. It covers all these different food items and asks about the source of these things. On that Wednesday evening I interviewed three cases, and two of the three cases told me they'd eaten Schwan's ice cream. That struck me as being unusual. Over the course of the next twenty-four hours we completed probably a dozen or so matched pairs, and as we interviewed them, they continually told us they had eaten Schwan's ice cream and the controls hadn't."

Within a span as brief as twenty-four hours the investigators had accumulated enough cases and controls to link the illnesses to eating the nationally distributed ice cream. By Thursday evening they had enough information to take the investigation to the next step. When Hedberg gave the FDA and the head of the food safety group of the Minnesota Department of Agriculture a heads-up call on Thursday afternoon based on preliminary trend analysis, he found out where he could reach them later that day if things continued to move in the direction of a convincing link to Schwan's.

"So I laid the groundwork that we might have something big afoot and when I called, it wasn't a total shock." He described the data and what the department thought the next course of action should be. They had already set up a plan, and now they put it in motion. They would first get together with the Minnesota Department of Agriculture and the FDA to look at the data, and then the FDA would contact the company. The agencies would meet with the company, review the health department's findings, and lay out a course of action.

As luck would have it, some of the key Schwan's executives were on airplanes flying off in different directions, but by Friday afternoon everyone had been contacted either by phone or in person, and that evening at a press conference they jointly announced a plant closure and a nationwide recall of Schwan's ice cream. "Initially it appeared to be a regional problem," says Hedberg, "but once we figured out the source of the problem in Minnesota, it was apparent that it was in all likelihood a nationwide outbreak, which ultimately it proved to be."

The CDC was involved from the beginning through its EIS officer assigned to Minnesota, Dr. Tom Hennessy, who worked with Michael Osterholm and Hedberg on the investigation. As it became clear how widespread the outbreak was, the CDC would lend its full assistance. The FDA would find *Salmonella enteritidis* in an unopened carton of ice cream on October 17, 1994, confirming the epidemiological investigation. The news was not just alarming to Schwan's, it was downright perplexing. *Salmonella enteritidis* was linked to eggs, but the company did not use eggs in its ice cream. How it got there would take an investigation to discover.

When the full impact of the outbreak was evaluated months later, the team would estimate that 224,000 people in the United States had developed infection with *Salmonella enteritidis* after eating Schwan's ice cream. The CDC itself found a 71-percent increase in the number of *Salmonella enteritidis* cases reported nationwide for the months of September and October. The Schwan's-associated cases raised the year's cases by 21 percent, making it the "largest common-vehicle outbreak of salmonellosis ever recognized in the United States," the investigators wrote.[41] What had gone wrong?

Schwan's ice cream is nationally distributed to all forty-eight contiguous states through a household delivery system. The primary production facility for the ice cream is in Marshall, Minnesota. The ice cream is made from premix purchased from two different suppliers; it is transported to the Marshall plant in tanker-trailers, transferred to storage silos, moved to vats where flavor is added, and finally moved to the freezers. Between two and nineteen tanker-trailer loads of premix were used in making each flavor.

When the team inspected the plant, they found no problems with the plant's equipment or its quality assurance programs. They found no explanation at the plant for the *Salmonella* contamination. But truck drivers like to "back-haul" goods on every trip. The rule: never travel empty. The tanker-trailers that carried the premix from supplier to plant were found to have back-hauled oils, molasses, corn syrup, pasteurized dairy products—and unpasteurized eggs. The trucking company had taken on new contracts that significantly increased the quantities of unpasteurized eggs its trucks were hauling from egg-breaking plants.

The investigators found a strong relationship between the ice cream that had made people sick and the number of loads of premix that had been transported by trucks that had just carried liquid, non-pasteurized eggs. The delivery firm was also found to have ignored written instructions on cleaning and sanitizing the tanker-trailers after each delivery of liquid eggs. FDA and Minnesota Department of Agriculture officials found soiled outlet-valve gaskets, poor record keeping, and tanker-trailers that weren't being regularly inspected. They also found egg residue on one trailer even after cleaning, and in the lining of five of the trailers they found cracks. One of these trailers had carried mix that had gone into the ice cream that had made people sick. To save time, investigators discovered, some drivers were apparently bypassing altogether the cleaning procedure after unloading eggs. When the contaminated ice cream was recalled, the outbreak stopped.

This massive outbreak was detected by a case-control study that involved only fifteen matched pairs of ills and wells. The microorganism wasn't found in the product until ten days later. The message was clear. Public health officials shouldn't wait for confirmation of microbial contamination before taking action when the epidemiological evidence points clearly to a certain product. The outbreak also pointed out the enormous savings when an outbreak can be detected early by surveillance, serotyping, and prompt investigation. In Maine, for instance, Dr. Kathleen Gensheimer, the state's epidemiologist, had noticed an increase in *Salmonella* cases during the same time period, but without the resources to investigate, she learned of the cause only after the Minnesota findings and the FDA announcement.

One more interesting fact emerged. If you look in the textbooks to see how many *Salmonella* microbes it takes to make a person sick, you are likely to find that it might take 100,000 organisms. But the experience in Minnesota suggests a very different scenario. In the Schwan's outbreak the products with the highest level of contamination contained only six organisms per half-cup of ice cream. Since they were frozen and eaten frozen, which wouldn't allow for any growth before they were consumed, that was likely to be the amount people who became ill had ingested. The contaminated cheese outbreak had, as well, revealed a startling fact that others would later confirm: it doesn't take much at all to make someone very sick. Apparently quality assurance programs that set high levels for *Salmonella* on the assumption that lower levels do not cause disease are way off and need to be completely reevaluated. But will they be?

The team also recommended that food-grade products be repasteurized after transportation or be transported in dedicated tanker-trailers. The trucking company shipping the premix had been hired by the company that supplied Schwan's. What had frustrated Schwan's was its lack of control over something that had been costly to both its bottom line and its reputation. The company wanted to make sure it never happened again. It now leases its own trailers and has added another step to the manufacturing process. The premix is repasteurized when it enters the plant, before it goes into the product.

The Schwan's story was the outbreak the Minnesota team and others had predicted. Everything had come together—the contaminated chickens, the bad eggs, the complex processing system that involved subcontracted transportation, human failure, mass-distribution, and cultural and lifestyle changes. It was an outbreak that should have prompted dramatic changes in the food system. Instead, it prompted small adjustments in the practices of one company. This outbreak was a dramatic warning, and there is nothing to prevent it from happening again.

8/*Campylobacter* and the Poultry Connection

I don't know which is more discouraging, literature or chickens.

E. B. WHITE,
Letter to James Thurber,
18 November 1938

If there were a contest for the most contaminated product Americans bring into their kitchens, poultry would win hands down. The contamination occurs when the bird's bacteria-laden intestinal contents come in contact with the flesh we eat.

It has been known probably from earliest human history that contaminating any meat with the contents of an animal's digestive system has the potential to cause disease in humans because the contents may harbor harmful bacteria. Even the healthy gut contains swarms of other bacteria, such as generic *E. coli*, which are easy to culture and easy to identify. Thus, a simple way to discover whether meat—or anything else—has been contaminated with fecal material is to test for *Escherichia coli* bacteria. When the USDA in 1995 did a baseline study to determine the prevailing contamination level of chickens, its studies found, astonishingly, "greater than 99 percent of broiler carcasses had detectable *E. coli*."[1]

Most members of the *E. coli* genus are not (with notable exceptions) harmful, but their presence is an indicator that other, more dangerous microbes may be hitching a ride along with them. The USDA study means that virtually every chicken is potentially infected with pathogens. If chicken were tap water, the supply would be cut off. Instead, chickens with heavy *E. coli* counts still receive the USDA's blessing of wholesomeness and safety.

One of the pathogens that may be hitching a ride is *Salmonella*, which most people have now heard of, although they may not realize what a serious hazard it poses. Different studies reveal levels of *Salmonella* contamination on poultry that range from 20 to 80 percent. Chickens are not unique in being contaminated. Studies have found from 2 to 29 percent of turkeys to be positive for *Salmonella*, and the new USDA performance standards, based on existing levels of contamination, identify ground turkey as the poultry product with the highest (49.9) average levels of *Salmonella*.[2] Another tagalong pathogen is *Campylobacter*, which can cause serious illness, is even more widespread, and yet is virtually unknown to consumers.

The story of *Campylobacter* and how it came to be present on poultry parallels the story of poultry itself and how it is raised and processed. Given our sense that modern medicine has gone well beyond the basics, forging on to science-fiction realms of transplants and gene splicing, it is also the surprising story of how long it took to identify an important and apparently long-standing cause of human disease. But it is an environmental story as well, revealing how we spread and recycle pathogens. Once again, the very complexity of the story is the story. We didn't create *Campylobacter*, but by a series of interrelated changes in the food system, we have created an environment that makes life easier for this pathogen.

The Farm Fantasy

The chicken breasts in my local supermarket come prepackaged from the processor. This is a good and bad thing: good, because further handling, which could cause or spread contamination, is eliminated at the level of the local supermarket; but bad because the package contains, in that absorbent pad at the bottom, the bacteria-laden water that would have drained off had the chickens been shipped in bulk and for which the store—and the consumer—now must pay. Nevertheless, packaging at the processing level has a distinct advantage for the consumer in that accountability is now on the package. The quality and reliability of a pre-packaged chicken, good or bad, can be directly linked to the company that produced it.

But it is the packaging itself that draws my attention. On the plastic that wraps the pink, fresh-looking breast meat is a charming design of a farm with a red barn, mature trees, and a pleasant, rutted dirt road. The reassuring image works. It is something we want to associate with our food.

That idea of the farm has always been something of an illusion. Producing food from animals was never as gentle an enterprise as it has been portrayed. It inevitably involves an element of brutality: Calves must be taken from their mothers for humans to reap the maximum amount of milk; chickens must endlessly produce potential offspring that are seldom allowed to hatch; and obviously the lives of animals are severely truncated—slaughtering has never been a pretty process—so that we might be nourished, however humanely that is accomplished. Ways of manipulating the environments of food animals so as to produce more food more cheaply have been known and practiced—although sometimes carefully guarded by those who possessed such information—since second-century Ptolemaic Egypt. According to Margaret Visser in *Much Depends on Dinner*, the technique of incubating chickens artificially was lost during Roman times but maintained by one small Egyptian village in the Nile Delta where techniques of hatching chickens throughout the year was a closely held secret. A native of the village was persuaded to divulge the secret to the grand duke of Tuscany in the early eighteenth century, and the life of the chicken has never been the same since.[3]

However naive it is, the intuitive appeal of the traditional idea of a farm, set in an uncluttered rural landscape, is well understood and appreciated by marketers, food producers, and processors. Farmer John's Slaughterhouse and Meatpacking Plant, a major meat processor in Los Angeles, is housed in a facility surrounded by high walls painted with a bright countryside mural that includes blue skies, fluffy clouds, rivers, mountains, healthy green trees, handsome fields, picturesque barns, farm animals, and a farmhouse where children play.

Since the windows of the facility are painted over with these scenes, it is impossible to see what goes on behind the walls, but the process of raising food animals today, then turning them into meat on a large scale, bears no relationship at all to these charming images.

That people cling to these illusions is uniformly dismissed as irrational, though exploitable, romantic nostalgia. But I think we reject too quickly this preference. Our instinctive sense of what is healthy and what is not healthy, what is safe and what is not safe, protected us for thousands of years before the FDA and the USDA assumed that responsibility. And human evolution surely selected for those whose instincts were finely honed. What we now experience is a growing anxiety, says Visser, "as we watch ways of life we have loved being killed off, apparently inexorably." The struggle over factory versus free-range chickens, or to use Visser's example, over butter versus margarine,

> represents the great oppositions articulated in our culture: the land versus the city, the farm (despite the extent to which dairy farming has become mechanized) versus the factory, independent versus corporately-controlled business, tradition versus not-necessarily-preferable novelty, nature versus human manipulation, labour-intensive versus machine-operated industry, uniqueness versus interchangeability.[4]

But consider that we might have loved this bucolic way of life, this direct relationship and sane approach to food production, for a reason. In fact, scientific evidence is growing that there is something protective—or at least less risky—about less intensive methods of animal production.

Here and there, over the past few years, in paragraphs buried deep in the medical literature, suggestions have been made that our intensive food production methods (based on efficient, industrial models) might be related to the rise in foodborne illness. Dr. Martin Blaser, a professor of medicine at Vanderbilt and a physician at the university's hospital, recently brought those concerns out in the open with an article in the *New England Journal of Medicine*. He linked the increase in foodborne disease not only to cultural changes but also to the startling transformation of the agriculture and food industry.

> In the not-so-distant past, most food was produced and consumed locally. In the 20th century, however, the production and distribution of food in the developed countries of the world have become increasingly industrialized. Small farms are being replaced by feedlots, local dairies supplanted by industrial plants, farmers' markets displaced by supermarkets, and local restaurants edged out by huge national chains. The relations among agricultural workers, food processors, and distributors have become increasingly complex and distant. The food chain and steps in food production are being varied in ways that stretch the imagination. From an economic standpoint, modern agribusiness offers many benefits, including the wide choices and apparently low costs of food available to the consumer. The development of national standards has improved food safety, yet certain microbial pathogens persist, the scale of foodborne transmission is increasing, and new hazards are being recognized.[5]

Health-food advocates and the organic farming movement have been saying for as long as they have been around—virtually from the onset of the agricultural revolution, in one form or another—that applying the industrial model to food production is not a good thing. But so successful has industrialized farming been at producing more food at less cost that to question its application in the developed world is widely viewed as heretical. Until recently the chorus of experts from the scientific, agricultural, and industrial establishments has maintained that the final product of intensive farming is nutritionally comparable to organically grown food—or nearly—and that in fact natural or organic farming is not safer but probably less safe in terms of microbes. Only recently has the link between intensive production and distribution and foodborne disease been noted by the medical community. But the relationship between microbes, agriculture, and sanitation is more complex than anyone really suspected. While poor sanitation conditions in developing countries set the stage for microbial diseases, some pathogens, paradoxically, actually appear to be more prevalent in developed countries where sanitation is good.

"Salmonellosis is rare," Blaser says, "in developing countries where sanitation is poor and diarrheal diseases are endemic, but where food production and consumption are local. In contrast, in the United States the reported incidence of salmonellosis has been increasing over the past 50 years"—during which time farming has become increasingly intensive. One reason for this increase is not difficult to spot. "*Salmonella* is an opportunistic organism," Blaser explains. "All microbes are opportunistic, but *Salmonella* is especially so. If a flock of chickens in a developing country gets *Salmonella*, it just stays localized in that small flock. Compare that to the U.S. where flocks are huge and the distribution of animal products is widespread."[6] Indeed, in the United States, there may be thirty thousand—even one hundred thousand—chickens in a flock, housed together in tight quarters with a common water source. It's no wonder that bacterial infections from *Salmonella* to *Campylobacter* spread quickly through such enormous numbers of birds, and that these infections can eventually reach the human population that consumes them.

Blaser is one of those scientists who grasps the larger picture of emerging pathogens and the role of culture and economics in their spread. He is also an expert on *Campylobacter*. Looking at all the elements that favor the spread of this one organism through the process of raising and preparing food animals for our tables can give us insights into the complex ecology of human foodborne diseases.

In many places *Campylobacter* infections are running at higher rates than those from *Salmonella*; it is one of the most frequently identified causes of acute infectious diarrhea in the United States and other developed nations. And yet most people, even many in the medical field, have never heard of it.[7] This widespread ignorance has unfortunate implications for public health.

Campylobacter fetus, then called *Vibrio fetus*, was discovered as early as 1913 when it was found to be associated with cattle and sheep who had aborted or were infertile. The organism was not proven to be a cause of foodborne disease until fairly recently. The late Elizabeth King at the CDC accomplished groundbreaking work on *Campylobacter* in the 1950s, recognizing important differences between the *vibrios* (of which cholera is one) and calling what would eventually be named *Campylobacter* "related *vibrios*." She felt that the bacteria were much less rare than people thought but seemingly impossible to isolate from stools. She noticed, however, that they were found in the bloodstream of people who reported having had diarrhea and that they were identical to those already described in the veterinary work as being found in chickens. Unfortunately, she never saw the outcome of her pioneering work.[8]

The responsibility for coming up with a name no one can spell or wants to try and pronounce apparently goes to the researchers M. Sebald and M. Veron, who in 1963 described the ways in which these tiny creatures differed from the other *vibrios* and proposed *Campylobacter*, meaning "curved rod" in Greek. Looking back through medical history, it seems clear that *Campylobacter* was what Theodor Escherich (of *Escherichia coli* fame) was describing in 1886 in the large-intestinal mucus of young children who had died of what he called "cholera infantum." His drawings show what clearly seems to be the elusive organism, although he could not manage to cultivate them and did not think they had caused the disease.

One reason *Campylobacter* wasn't identified sooner is that the organism is what bacteriologists call "fastidious," or hard to find, because it can be overwhelmed in a culture by the growth of more aggressive bacteria such as *Salmonella*. Martin Skirrow, honorary emeritus consultant microbiologist at Gloucester Public Health Laboratory in Great Britain, is the person who gets most of the credit for making *Campylobacter* routine to detect, and he was among the first to make clear its role as a major cause of human diarrheal disease. Then at the Public Health Labs in Worcester, England, he remembers being called in to look at an unusual organism in a blood culture from a one-month-old infant. It was like nothing he had ever seen, but he remembered a series of case reports published in the *British Medical Journal* of three similar blood cultures. That report cited articles from the *Journal of Pediatrics* and the *Journal of Infectious Diseases* by Jean-Paul Butzler of Brussels. Butzler had first isolated the organism in a nurse in his lab—and then he claimed to have isolated the organism from 5 percent of children with diarrhea in a Moroccan community in Brussels.

To find it in so many children was startling to Skirrow. "Here was this organism," he says, "of which there were only about thirty or forty reports in the world literature ever, so I thought, either this work is a load of rubbish, which is what I rather expected—I was rather skeptical—or else it's extremely important." Skirrow then set about looking for it, using Butzler's techniques of filtration and a culture medium that contained antibiotics to which the more aggressive bacteria were sensitive and to which *Campylobacter* was resistant. "Within almost days, I was getting positives."[9]

"I more or less reproduced what Butzler did," says Skirrow modestly. But it was perhaps what Skirrow did next that made his work so memorable. He conducted a study in a nearby medical practice that found the organism to be present in 7 percent of the patients with diarrhea and in none of the controls. Being in a public health lab in a rural area, Skirrow was able to follow up people who tested positive. He found a connection between disease and close association with animals or animal foods. "Three patients—two young farmers and a child—were apparently infected from chickens. In each case the feces of the birds with which they were in contact contained organisms that were indistinguishable from their own." Similarly, three more patients—two butcher's assistants and a housewife—were presumed to have been infected while handling dressed chickens found to be contaminated with *Campylobacter* that matched their own strains. Dogs with diarrhea were thought to have transmitted the infection in three of the fifty-seven cases he found. Skirrow suspected that they picked up the organism from eating chickens.

As Skirrow became more interested in the organism, he felt that he needed to meet Butzler. A microbiological gathering provided the opportunity. They discovered that they had both tried to interest others in setting up similar techniques to look for the organism—"We weren't trying to hide our lights under a bushel," Skirrow says. Yet while everyone said, "Yes, it sounds very interesting," no one was willing to change his or her lab techniques. This proves to be a critical point in the widespread failure to identify a number of emerging enteric pathogens. Getting labs to change their routines, to try new culture media, or to begin looking for bugs they haven't looked for before—or even to stop looking for organisms that are no longer important causes of disease—is difficult, and as a result, important pathogens and even widespread outbreaks of disease are routinely missed. The problem in this case was the filtration step, which made the process tedious. (It was a method veterinarians had long used, and Butzler had gone to veterinarians when he wanted to isolate *Campylobacter* in humans, crossing that invisible line that once divided human and animal researchers.) Each went back to his lab to look for a way around it. When Skirrow announced the new selective medium, the door was opened for easy culturing of the bug. His 1977 article about it in the *British Medical Journal* would change Martin Blaser's life.

Blaser had finished a residency in internal medicine at the University of Colorado Medical Center and decided, more or less at the last minute, to get a further specialty under his belt. By chance, an opening came up in infectious disease. He began the residency on July 1, and on July 10 he was called to see a patient at Colorado General Hospital who had meningitis and bacteremia—the presence of bacteria in the blood, a place where it should not be. In general, Blaser says, bacteremia "is a bad thing." The lab had isolated *Campylobacter fetus*, "an organism I knew nothing about."

"I read about it, and we took care of it," he remembers. But he was intrigued.

About the same time Blaser was assigned to present a conference topic and was looking for a subject; he decided to pick something obscure. One evening in the hospital library he came across an article in the *British Medical Journal* called "*Campylobacter Enteritis*: A 'New' Disease." It was by Martin Skirrow. He realized that *Campylobacter enteritis* was the cousin of what had infected his patient. Blaser made *Campylobacter* his subject.

After Blaser gave his lecture, a Chinese-American microbiologist, Wen-Lan Wang, came up to him and said that she was interested in *Campylobacter* as well. She'd isolated it from several patients in the past. The two decided to run a diarrhea study of patients in Denver to see whether they could find the bacteria. The search began on March 1, 1978, and by March 15 they had their first positive. The results were truly startling. This virtually unknown pathogen was actually a major cause of serious diarrhea. "That was the first real study in the United States, and we just kept getting positives. We found that in the Denver area it was more common than *Salmonella* and *Shigella* combined."

It was clear now that *Campylobacter* had been there all along, an unseen, unidentified cause of disease, slipping through the noose of medical science even as late as the 1970s, making fools of all the disease experts, the fancy techniques, the high-tech equipment. Revealing that even now—perhaps especially now—we seldom see what we're not looking for.

Blaser first saw the *Campylobacter* bacteria in 1977 and found them beautiful. They are long, curved organisms, and under a phase-contrast or dark-field microscope they can be seen to have a flagella at each end, allowing them to be highly motile. "They move around a lot; they dart this way and that way," says Blaser, moving his hands back and forth. Over the last twenty years, from the many research projects he and others have conducted, a broader picture of the organism and what it can do has emerged.

> We know the clinical and epidemiological features of the infection. We know where it comes from. We know how people get it and what the risk factors are. When we talk about *Campylobacter* and human disease, we are mainly talking about *Campylobacter jejuni* and *Campylobacter coli*. These organisms live in the intestinal tract of animals—the animals we use for food, such as cattle, sheep, goats, pigs, and poultry—and also, on occasion, in domestic dogs and cats, wild birds, rodents, and primates. Humans are merely accidental hosts.

Just how accidental had been brought home to Martin Skirrow when he investigated a case of *Campylobacter* illness in a farmer near the village of Pershore, southeast of Worcester, England. The farmer was helping a ewe give birth, and when the fetus emerged unresponsive, he had given it the "kiss of life." It was useless to the dead lamb, and the farmer had come down with the *Campylobacter* infection that had probably caused the unsuccessful birth in the first place.

When Blaser's fellowship in Denver ended in 1978, he joined the CDC's Enteric Diseases Branch. He was hardly the only person interested in *Campylobacter*. Martin Skirrow's culture medium and his study revealing the extent of disease caused by this pathogen had excited other researchers around the world, and it was already clear that the bacteria were indeed a major cause of diarrheal disease almost everywhere they looked. They were looking, however, mainly in the developed world. The studies that had been done in Africa and Brazil had not established precisely what role *Campylobacter* played in causing disease there, although one researcher had shown that it was frequently isolated from healthy children in South Africa. That indicated that the bacteria might be acting differently in different environments.

The CDC sent Blaser to Bangladesh. There he met up with Roger Glass, already on assignment in Dacca from the CDC. Because Bangladesh is a developing country, diarrheal diseases are endemic, especially among children. If research could reveal the extent of infection with a particular pathogen such as *Campylobacter*, they might be able to institute specific measures to control it.

Working with Bangladesh's International Center for Diarrheal Disease Research, Blaser and Glass set up studies among three populations. They did a random sampling of patients at a clinic known in the community as "the Diarrhea Hospital"; their second group was composed of patients who had specifically reported having both blood and mucus in their stools; and the third group consisted of healthy children for comparison.

One of the challenges was to find a way to isolate the cranky pathogen under the lab conditions typical in the developing world, where high-tech equipment was rare. Because bottled gas mixtures and vacuum jars were not available to provide the low level of oxygen the bacteria preferred, they used candle jars; the candle uses up the oxygen and is an incubator, creating the 42-degree Celsius temperature the organism prefers. They found it worked well enough to be used satisfactorily.

Blaser and Glass's study found, to their surprise, that *Campylobacter* infection was very common among young healthy children in the area but decreased quickly as the children grew older. That suggested, says Blaser, that immunity was developing. By the time the children in this environment were around five, he says, they were completely immune. The situation would be quite different in the developed world. With better overall sanitation, the sudden introduction of a heavily contaminated food product might be a serious challenge to an unsuspecting intestinal tract that had no previous experience with the bug.

After his stint at the CDC, Blaser returned to Denver, where his research colleagues Marc LaForce and Wen-Lan Wang remained. With them, he became deeply immersed in research. The interest in *Campylobacter* had been growing everywhere. In 1979 Dutch scientists had reported what they called "an explosive outbreak of *Campylobacter enteritis* in soldiers." It happened during a survival-training exercise for 123 young Dutch military cadets. They were given live chickens, which they killed and cooked over a wood fire for their evening meal. During the following week, eighty-nine of them became ill with symptoms of classic campylobacteriosis. The outbreak may have occurred, the authors suggested, because the chicken probably wasn't cooked thoroughly; nor, they suggested rather dryly, was sanitation as good as it might have been during the preparation.[10] The potential for an army campaign to be seriously hampered by foodborne disease was all too obvious. Diarrhea had changed the course of history before, and it might well do it again.

This was, in fact, one of the earliest episodes linking the pathogen to chicken. Another military episode happened in Holland not long afterward. This time it was soldiers in barracks, and the culprit seemed to be raw hamburger meat.[11] And other outbreaks were being detected and identified. As often happens, scattered outbreaks can present a confusing picture early on, as knowledge about a newly identified pathogen is developing.

In April 1982 Dr. Gregory R. Istre was on assignment from the CDC with the Colorado Department of Health. The department received a report that eleven of fifteen members and friends of an extended family had become ill between April 20 and 24 and that C. *jejuni* had been isolated from the stool specimen of one of them. The group had had a party a few days earlier and cooked out. The health department began an investigation on April 28, calling on Dr. Blaser for assistance. He was back at the University of Colorado School of Medicine by then but also working for the Veterans Administration Medical Center as chief of the infectious disease section. With Istre, Pamela Shillam and Dr. Richard Hopkins of the Colorado Department of Health, he began an investigation.

The patients had diarrhea, abdominal cramps, and fever, and some of them experienced headaches, weakness, and lethargy. Three of the eleven had grossly bloody stools. *Campylobacter* was isolated from one of the ill cases who had not received treatment with erythromycin (which can destroy the evidence in the intestines). Their illness lasted six days on average. Clearly the party was the problem, since that was the only time they had all been together. When the ill were interviewed, they remembered eating barbecued chicken, cake, and ice cream. The article about the outbreak described how the meal was prepared:

> Whole raw chickens had been purchased at a local grocery store and cut in half before being placed on a barbecue grill. A homemade sauce made of a commercial brand of butter, salt, and garlic salt was

> brushed over the chickens as they cooked over the coals for approximately 45–60 minutes. The chickens were turned with tongs and never speared or punctured. The ice cream had been bought at a grocery store; the cake was made at home. Raw eggs were used in the preparation of the cake mix, which was then baked for 35–40 minutes at 375 degrees F. No raw milk was used in the cake mix or at the party. None of the persons involved in preparing or serving the food had had any gastrointestinal symptoms before the party.[12]

What the investigators concluded from the study was that the chicken was the most likely source of the *Campylobacter*. Those who had eaten the skin or whose chicken had been clearly undercooked, but who had eaten it anyway, were also sicker. What they suspected was that the brush used to apply the sauce had continued to transfer *Campylobacter* to the chicken as it cooked. And then it had been used to apply more sauce even after cooking. The *Campylobacter* burden had been amplified with each swipe. Although *Campylobacter* is found elsewhere, its presence on chicken is not unexpected.

The Poultry Connection

The optimal growth temperature for C. *jejuni* is 42 degrees Celsius (about 107 degrees Fahrenheit). Coincidentally, that is the body temperature for birds. *Campylobacter* is a commensal in birds—meaning that it seems to live quite happily in them without causing disease. The bacteria present a very serious threat to humans, however, with the dramatic increase in chicken consumption in this country. Research has determined that *Campylobacter* is never found in the absence of fecal coliforms, the common bacterial residents of the gut.[13] And as Martin Skirrow and others have pointed out, humans come in contact with *Campylobacter* on animal foods as a result of fecal contamination during processing. Studies have shown that the majority of broiler carcasses and parts sold in U.S. retail stores are contaminated with *Campylobacter*. In fact, Blaser assumes that they all are. Like most bacteria, they are destroyed by the heat of thorough cooking, if the rest of the kitchen hasn't already been contaminated in the preparation process.

Campylobacter also has an advantage in that it survives storage at refrigerator temperatures even better than at room temperature, and it likes an environment with only a small amount of oxygen—which sounds exactly like the conditions in plastic-wrapped packages of chicken in the refrigerated meat section of the grocery store.

While living chickens may have no trouble with *Campylobacter*, in humans it burrows into the mucosal layer of the intestines, where it finds a happy home—not ideal conditions, it would prefer to be a bit warmer if it could have its choice—but it can manage nicely enough. Nestling in, it can find the microaerophilic situation it prefers—about 5-percent oxygen is ideal. It causes a disease characterized by watery diarrhea, sometimes bloody, usually accompanied by fever, malaise, and abdominal pain. The diarrhea begins within a week after exposure and usually lasts about a week. If that were all it did, it would be unpleasant enough, but up to 20 percent of patients may have a relapse or a prolonged or severe illness.[14] It can produce symptoms that result in unnecessary appendectomies, and quite surprising to most people, it can cause reactive arthritis; Reiter's syndrome, characterized by arthritis, urethritis, and conjunctivitis; and Guillain-Barré syndrome, a life-threatening disorder of progressive paralysis. In fact, recent studies indicate that from 38 percent[15] to 46 percent[16] of cases of Guillain-Barré syndrome follow recognized infection with *Campylobacter jejuni*. It has also been associated on occasion with meningitis, convulsions, and bacteremia, and, rarely, miscarriage.[17]

Clearly a chance encounter with *Campylobacter* can leave its victims with debilitating souvenirs that substantially alter their lives. While it doesn't often cause death, it certainly can. The CDC estimates that between 200 and 730 people die of *Campylobacter* infection in the United States each year, and it considers that figure an underestimate.[18] When added to the illnesses and deaths attributed to *Salmonella*, the tendency of health-conscious Americans to make chicken the centerpiece of their diets seems ironic. As *Time* magazine put it, "The good news about chicken is that thanks to modern processing techniques, it costs only about a third of what it did two decades ago. The bad news is that an uncooked chicken has become one of the most dangerous items in the American home."[19]

C. jejuni can be cultured from most chicken and turkey carcasses sold at retail outlets not only in the United States but in Britain, the

Netherlands, Canada, South Africa, and Australia, says Skirrow. And interestingly, in studies of families in which everyone had eaten chicken, it was the cook who generally got sick. All too often, that is where the blame has gone.

Skirrow says that contaminated meat and offal can cause infection in any of three ways: The contaminated product itself may be eaten raw or undercooked; the person preparing the food may become self-infected through handling the raw product; or other foods that are to be eaten raw or without further cooking may become cross-contaminated from the raw product. With *Campylobacter*, "evidence indicates that all three methods can operate."

Thoroughly cooked chicken is perfectly safe—for everybody except the person who prepares it apparently. The difficulty for those not trained in safe food-handling techniques—which includes most people—is in trying to avoid infecting themselves or cross-contaminating other foods in the kitchen.

It is assumed that *Salmonella* has to multiply before causing infection. *Campylobacter* does not, and the evidence is that the infective dose is quite small.[20] Contaminated food products come, it would seem, with plenty of bacteria—more than enough to make you sick. The cook doesn't have to do anything wrong to make contaminated food worse. But it's up to you, the consumer, to kill them before you eat the food. The question most people might ask is, How did the bacteria get there?

Awash in Pathogens

If there is one thing that significantly transformed the chicken industry it must have been the ability to take a number of chicks, provide them with exactly the same conditions for exactly the same amount of time, and produce year after year birds of exactly the same size. The consumer might wonder why years ago chickens came in all sizes but now there are virtually only two: roasters, which are larger, and broilers. The answer is that every aspect of a chicken's life is organized specifically to produce a uniform product that grows as fast as possible while consuming as little energy as possible. Once the processing plants could be certain that chickens were going to be approximately the same size, they could replace with machines many of the workers who had been needed to cope with the processing of birds of different sizes.

After the chicken is hung, stunned, bled, and killed, it is scalded and its feathers are plucked with mechanized rubber "fingers" that imitate the plucking motion of real hands. These rubber fingers quickly become worn and cracked, collecting and spreading the pathogens routinely found on the chicken feathers. They contribute to contamination by pressing the bacteria into the bird and, if not replaced often, by carrying pathogens from bird to bird. Then the bird is singed, rinsed, and its feet are removed. Some of these procedures can remove pathogens—which can then be reintroduced by the rest of the process, but they also can open the door to more thorough impregnation of bacteria on the product.

After the chicken is "vented" (opened), it is gutted, also by machine. A metal hook reaches up inside the bird and pulls out the intestines. The trouble is, there are going to be variations even among chickens of almost exactly the same size. Unlike human fingers, the hook has no feelings, no sensitivity. It can't adjust for intestines that cling more tightly to the cavity or are irregular in some other way. Often the intestines break and the contaminated contents spill out. These birds should be removed from the process, but often they are not. On they go, contaminating other birds along the way.

The chickens are rinsed twice more, but then they are dragged through a bath of chill water, where they remain for an hour. The previous rinses do not remove all the bacteria; it has been demonstrated that *Campylobacter* in particular burrows into the skin and nestles in the now-empty feather follicles and avoids being washed away. How dangerous is the chill bath? A USDA study revealed that at one federally inspected slaughter establishment 58 percent of the chickens were contaminated with *Salmonella* before they were eviscerated. That dropped to 48 percent after the two washes. But after an hour in the chill water, contamination increased to 72 percent. The chill bath is clearly a point at which bacteria can be spread from bird to bird.[21] There is little reason to think that *Campylobacter* contamination, especially given the high percentage of chickens at the end of the line that test positive, acts any differently from *Salmonella* during the process. In fact, *Campylobacter* can survive processing—even applications of chlorine and other bactericides—by hiding in feather follicles.[22]

Time quoted a former USDA microbiologist, Gerald Kuester, on processed chicken: the "final product is no different than if you stuck it in the toilet and ate it."[23] That sounds like gross hyperbole. In fact, the reality is worse. A recent study by the University of Arizona found more coliform bacteria in the kitchen than on the rim of the toilet. The report of this study in *Science News* discreetly did not mention where this gross contamination came from, but the only reasonable explanation is that it arrives as a bonus on the animal foods people bring into their kitchens. The bathroom is cleaner because people are not washing their chickens in the toilet.

Why does poultry spend so long in the chill bath? The "fecal soup" process adds 8 percent water weight. That extra 24 percent of contamination the chill water adds can be credited to pure greed. Thus, the industry has looked for techniques that allow it to keep the water-chilling process while producing a cleaner bird. Continually adding chlorine, the USDA study demonstrated, did not actually reduce the *Salmonella* but held the increase to 3 percent over the pre-chill count.[24] The chickens could be air-chilled, which has been shown to reduce bacterial levels of *Campylobacter*, in particular, or chilled in individually sealed plastic bags, but that would prevent them from absorbing that all-important (to the industry) water.[25] *Time* reported that giving up the added water weight would cost Tyson, a major chicken producer, about $40 million in annual gross profits. But one agricultural specialist in the USDA wonders whether consumers will continue to tolerate chlorine on their chickens and whether it is safe for the workers who have to handle it.

Chilling takes place before the birds are cut up. As conveyor belts carry them to the cutters, more contamination can occur. Workers report picking up chicken parts off the floor and sending them along to be frozen and shipped to consumers.

Chicken Cities

A fair question might be why chickens arrive at the processing plant so thoroughly contaminated with *Salmonella* and *Campylobacter* in the first place. The life of a chicken destined for the dinner table is carefully organized from its conception to its arrival in the supermarket. Nothing is left to chance. The chicks are descended from top breeding stock and carefully selected for the characteristics that enhance the economics of poultry rearing. Only a few breeds are suitable for each purpose, and in general the growers stick to one or two breeds. The result: a lowering of the diversity that would guarantee that some would resist certain diseases. Broilers arrive at a grower's plant as baby chicks, free of *Campylobacter;* while the bacteria has an affinity for poultry, the relationship is not automatic. These designer birds—they put on meat in all the right places, such as the breast, because Americans prefer the white breast meat—grow quickly and consume less feed than they once did (down from ten weeks to maturity on ten pounds of feed in 1960 to six and a half weeks today on eight pounds of feed). Many receive antibiotics in their feed to ward off the diseases encouraged by close confinement. Bacitracin, one commonly used, has the added advantage of encouraging growth, although it is not officially classified as a growth enhancer.

Broilers are raised in close quarters but not, as is often thought, in cages. (That fate is reserved for laying hens.) A grower with one house would be an oddity; most have four or five houses. With 50,000 or more birds per house, an average grower can produce five and a half flocks a year—or well over one million birds for the insatiable U.S. market. North Carolina, the fourth-largest poultry-producing state, grows 675 million birds a year. The birds are loose in these houses; they can move around at will, although doing so becomes more difficult as they grow. Raised by growers in 40' x 400' houses where 25,000 may live together in fairly close confinement, each full-grown bird has about eight square inches. Their increasing inability to move about freely as they become more cramped has the same effect on chickens as on humans. They gain weight, they begin to feel stressed, and they subsequently become increasingly vulnerable to disease. Suddenly they become infected with *Campylobacter*. It can happen almost overnight, researchers say. One day the flock is clean; three days later all the birds are infected. There is worldwide interest in answering the question of how they get the infection.

In 1994 a group of experts from around the world gathered at Bilthaven in the Netherlands, at a meeting jointly sponsored by that country's National Institute of Public Health and Environmental Protection (RIVM) and the World Health Organization, to discuss ways of controlling *Campylobacter* in poultry. The conference report began with what is known: Flocks are more likely to become contaminated when a breakdown in biosecurity or hygiene control on the farm allows contamination to enter the environment. There are any number of possible reservoirs in nature and many different ways a poultry flock can be exposed. Nevertheless, most of the factors that contribute to infection are made worse by the nature of factory farming: the limited diversity of the stock; the crowded mass of animals; the necessity of moving animals to slaughter under stressful conditions; the common sources of food and water.

Unlike *Salmonella*, *Campylobacter* is not believed to be transferred vertically—that is, from hen to egg to chick. But chicks are highly susceptible to *Campylobacter*, and once one chick is infected, it sheds billions of organisms that can infect the other birds quickly. Young birds can be infected when they enter a contaminated facility: The infection from the previous flock can hide in out-of-the-way places, such as rafters and lighting fixtures. While contaminated feed is not as much of a problem with *Campylobacter* as with *Salmonella* (*Campylobacter* prefers more moisture than most feed contains), drinking water is a big problem says Frank Jones, a poultry specialist with the Department of Poultry Science at North Carolina State University in Raleigh, North Carolina. When water is contaminated, or when it is clean but the delivery systems become contaminated—say by an infected chicken walking in the water trough—the flock quickly becomes infected. Visitors and farmworkers can also bring in the bacteria or spread it from one house to another. And finally, the transport vehicles and the transportation system can encourage spread. More bacteria have been found on chickens after being transported, whether they were alive or slaughtered. In fact, the common practice of withdrawing food from animals that are to be slaughtered, the lack of food and water during transport, and other causes of stress, such as heat and cold and antibiotic therapy, are also thought to contribute to the shedding of *Campylobacter*.[26] The humane treatment of food animals, surprisingly perhaps to some, seems to be a factor in making them safer for consumption, and producers in some countries are applying this wisdom.

One potential source of contamination is litter—the bedding material on the floor of the chicken house. It is usually wood shavings, rice hulls, or some other material available locally, depending on the agriculture in the area. It quickly becomes mixed with chicken manure. In Europe it is typically cleaned between every flock, but in the United States it normally stays around a long time—it remains when a flock is sent to be slaughtered, and it is there when the new birds arrive. It might stay on the floor a year or even two, building up in layers, before it is scooped out and taken away. European growers visiting U.S. production facilities inevitably find this practice shocking.

Growers in the United States don't routinely clean the litter out of the houses between flocks, says Jones, because "we didn't start off doing that. We haven't ever done it." Most of the houses in the United States have dirt floors, not concrete. Each time they are cleaned, some of the dirt is scraped up and eventually would have to be replaced with more dirt or rocks. To install concrete floors in existing houses would be an expensive option. "We're really locked into the system we've got. . . . It would cost dearly to change it now," says Jones.[27]

The U.S. practice is clearly out of sync with accepted standards elsewhere and continues only because consumers are not aware of the consequences and have not demanded that it be changed. The Bilthaven report concluded that, while other factors contributed to the infection in animals, the stocking principle of "all in, all out" needed to be encouraged and that "all the litter should be removed prior to thorough cleaning and disinfection of both premises and equipment," followed by a shutdown time.[28]

While flocks of broiler chickens all over the world are thoroughly infected with *Campylobacter*, there are a few exceptions, such as Norway and Sweden, where infection is, by comparison, very low. What do they do that other countries do not?

Sweden shows the impact consumers can have on food production. When Swedes learned in 1987 through media reports that their chickens were contaminated with *Campylobacter*, consumption of chickens, already low at a mere seven and a half kilograms per person per year, dropped at once by 40 percent. The industry itself instituted a control program based on what the growers already knew about controlling *Salmonella*, even though the bacteria have somewhat different lifestyles. The Swedish *Salmonella* program begins with the breeding stock, making sure it is free of infection. Feed is heated to ensure that it is *Salmonella*-free. Hygienic measures include sealing the chicken houses against rats, mice, and birds, and strict requirements are set up for anyone entering or leaving the houses. When the animals are slaughtered, the houses are emptied, cleaned, disinfected, and sealed until the new chickens arrive. Contaminated birds are not taken to slaughter, avoiding the problem of cross-contaminating other birds during transport, slaughter, or processing, and flocks found positive for *Salmonella* are destroyed. These practices have virtually eliminated the *Salmonella* problem in Sweden. Although most birds are still water-chilled, the use of chlorine is prohibited in Sweden, and the practice of air chilling is becoming more widespread.[29] Dr. Eva Berndtson of the Swedish University of Agricultural Sciences in Uppsala says that Sweden has other advantages: "a good location without large, dirty rivers through farm land, and generally not too dense farming."

When the Swedish public became aware of the presence of *Campylobacter*, the measures were tightened further, especially with regard to biosecurity. To avoid further stress chickens are now transported in clean cages and trucks that are heated or cooled, as needed, and given adequate ventilation and water. But *Campylobacter* is difficult to eliminate; although the contamination rate has been brought down to 12 percent, the bacteria are still present on some birds going to slaughter. Another factor in Sweden's low rate of infection may well be the prohibition against using antibiotics to promote growth or to prevent disease. (They are used, of course, to treat disease where appropriate.)[30]

Certain antibiotic use has clearly created resistant strains. Between 1982 and 1989, 883 strains of *Campylobacter* were screened for quinolone resistance by a group of Dutch researchers at laboratories in the Netherlands. When the seven-year period began, no strains were resistant. At the end of the period, 14 percent of the *Campylobacter* strains from animals were resistant. (And during the same period the prevalence of resistant strains in man increased from 0 percent to 11 percent.) Their explanation was that, since enrofloxacin was widely used in poultry and the almost exclusive transmission route of *Campylobacter* in the Netherlands was from chicken to man, the resistance therefore was mainly due to the use of enrofloxacin in the poultry industry.[31]

Some experts say that the practice of feeding birds antibiotics regularly simply to promote growth or as a prophylactic to ward off disease is waning in the United States, but major producers such as Purdue still use bacitracin.[32] Nevertheless, is there a chance that routine feeding of antibiotics has created an opportunity for *Campylobacter* by killing off the bacterial competition in chicken intestines?

Martin Blaser says that *Campylobacter* is naturally resistant to bacitracin, and so, by killing other bacteria in the chicks, the antimicrobial is actually selecting for *Campylobacter*.[33] Anita Rampling noted something else: "If you want to do experiments with *Campylobacter* and *Salmonella*, one of the ways of making these experiments work in animals is to give them antibiotics and then to give them these bugs and then you can produce invasive infection." But Rampling sees antibiotic use and its possible role in favoring *Campylobacter* infection as only part of the problem: "The whole system of agriculture and medicine doesn't bode well, I'm afraid. I think the use of antibiotics is a problem, but also the ever-increasing intensification and the tendency to breed hybrids so that everyone has the same strain—that's causing trouble."[34]

The High Cost of Cheap

Could we change the way we produce and process chickens and substantially lower the contamination rate? The EU is working on standards for "extensive" (organic) farming that would, among many changes, reduce the number of chickens in flocks to forty-eight hundred.[35] With a smaller number of birds, certainly an infection could be controlled more easily and more cheaply. Ironically, the conditions of intensive rearing of chickens not only set an efficient stage for the spread of *Campylobacter* and other infections but create the most effective conditions for their reduction. When birds are raised in a closed, "biosecure" environment, it is possible to make certain that food and water are clean and that the possibility for outside contamination is reduced. Researchers are looking at why some breeds seem resistant to *Campylobacter*, with the thought of even further selective breeding—or perhaps genetic engineering—for resistant characteristics. Also under consideration as means of controlling the bacteria are vaccination or competitive exclusion—dosing the bird with microbes that compete with *Campylobacter* in its intestines. Inevitably, each high-tech solution seems to demand yet another high-tech solution.

At the USDA research facility, Dr. Norman Stern, by contrast, is focusing his work on reducing the amount of *Campylobacter* in chicken rather than eradicating it. What he wants is to lower the overall bacterial load on any one poultry carcass. All too often, however, there is little communication between the veterinary researchers and the medical community. Almost every day now what was considered the infectious dose of microbes is being reconsidered and lowered. Skirrow has suggested that it might be quite low indeed, especially if the infected person is old, young, a cancer patient, or immune-compromised by AIDS or some other illness. AIDS patients have an especially high rate of infection with *Campylobacter* at forty to one hundred times that of the non-HIV population. What Stern is proposing may not be a safe option at all.

Other researchers think *Campylobacter* could be eliminated—at a cost. But there is no incentive, economic or otherwise, to reduce the level of a pathogen that the consumer has never heard of. People will not pay more for a *Campylobacter*-reduced chicken if they don't know that they are at risk from the bacteria. In fact, Stern agrees that the only thing standing between the consumer and *Campylobacter*-free chickens is will and economics. Even simply cleaning out the litter frequently would have its costs. Chickens in Europe do cost more. Berndtson says that she can buy fresh Swedish chickens for about SKr 30 a kilogram, which, at SKr 6.5 to the dollar, works out to about $2.20/pound, as compared to the $1.00/pound broilers in U.S. supermarkets. In fact, all food in Europe costs more because of higher labor costs and other factors.

"But we are able to produce a great deal more chicken than they are," Stern responds. Relatively high production in the United States allows people to buy chicken more often. "It's not bad for people to know there are those trade-offs." He doubts that people would pay more for clean chickens.

Actually, consumers are seldom given the choice. Chicken mass marketing in the United States seems to avoid the question of microbial safety altogether. (The safe-handling labels on raw meat products are a government mandate.)

Swedish techniques can be applied elsewhere with good results. The Danish had a considerable problem with bacterial contamination of poultry up until recently, when the largest Danish poultry company, Danpo, was bought out by the Swedish giant Kronfagel. The process of producing all *Salmonella*-free chickens takes time, but the company reports that its chickens are now much less contaminated with *Salmonella* and have reduced levels of *Campylobacter*, yet cost virtually the same. The company has also dedicated one slaughterhouse to processing only *Salmonella*-free chickens, which are labeled as such, a purchase option that American consumers are not given.

Consumers will pay more for chickens when there is a clear difference in what is offered—even though that difference may be quality, not safety. They may pay more for a chicken because it simply tastes better—not a frivolous matter. Chickens generally taste of what they are fed. Since protein makes them grow faster, they are no longer simply fed grain but given feed mixtures that contain rendered animal protein (often made up of their dead ancestors) or sometimes fish meal—which gives a fishy taste to the meat. Mimi Sheraton, writing about her mother's preparation of that Jewish classic, chicken soup, wrote in 1979:

> Only in recent years did she realize how difficult it is to make really good chicken soup with the chickens one has to work with today. Naturally fed chickens, which were allowed to scratch and which were then killed to order and bled to death—the kosher method of killing—did actually produce lighter, clearer, more golden soup of more delicate flavor. Chickens are not nearly as good as their ancestors.[36]

Perhaps flavor, not safety, is the reason, but demand is strong for the chickens produced by Salvadore Iacono in East Hampton, Long Island. He feeds them grain, uses no antibiotics or chemicals, and thinks sanitation is the key to healthy, disease-free birds. And one other thing. "We change the litter every time between chicks. That's where the real work is. We work hard at it. We take out the old litter, scrub it down, disinfect it, and air it out for a while."[37] His chickens sell for $2.20 a pound.

It will take education to make consumers aware of the hidden costs of cheap food. As Blaser wrote in his opinion piece "How Safe Is Our Food?" for the *New England Journal of Medicine*, we seriously need to question whether, when the enormous costs of foodborne disease are taken into account, our food is actually cheap. A Chinese proverb expresses it well. "Cheap is not cheap; expensive is not expensive." It takes only one round of foodborne disease in a family involving a hospital stay that could run into thousands of dollars to make it very clear that saving $1.20 a pound on chicken isn't such a saving after all.

Another cost of contaminated chickens is in reduced sales abroad for American producers. If American consumers are not aware of the state of our poultry, potential importers are. When in the spring of 1996 Russia refused to take our chickens, the refusal was brushed aside by the poultry industry as simply a ploy and a bargaining chip in a trade squabble. The chicken exports were important to the U.S. market. Since most Americans prefer white meat and also buy their chicken in parts, a great deal of dark meat is left going begging. The Russians, who prefer dark meat, are obvious customers. The convenient swap created a $500 million market for U.S. poultry producers. Some protectionism was involved in the resistance to buy American. Russians had raised tariffs on imported chickens significantly. The country didn't want America's cheaper chickens undercutting its own production. But once again, cheap is not cheap if it contributes to foodborne disease, and when the Russians complained about the safety of U.S. poultry, they knew exactly what they were talking about. The announcement by Agriculture Secretary Dan Glickman on March 26 that poultry export to Russia would resume represented a month of hard work on the part of his staff. It wasn't a tariff trade-off that clinched the deal, but an agreement by the USDA to carry out special reviews of U.S. poultry production facilities for the Russians. The reviews would be conducted jointly by U.S. and Russian veterinarians for a time. Then the FSIS would take over "to insure that facilities exporting poultry to Russia continue to meet the established criteria."[38]

In other words, the Russians demanded new and higher standards for the chicken they would import—their standards—and the USDA was going to see to it that they got them. The Russians also asked for more information on the U.S. veterinary certificate to provide "an additional assurance that U.S. poultry is safe and wholesome," and they insisted on the right to spot-check 10–15 percent of the U.S. poultry facilities annually. What American consumers might reasonably wonder is why they weren't important enough to get chicken of the same quality and the same right to inspection as the USDA granted the Russians. There's not much satisfaction in winning the cold war if the Russians get safer chickens than we do.

A Widening Environmental Problem

Campylobacter, unfortunately, isn't just in chicken. If there is one food product that stirs near-violent emotion in enteric disease specialists, it is raw milk. That is because it is implicated in so many outbreaks of foodborne disease. Many health advocates still promote its consumption, saying that milk is more nutritious before it is pasteurized. It certainly tastes better, although few people remain who know that, as milk has been regularly pasteurized for many years. And certain nutrients are harmed by the process—about 10 percent of milk's nutritional value is lost in pasteurization. But food scientists say that the safety advantage far outweighs the disadvantage of nutritional loss.

Raw milk was not always so dangerous. When my cousins were young, they spent every summer at a family camp, a log house with bunkhouses nearby, in the Virginia hills that push up against the Blue Ridge Mountains. They invited their friends and spent their time riding ponies bareback over the rough terrain. Each summer their father would buy a cow, which provided milk for the cooking and drinking needs of what always became an extended family. He would sell it back to the farmer in the fall. No one remembers ever getting sick.

Today raw milk is clearly a risky product. Whether this is because something in the way we raise dairy cows has changed over the years (and certainly there have been changes in what they eat, what drugs they receive, and how they are milked) or because the human population has become more susceptible as general sanitation improves, no one has bothered to find out. But widespread outbreaks of milk-borne disease followed the industrialization in the 1800s of the diary industry and the subsequent mass-production and distribution of milk and dairy products.[39]

For whatever reason, the number of outbreaks today associated with raw milk is impressive. Some have involved *Salmonella*. Very recently they have been associated with the virulent new pathogen *E. coli* O157:H7. Sadly, the outbreaks often involve schoolchildren who visit a farm and are given raw milk by their teachers.

It is memories such as those of my cousins that lead people to drink raw milk. Telling people who have good memories of a product that once gave no trouble that it is now dangerous does no good unless one explains that something has happened to make it dangerous. But such explanations implicate our systems of production, so too often vital information is not disseminated and people continue to do what their experience has led them to think is appropriate.

In October 1982 a group of college freshman fraternity pledges took a break on a dairy farm owned by the parents of one of the students. They had brought coolers full of food prepared by their fraternity kitchen: ham, potato salad, and hamburger. Some of the same food had been left behind with the fraternity members who didn't go on the trip. Arriving early, they ate breakfast at the farm: cereal, eggs, and milk from the farm's cows, taken from the refrigerated bulk tank before being sent for pasteurization.

The group of thirty-one spent the day relaxing outdoors and in a hot tub, among other things. They returned to the fraternity house that evening. Over the next week, nineteen of them developed acute gastrointestinal illness. All the sick students had infection with *Campylobacter jejuni*. But there was a very interesting aspect to the outbreak.

The health department in Corvallis, Oregon, naturally called on Martin Blaser to help them with the investigation. He, after all, had experience both with the CDC and with investigating *Campylobacter*. He and Dr. Elizabeth Sazie from the Benton County Health Department conducted a classic case-control study.[40] They looked at the water and sewer arrangements at the fraternity house and in the kitchen and found nothing remarkable. The sanitary arrangements at the dairy farm were equally satisfactory. The food shared by both those who had gone on the trip and those who had stayed behind seemed highly unlikely to be a problem since none of those who had not gone to the farm became ill. The culprit looked like the milk, but there was one hitch: Some of those who had drunk milk did not become ill; neither did the farm family or the farm employees. It turned out that four of the pledges who drank the raw milk without becoming infected had, in Blaser's words, "substantial prior raw milk consumption." The six people on the farm drank raw milk regularly as well, and they were immune to *Campylobacter*.

Campylobacter, once it was spotted as a pathogen, began showing up everywhere. The Eden-like quality of the Grand Tetons lulls visitors into careless behavior. You have been hiking for miles and are hot and thirsty. The stream looks cool and clear and sweet. The threats of civilization and pollution seem remote. You cup your hands and drink.

In 1980 a number of patients began showing up at the hospital in Jackson, Wyoming, with acute diarrheal disease. The CDC conducted an investigation and the next summer ran a follow-up study, testing and interviewing everyone with acute gastrointestinal illness. The *Campylobacter* cases were matched with a group of controls and then interviewed. At the same time investigators collected water samples from more than one hundred sites in the Grand Teton National Park.

Of the people with diarrhea, 23 percent had *Campylobacter* infection. Of those with *Campylobacter* in 1980, 60 percent had drunk from mountain streams in the two weeks before the onset of illness, compared to 25 percent of the controls. In 1981, 92 percent of the patients had been in the backcountry during the month before their illness, and 65 percent had drunk untreated surface water. *Campylobacter* was isolated from 2 percent of water specimens and from 3 percent of specimens taken from animals.[41]

Other studies have confirmed that animals that carry *Campylobacter* in the wild can contaminate untreated surface water; as clear as streams appear to be, drinking from them comes with a risk attached. But what of municipal water systems? Are they protected from *Campylobacter* contamination?

Greenville, Florida, is a tiny town in Madison County, not far south of the Georgia border. It's a farm community. According to Dot Pridgeon, the assistant clerk,

> there are two lumber mills, one grocery store, one drugstore, two branch banks, a middle school, a primary school, six or seven churches, a three-man police department, a city manager, and a city clerk. The main crops are peanuts, cotton, tobacco, and corn. A little

> over five miles out there is a chicken farm, and further out still a hog farm. The downtown is boarded up. It looks deserted. It does not look good. The area around Greenville is flat, hot, and sultry.[42]

In May 1983 the county public health unit received reports of a large outbreak of gastroenteritis in the small town, which then had 1,096 residents. It was alarming. There were hundreds of cases. "That was a bad time. Some of them were desperately ill," says Pridgeon, who remembers it all well. She escaped illness. She was one of the very few.

When the number of people sick climbed to astonishing numbers, the health department stepped in. Water samples were taken from the town's water supply and total coliform counts, which are looked at as indicators of contamination, were high. Eventually 865 people would become ill. The culprit was *Campylobacter*.

The water plant was found by researchers to be completely inadequate, and the prechlorinator had failed not once but twice. When the investigators took a closer look at the system, they saw birds perched on the settling tank aerator, and they found bird droppings on the grates over the open tank. The obvious next course was to trap birds and look for *Campylobacter*. They found it in 37 percent of the birds. It was not the outbreak strain, but the tower was on a migration path and the birds that had done the infecting were long gone by that time. They might have arrived harboring the bacteria, or they might have picked it up at a nearby chicken or hog farm.

It's becoming obvious that *Campylobacter* is now commonplace in the environment. Poultry waste, presumably heavily contaminated with *Campylobacter* if the level of contamination of finished poultry is any indication, is often dumped to rot in the open air. *USA Today* reported in August 1994 that in West Virginia fifty tons of rotting, maggot-infected, stinking chicken manure were dumped in Hardy County near the shop of Linda Scott, where it stayed for months, washing down into the creeks when it rained. There is no reason to believe that this is a unique occurrence. The poultry industry is both growing and taking a toll on the environment. In Arkansas, which has an enormous poultry industry, the paper reported that "Flies and buzzards used to feast on untreated chicken waste dumped into streams and public sewers and on carcasses thrown into smelly pits. Tons of chicken litter seeped from fields, at one time contaminating 90 percent of Arkansas' water supply with disease. The situation has improved but some problems remain."[43]

Hogs are a growth industry in North Carolina, where they now outnumber people seventeen to one. Hog waste, stored in open lagoons, has become an extremely serious problem. Breaks in embankments have sent foul waste running into fields and streams, where it kills wildlife and contaminates water supplies. Hogs also harbor *Campylobacter*. That birds would find attractive the open lagoons of waste and piles of maggot-infested litter, with its content of spilled and undigested chicken feed, goes without saying. It's little wonder that birds carry *Campylobacter*. It is now considered to be normal flora in many wild birds. But need it be? Birds in some areas have tested negative for the bacteria.

The problem of *Campylobacter* persists even when the effluent from animal production is treated. When Dutch researchers looked at the problem of *Campylobacter* in poultry waste water, they found it to be "an important environmental source" of *Campylobacter* contamination—and this was after the effluent had gone through a sewage treatment plant. Thus, both humans ill with *Campylobacter* infection and healthy carrier animals "provide a constant flow of this pathogen into the environment," they concluded.[44] Keeping the bacteria out of water supplies is likely to be an ongoing problem. There have been waterborne outbreaks of *Campylobacter* in Vermont, Sweden, British Columbia, and Wales.

There are forty thousand confirmed cases of *Campylobacter* infection a year in England, and studies show that the number of actual cases may be at least twenty times that many. Some are from sources that can be identified, but other infections—especially those that seem to be sporadic and not connected with an established outbreak—have remained a mystery. Professor Stephen Palmer of the Welsh unit of the British Center for Disease Surveillance and Control noticed a seasonal fluctuation in the number of sporadic cases. "The peak is late spring," he told a BBC radio interviewer in February 1995. "By the middle of May we notice a very sharp increase which persists through June, and then begins to gradually decline throughout the rest of the year. And that's quite a different peak from *Salmonella* food poisoning, which peaks in August and [is] associate[d] with warmer temperatures where the organism grows on food."

Palmer investigated a small outbreak in a hospital in Mid Glamorgan and discovered that the milk supply to the ward where the infected patients were had been attacked by local magpies. Home delivery of milk is still an option in England, and titmice frequently peck a hole in the silver bottle caps to drink the milk. But the magpies hadn't been just pecking at the caps: People had noticed they were taking the whole cap off. That kind of behavior is more typical of jackdaws and magpies, and magpie numbers in particular have increased over the last few years. Those ill with *Campylobacter* infection were questioned and their responses compared to a matched control group of wells. "We found a huge increased risk of getting *Campylobacter* if people had drunk milk from a bottle which had had the top removed, and the more likely the bottles were to be attacked in those households, the greater the risk," said Palmer.

In Newcastle an investigation went even further. Milk bottles that had been attacked were collected. *Campylobacter* was cultured from a high proportion of them. The jackdaws were also trapped and the same strains of *Campylobacter* could be cultured from the birds' beaks as well.

Recently Palmer and his colleagues did another study, working with twenty-one local environmental health departments throughout Wales. Of 551 *Campylobacter*-infected patients, ninety-three, or 17 percent, had drunk milk from a bottle that had been attacked by a bird the week before they became ill. While that fact by no means explains all the cases of *Campylobacter*, or even all the seasonal increase, it explained half the seasonal increase in May. "We're not saying it's the only factor, but there does seem to be some ecological phenomena happening around late spring, and jackdaw and magpie behavior seems part of that," says Palmer.

What Palmer's team did not try to explain was why the jackdaw and magpie populations had increased or why they were so thoroughly infested with *Campylobacter*. Martin Skirrow believes it is probably because the birds probe cowpats for invertebrates and in doing so contaminate their beaks, then transfer that contamination to the milk. But there are other possible reasons for the seasonal variation in temperate climates. After the winter (when presumably they receive a different sort of feed), dairy cows carry a higher percentage of *Campylobacter*. That is a hint that what we are feeding animals may have a direct impact on the microbial colonies that establish themselves in the animals. The routine addition of extra protein in animal feed has had an unexpected consequence: manure now has a higher nitrogen content than it once did.[45] We do not yet understand how this affects the colonies of microbes they harbor.

My neighbor across the street was given two little black hens. Her husband built them a house in the backyard, into which they went each evening. During the day for several years they clucked and pecked almost in tandem about her yard. They were very happy hens, tame, friendly, productive, and obviously very healthy. Then one spring a hungry fox nabbed them for supper, leaving behind a sad scattering of shiny black feathers. I miss the satisfying sight of two handsome, independent creatures going efficiently about their business. What I do not know was whether they were *Salmonella-* or *Campylobacter*-free. I have seen no studies to show that chickens living as my neighbor's did are healthier. They seemed so, but E. B. White might have disagreed. In 1938 he wrote complainingly to James Thurber of his attempts to raise poultry on his North Brooklin, Maine, farm; he listed the problems with his birds: "Roup, favus, thrush, range paralysis, the spiral stomach worm, the incessant shaking of the head." Keeping them healthy was no simple matter. His neighbor, when White asked him for advice, said: "God, I dunno nuthin' 'bout chickens. I just feed 'em, and if they do good I take the money; if they sicken, I dump 'em. That's all I know 'bout chickens."[46] Clearly to "dump 'em" is not an alternative the poultry industry wants to consider.

No single solution will rid the food supply and the environment of *Campylobacter*. Any effective approach will require a complete overhaul of well-established patterns of production, processing, and distribution—not to mention a revolution in our thinking about our relationship to the natural world. If one thing is obvious it is that before intensive farming, 25,000 to 125,000 birds didn't become infected in a matter of days as they do now; moreover, before mass-processing and mass-distribution, infected animal foods weren't shipped around the world. We have created a beautiful niche and an efficient means of distribution for an opportunistic microbe. From the microbe's perspective, it's got to be heaven. Unfortunately, any thought of applying standards to the animal production environment has been greeted reflexively by agribusiness; at the 1996 Republican convention delegates were spotted wearing buttons opposing government regulation of the farm.

But animal waste and how we manage it may be at the root of many of our problems with emerging pathogens. Chicken litter, which contains some nutrients, is stored for a time and then often fed to cattle. The consumer will be relieved to know that feeding chicken litter to milk-producing dairy cattle is not permitted because of the potential residues of antibiotics and heavy metals.[47] *Campylobacter* does not survive well in adverse conditions such as stored litter might endure, but it does change into coccoid forms that are not culturable. Some researchers think this stage of the *Campylobacter* is part of its dying process; others believe these damaged cells can revive in the intestinal tract. Still another group of researchers has demonstrated that these cells may indeed be viable under the right circumstances.

At the CDC, Robert Tauxe is rapidly coming to the conclusion that controlling foodborne disease is going to have to focus on controlling animal waste. He points to two changes in human history that have had a serious impact on disease. The first, about ten thousand years ago, was our domestication of animals. The second came in the nineteenth century when we seriously began to urbanize. At that point the necessity of providing clean water and food and controlling human waste to decrease the potential for the spread of disease became obvious. Now animals live in vast "cities," and it has become equally obvious as the foodborne disease outbreaks increase that ensuring the safety of what food animals eat and drink and how their waste is disposed of must be tackled with the same vigor.

9/The Hamburger Bacteria

> The microcosm is still evolving around us and within us.
>
> LYNN MARGULIS AND DORION SAGAN, *Microcosmos*, 1986

In February 1992 Mary Heersink, the wife of an Alabama physician and mother of four, was driving two of her sons to Florida. It was a wonderful, balmy Saturday in early spring, and she'd taken the boys out of school so that they could do something special.

Damion, eleven at the time, had returned from a Boy Scout camping trip the weekend before and was looking forward to attending space camp. His younger brother, Sebastian, was headed for a tennis camp. Damion, she remembers now, didn't seem himself. He was pale and quiet. She thought he might be coming down with flu, and if it developed further, she thought she would keep him in the hotel room with her for a day or so until he recovered, then send him on to camp. "He was very groggy and hard to awaken. He seemed mentally detached," she says now.[1]

Along the way he became ill with stomach cramps, then fierce diarrhea. She wasn't overly worried at first. Then it became bloody diarrhea. In fact, it was more blood than diarrhea, she says. "It looked like pure clotted blood, and it just came constantly. At that point you know you have to get to a hospital."

She made it to Tampa, her destination, quickly checked into the hotel facility adjacent to the tennis camp, where she left Sebastian, and took Damion to the University Community Hospital. The name reassured her. An association with a university was a good sign.

Mary Heersink had managed, while checking in, to contact her husband Marnix, an ophthalmologist, who was attending a continuing education course on pediatric medicine in Birmingham that weekend. Discussing the symptoms over the phone, they concluded it was just a vicious case of the "flu." They assumed that a trip to the hospital could secure some kind of medication that would make Damion comfortable, then Mary would turn around and get him home.

It wouldn't work out that way.

The hospital was busy and crowded. "It was like a war zone in there," she says of the emergency room. Damion's problem, acute diarrhea, didn't seem that compelling. He wouldn't be seen before 10:30 that night, five hours after they had come in. Within hours they were getting the results of his tests. "They were extremely alarming," she remembers. "His body was destroying blood platelets [small disks, smaller than cells, that promote coagulation] and shredding red blood cells." A gastro-graph showed that his intestine was edematous—"very puffy, very abnormal-looking." He was hospitalized at that point, but his condition worsened rapidly. "He went from bloody diarrhea to hallucinating, platelets being destroyed, kidneys malfunctioning. He was just plummeting," she says.

It would take an astute gastroenterologist to diagnose his condition as hemolytic uremic syndrome; this doctor insisted that Damion be transferred to St. Joseph's Children's Hospital in Tampa. (University Community Hospital had no connection at all to any university, Mary would later discover.) That in itself took hours. The confusion was almost intolerably frustrating. "There were mix-ups. They kept bumping my son. And it was so urgent," Mary remembers.

Neither she nor her physician husband had heard of hemolytic uremic syndrome. Still in Birmingham, Marnix Heersink tried looking it up in the library of the medical school library and found almost nothing. Both he and Mary were misled by the term "syndrome." It sounded better than some of the other possible illnesses that had been suggested. She would subsequently find out how horrifying HUS is—

an illness that would attack first one then another of the chief organs in Damion's body in what is described as a cascading pattern.

The doctors had assembled a team to deal with his case. As gently as they could, they told Mary what the mortality rates were and what she could expect. She was told that the course of the disease might well be five or six weeks; that it would be like a horrible roller-coaster ride as the disease created an anarchical situation in the body. There would be swelling of each organ affected. There would be clots. There could be strokes. Mary relayed this to her husband over the phone, and he booked a plane.

Marnix Heersink, born in Holland, is tall, handsome, and scholarly, with a ready smile that belies his otherwise serious demeanor. The name of his son's syndrome had told him one thing—it was as much a blood disease (hemolytic) as a kidney (uremic) disease. He thought of a fellow student at Western Ontario, where he'd done his medical training, who'd gone on to specialize in blood diseases, and although it was a wild shot, he contacted his friend's office in Canada. He explained to the nurse why he was calling, but when told that his friend was on a two-week vacation out of the country, he didn't bother to leave a number.

Damion was now strapped to his bed in intensive care and on kidney dialysis. A peritoneal catheter had been inserted in a surgical procedure into his abdominal cavity to take on the work of the failing kidneys. He could have no food or drink, and no tranquilizers that might disguise his neurological condition. His platelet count had dropped to precipitous levels. He might hemorrhage internally, they were told. Three days before he had been a strong, healthy, active boy. The Heersinks were frantic.

No one knows how many people get HUS each year, but estimates are that the disease kills as many as five hundred, most of them children. If the mortality rate is figured at around 5 percent, that means there must be at least eight thousand cases, but the CDC assumes that many go undiagnosed. What was an obscure disease just fifteen years ago is now the leading cause of renal failure in U.S. and Canadian children. And its numbers are increasing.

The doctors at St. Joseph's suspected at once where it had come from. Damion knew, too. He was in and out of consciousness in the early days of his illness, but in a moment of lucidity he told his father, "I know what made me sick."

On the Boy Scout camping trip a few days earlier, one of the projects of the group of ten-, eleven-, and twelve-year-olds had been the preparation of food. Hamburger was on the menu. Damion, who had little practical experience with meat because his health-conscious family seldom ate it, had formed the patties and a corner of one had fallen off. While the burgers were cooking, this small piece of ground beef was not. The morsel lay on the platter, exposed to the air and discoloring. When the burgers were returned to the tray, it was beside them, grayish-brown and looking for all the world like a piece of cooked meat to Damion. He popped it into his mouth. He realized his mistake at once, but not wanting to look foolish in front of his friends, he swallowed it quickly. In all likelihood that indeed was what made him sick.

Hamburger meat is the chief culprit in infections caused by the bacterium *E. coli* O157:H7. Ground meat is especially vulnerable because the grinding spreads the bacteria, a product of fecal contamination of meat, throughout the product. The most unfortunate of those who begin with the bloody diarrhea that characterizes the infection will develop HUS. Researchers think that 85 percent of all HUS cases in the United States are caused by *E. coli* O157:H7.

As Damion lay ill, growing worse by the moment, Dr. John Kelton, Marnix Heersink's friend, called his office at McMaster University for his messages and learned that his friend's son was in trouble. He had no idea where the Heersinks were. By calling friends and relatives, he was able to track them down. He told them, to their astonishment, that he was on his first vacation in years, in Florida, and that HUS was his specialty. He would come on board—via teleconferencing and the fax machine—from his vacation site on the eastern coast.

Kelton suggested an aggressive treatment of the blood using plasma exchange. Used in France and Canada with good results, the technique is less frequently employed in the United States, where nephrologists have claimed HUS as their disease. It's a sophisticated and costly treatment, and the St. Joseph's team had some reservations. But Kelton persuaded them of its effectiveness. And Damion was getting worse. It was clear they had to do something.

The boy was infused with eight units of fresh frozen plasma, which was exchanged for his own plasma. "What was coming out was just evil-looking," says Mary. "Foul and viscous-looking, discolored; full of shredded blood cells and toxins, which I couldn't see, of course, but were there. It just looked bad. And there was a strong, alien, most offensive odor. It was coming from everywhere; his hair, his skin, everything."

Damion underwent eleven plasma exchanges over the next two to three weeks, each lasting two to three hours each. After the exchanges, she noticed that the smell disappeared. A period of unwarranted optimism can occur at this point. Damion was suddenly improving. But the toxins had already done their damage and complications could set in. They could kill.

What would follow was a horrifying series of events. Damion's heart enlarged. His lungs filled with fluid, and he had to go on a respirator. Three times he underwent pericardialcentesis: With Damion awake under a light local anesthetic, the doctor inserted a needle just inside the pericardium, the sac enclosing the heart, to drain off the fluid. Each time a liter of fluid was drawn off. Then it would begin again. His heart rhythms began to be affected. At one point "he became convulsive. He was turning blue. His blood pressure plummeted to 40/20. He was dying. We had to leave the room, and they brought in the crash cart," says Mary.

The doctors patched him up and told her they would have to use more drastic surgical procedures to drain off the fluid. The idea was to cut a little window in the pericardium to release the fluid. When the surgeons got inside, they found that the pericardium was in such bad shape, she says, "so nasty-looking, so ragged," that "it had to be stripped off and discarded." No one really knows what the long-term implications are of having no pericardium, but the doctors told the Heersinks that any number of people were walking around without them. Damion could have a normal life, they said. They didn't have to add: if he survived.

Even then, Damion's course would not be smooth. The first sip of fluid by mouth was followed by agonizing pain. He had perforated his intestines and would have to go back to surgery. This time he was "split," as his mother says, "from neck to pubis, the way you eviscerate a cow." They found a small hole that unfortunately had spilled its contents into the body cavity. The doctors washed Damion's intestines and packed them back into his ravaged body. Somehow he would avoid massive infection, but there were abscess pockets in his abdomen, and a few days later he would go back into surgery. Incredibly, after a total of seven surgeries, Damion finally began a genuine recovery. And this time there would be no further setbacks. He had lost twenty-five pounds and had "absolutely no muscle tissue." When he was finally able to sit up on the bed, Mary "cried and cried to see the back of him look like that, so bony, so frail."

The mortality rate for children in his category of HUS is 5–10 percent. If a child has a complicated case, as Damion did, there is a one-in-five chance of death and a 74 percent chance of a major physical impairment. Children who have survived are often left with "souvenirs" such as diabetes, blindness, a colostomy, or kidney failure that requires dialysis or a new kidney. Damion was lucky, from beginning to end. His intestines had to relearn how to process food, and he had to learn to stand again, but finally, on his twelfth birthday, April 11, after seven weeks in intensive care, he was allowed to leave the hospital. "Then we spent a year putting him back together," says Mary.

Time after time during the ordeal she'd thought that her oldest son, her precious fair-haired boy, might be taken from her. Now he was home, recuperating slowly, but while caring for him, getting his body to work again, she found that the questions wouldn't go away. Her doctor father and brother-in-law, both ophthalmologists like Marnix, had bombarded her with questions when Damion was in the hospital. "What is the etiology of this syndrome?" they had sensibly wanted to know. What, in other words, was the cause? The members of this family were not merely inquisitive, they were confident people who demanded answers and expected to get them. The disease made no sense in the context of twentieth-century hygiene. Where had it come from? She remembered that Dr. Kelton's first question when he heard of Damion's illness was, "So when did he eat the hamburger?"

She was puzzled. There is hardly an American who does not think of hamburger as the National Food. No one had ever said ground meat did anything more ominous than raise cholesterol. Her brother-in-law had sent her articles from medical journals about HUS while Damion was still in the hospital, and she'd read about the hamburger connection again—and about a bacterium with an impossibly long name. What was it? Why was it in meat? Why had no one ever heard of it before?

At home, when she could spare a few minutes away from Damion and the rest of her family, she'd gone to her local medical center library and done more research. One name stood out as she went through the documents: P. Griffin at the CDC. Dr. Patricia Griffin had made the new bacterium her specialty. Boldly, Mary picked up the phone and called her. And Dr. Griffin listened. What Mary had already learned surprised her. Where there were gaps in her new knowledge, Griffin filled in the blanks. Coupled with what Mary already knew, the larger picture was horrifying. What was the matter with meat inspection that it allowed meat to be sold that was capable of putting people through what Damion had been through? Why wasn't anybody doing anything about it?

The questions didn't go away. Even as Damion improved, when she wasn't driving her three other children to their soccer games and school functions, she accumulated more information. She read reports from the General Accounting Office about the weaknesses of meat inspection procedures; she contacted a consumer group called the Government Accountability Project that sent her reports and the testimony of whistle-blowing meat inspectors. The more she learned about meat inspection policies, the more unacceptable the situation became.

Then one day in early 1993 Marnix Heersink came home with *USA Today* tucked under his arm. "Look," he said, "it's happening again." Together they read the first reports of what would eventually be known as the Jack-in-the-Box outbreak. The American public would read with them and learn what everyone should have learned a decade before, when the story actually began.

American Nightmare

Vegetarians of long-standing, who long ago stopped dreaming of steak and roast beef, tell me that in weak moments, with their defenses down, the memory of hamburger still returns to torment and test their will. The marriage of soft roll to juicy meat, adorned with the perfect blend of condiments, hot, dripping, and flavorful, has prospered and endured and proliferated beyond anyone's imagining. From its uncertain origins in the 1800s—two restaurateurs lay claim to its invention—the hamburger has become the most American of foods. But far more than that, it has become a cultural icon, a symbol of the national character. The hamburger is the backyard barbecue, the camping trip, the kitchen standby. Especially in its fast-food form, it is the American dream personified: modest, predictable, preseasoned, prewrapped, standardized, quality-controlled, and ready-to-go. It is more than a meal. It is a food wrapped in political and cultural mystique.

It was probably J. Walter Anderson, a short-order cook and future founder of the White Castle hamburger chain, who thought to place a ground beef patty not between bread, as had already been done, but between buns. Inspired, he added shredded onions. The bun, popped over the cooking burger, absorbed the flavors of the meat and onions. It was a hit. It would be refined over the years, but he had created, in its deceptive simplicity, the culinary equivalent of blue jeans. The only problem was getting the middle class to eat it. It was 1921, some fifteen years after Upton Sinclair had published *The Jungle*. That novel, with its graphic and horrifying descriptions of the conditions in slaughterhouses, had destroyed whatever illusions Americans may have had about the safety of their meat and prompted the first federal meat inspection laws. Consumer confidence had yet to be completely restored. One could cook foods carefully and thoroughly at home, but eating out was risky business. Anderson catered to this skepticism and apprehension by grinding his meat fresh and cooking his hamburgers in full view in an all-white, hospital-like atmosphere. Even the name of his establishment, White Castle, was chosen to signify purity and cleanliness, says Jeffrey Tennyson, author of *Hamburger Heaven*, a history of the burger.[2] But mothers still needed more convincing. Anderson sponsored scientific studies to demonstrate that his burgers were actually good for you.

Not long afterward, Billy Ingram, Anderson's investor, partner, and successor as White Castle president, introduced the frozen hamburger. It could be slapped, still frozen, on the grill; five critically placed holes in the frozen patty would allow the steam to escape and hasten the cooking. White Castle had invented fast food. Ingram, in a masterful triumph of public relations over reason, managed to convince customers that this dubious innovation was an advancement and an added assurance of quality. The flash freezing, the White Castle ads said, would guarantee that the beef patties were "at the peak of their perfection." Turning flaws into assets would become a standard PR approach.

The chain spawned a multitude of look- and sound-alikes, all eager to capitalize on a proven model in the growing burger market. The most successful competitor, White Tower, emphasized safety by hiring a team of "Towerettes," young women in nurselike uniforms posted for reassurance at every White Tower restaurant. By the time the multitudes of copycats had mutated into the ubiquitous drive-ins of the 1950s, with their gaudy, futuristic architecture, and then into the even faster-food franchises of the 1960s, the safety of the hamburger was taken for granted.[3]

Standing sentinel behind the reputations of the fast-food franchises was the U.S. Department of Agriculture, which inspected and approved the meat they used. Shoring up the agency's reputation was the assurance, repeated endlessly, that "America has the safest food supply in the world." From the 1950s to the 1980s that guarantee fit smoothly into the image America had of itself, an image of technological superiority, the highest standards, and an enviable lifestyle. It had become a part of the American mythology—often quoted and never questioned.

The era of the Great Hamburger Wars began in the late 1970s. Successful chains had toned down the flashy images of their recent past for something down-home and comfortable. "McDonald's and the other chains," says Tennyson, "decided to make a concerted effort to capture the family market." As a McDonald's spokesperson said in 1973, "That meant going for the kids. We decided to use television, so we created Ronald McDonald."

Kid appeal caught on, and the chains outdid themselves creating attractions for the younger set. All the while competition was growing as stronger chains gobbled up smaller ones and tough new contenders entered the arena. A fierce advertising war began in late 1981 between McDonald's, Burger King, and Wendy's. In one of the ads in the battle that followed, McDonald's described the intense urge for one of its burgers as "the Big Mac Attack." It would turn out to be an unfortunate slogan.[4]

Medford, Oregon, is a town of some forty-two thousand inhabitants in Jackson County, in the southern part of the state, just to the west of the Cascade Range. It is set in the Rogue Valley, a bowl almost, surrounded by hills, forests, and mountains, with a gentle climate that makes it ideal for growing fruit. For several decades now the area has been a draw for emigrants from California searching for a slower, more pastoral life. With the decline of lumbering, Medford has seen an influx of retirees, artists, writers, investors, and consultants who work from their homes, as well as those looking for an alternative lifestyle.

In February 1982 Dr. Richard Hebert, an internist on call at the Rogue Valley Medical Center, saw several patients with bloody diarrhea. It seemed unusual, something more than ordinary diarrhea. The cramping was fierce, like childbirth, said one female patient, and the diarrhea was often more like pure blood. It could easily be mistaken for hemorrhage. When a nurse mentioned that there were other similar cases, Dr. Hebert pulled their charts. He was struck by the similarity of their symptoms. The other area hospital, Providence Hospital and Medical Center, had patients with similar symptoms, he discovered. He also noticed that the patients had the same phone prefix in White City, a suburb of Medford. For lack of a better name, he and the staff began calling it 826 illness. Some of the patients had gone to physicians who had identified their symptoms as appendicitis. Some had even undergone surgery. No infected appendixes had been found.

Dr. Hebert had stool samples taken, and he contacted the Jackson County health officer, Dr. Taira Fukushima, who thought it might be *Yersinia enterocolitica*. When the cultures were negative for *Yersinia*, as well as *Salmonella*, *Shigella*, *Campylobacter*, and all of the several parasites that might cause such symptoms, Dr. Fukushima put in a call to the state epidemiologist's office, and from there, Dr. Steve Helgerson, an EIS officer on assignment to the Oregon Health Department, contacted the CDC. From then on, Dr. Hebert remembers, he simply "rode the wave. They were on this like white on rice," he says.[5]

Dr. Lee Riley, thirty-one years old and in his first year as an officer in the Epidemic Intelligence Service, took the call. Riley was the officer in charge of *Campylobacter* and *Salmonella*, and even though neither had been identified in the Medford cases, he suspected they eventually would be. In these early conversations with the state health department, he was struck by the fact that the victims showed little or no fever. Fever is an indicator of infectious disease, a sign that the body is doing battle with an invader. Its absence was enough to catch his attention. The health department had a number of questions, which he tried to answer, and then he made some suggestions.

The CDC heard nothing more from the Oregon Health Department for two weeks. Then it called again. There were more cases, and the department was no closer to finding a cause. Would the CDC come out and help? Riley packed his bags and got on a plane to Portland, Oregon. He linked up with Helgerson, and together they drove to Medford, discussing the case on the way. "We didn't know what was going on," Riley remembers. "We didn't even know if this was an infectious process. It could have been chemical, it could have been poison, it could have been anything."[6]

What would follow was a classic investigation. First came a "quick and dirty" case study to identify a set of hypotheses. They asked a good many questions just to find out what questions they needed to ask. This helped them to devise another set of questions that, together with the use of carefully selected controls, might lead them to the source of the disease. Although word of the outbreak was creeping out in the community, the first press coverage did not appear until March 17 in the *Medford Mail Tribune*. Allen Hallmark reported that the CDC had briefed the local board of health, but the names of the team members were not released. By the next day's edition Hallmark had found and talked to the investigators. The absence of fever, Riley told Hallmark, meant that it would be unusual if a bacterial agent were responsible.

Dr. Hebert, however, had always assumed it was an infectious disease, and gradually the cautious EIS officers were coming to the same conclusion. In some families more than one person was ill. And, Riley remembers, "One thing that kept coming up in the interviews was that the patients had eaten at one of the fast-food restaurants before they got sick." The restaurant was McDonald's. The thing was, eating at McDonald's wasn't that unusual. Given the popularity of the chain, it was almost expected. Nevertheless, they decided to include it as one of the questions in the case-control study.

In the meantime, Riley had been in constant contact with the Enteric Diseases Branch in Atlanta; as he and Helgerson were setting up the study, a team of epidemiologists and laboratory people had been just as busy looking for the cause of the outbreak in the samples. Joy Wells is the chief of epidemiological investigations for the Foodborne and Diarrheal Division laboratory. When word of the outbreak first came in, she was out of town at the annual meeting of the American Society of Microbiologists. As soon as she got back, she and the rest of the lab got straight to work. "The state epidemiologists had already looked for the routine pathogens, so we looked for those, but also for the unusual. We identified all the aerobic bacteria from the ills, and then, later, from the controls. We do everything very methodically," says Wells.[7]

Sometimes in this process something can appear that confuses the issue. They isolated *Klebseilla oxitoka* from some of the ills. "When we saw this in the stool samples from the patients, we got very excited," remembers Riley, who heard about the identification by phone. *Klebseilla oxitoka* is a bacterium often found in water where wood wastes from wood products mills have been deposited. That would make sense in Oregon.

Wells was not convinced. "What happens a lot of times, and what leads people astray," she says, "is that if someone has diarrhea and they isolate an organism and it's the predominant organism, they assume it's the pathogen. What happens when people have diarrhea is that the pathogen may not stay around very long, and there are organisms that are just good colonizers. When we did a further analysis of the *Klebseilla* in the ills, we found they were different strains." That would mean they came from different sources, she explained. "We also isolated *Klebseilla* from the controls," Wells says. At that point they lost interest in it.

The CDC had only six specimens from the Oregon patients. But in three of them they isolated something else—a rare serotype of *Escherichia coli*. It had the unwieldy name of *E. coli* O157:H7. Wells suspected at once that it might be the culprit. Her colleagues didn't agree, but she held out and just kept pursuing the unusual *E. coli* serotype. "We went back in our culture collection, and we found an O157:H7. It was from California. It had been isolated in 1975 from a patient with bloody diarrhea." That was a critical discovery. "This was a very unusual organism," Wells says. "You would not expect to find it in three out of six people. And we didn't find it in the controls," she explains.

The California patient from whom *E. coli* O157:H7 had been isolated had been ill with hemolytic uremic syndrome, but the significance of that wasn't obvious at the time. Joy Wells just stuck with what she was doing, a methodical retracing of a path worn smooth by the countless microbiologists who had come before her. Her strength was her refusal to be led off in the wrong direction. "We felt O157 might be the agent, but we didn't feel completely comfortable saying it was," says Wells. "We had no mechanism."

A "mechanism" is the method by which something causes a disease. Scientists don't feel confident saying something is responsible for an illness until they know how. At this point there was only an association between the outbreak of a disease characterized by bloody diarrhea and the rare serotype of *E. coli*. Often *E. coli* causes no trouble at all; sometimes it can even be helpful. The serotypes of *E. coli* that cause illness are divided into three groups, according to how they operate: enteropathogenic, enteroinvasive, and enterotoxigenic. *E. coli* O157 did not fit into any of these categories. Some serotypes are responsible for traveler's diarrhea, but that is a mild disease compared to what was happening in Oregon. These people had been very, very ill.

Riley completed his information gathering and headed back to the CDC. One thing became abundantly clear when the data were analyzed: Eating at McDonald's was strongly associated with the illness. But the fast-food chain served only USDA-approved meat. An association with foodborne illness, however tenuous, can have disastrous consequences for an eating establishment, which may or may not be implicated in the final analysis. There was no conclusive proof at this point, and that proof might never be forthcoming. The outbreak might just go into that thankfully small file of those that never really get solved. The epidemiologists tried their best to keep it quiet.

But Peggyann Hutchinson, a thirty-eight-year veteran of the *Medford Mail Tribune*, says that the fast-food association was no secret in the city. "We didn't hide it here. Everybody in town knew."[8] If the residents of Medford knew, it was because word travels fast in a small town. The coverage of the outbreak in the *Mail Tribune* ended on March 23 without a mention of either the McDonald's association or O157:H7.

It might have ended like that if three months later Bryce, Marcy, and Lance Turner of Traverse City, Michigan, hadn't been taken by their father to a fast-food restaurant one evening in June 1982 when their mother was working. Almost a week later the children began getting sick. "The kids had really bad stomachaches and really bad diarrhea," their mother Lynn says now.[9] Bryce, then eleven, was the worst off, and he was getting sicker. "His stools were pure blood." They took him to the hospital, where he was admitted. "I just remember him being hospitalized and IVs going and it was just very, very serious," Lynn Turner says now, trying to recall more than a decade later just what happened.[10]

At the hospital the doctors asked questions but didn't seem that surprised. They'd seen quite a few cases like Bryce's in the past few days. But they were certainly worried: These were very sick people. The doctors had strong suspicions—the link to McDonald's seemed to pop up often in answer to their questions—but they could find nothing in the lab test to explain the disease. Clearly they were in the midst of an outbreak of some kind. Clearly they needed help.

Dr. Robert Remis was then an EIS officer working with the Michigan Department of Public Health, and he remembers getting a call from the Tri-City Health Department in Traverse City asking for assistance. He and his boss, Dr. Harry McGee, knew this could be an important outbreak. Remis thought at once of the Oregon outbreak of bloody diarrhea he'd heard about through the CDC network. To complete the link to Oregon, the Tri-City Health Department had already made the association with McDonald's.

Remis had little doubt that his hunch was correct. "There is a dictum in our trade," he says. "If you go into an outbreak as an outside investigator and they don't tell you within the first fifteen minutes what the cause is, you'll never find it. Basically, what we really do is confirm what people already know."[10]

Before leaving for Traverse City, he called Riley. In the intervening months the CDC had been studying *E. coli* O157 further. It looked interesting enough to justify tests on animal models. Collaborating with the CDC veterinarian Morris Potter, the CDC investigators began with Rhesus monkeys, rabbits, and infant mice. They found nothing at first: The bacteria produced none of the standard responses they might have expected. That was discouraging. It's almost a tenet of modern medicine that what causes an illness in humans will cause an illness in animals. In fact, preventing illness in animals has long been a public health strategy in preventing illness in humans. Finally, in a test on infant rabbits, the bacteria did cause diarrhea—but not bloody diarrhea. All they could do at that point was watch and wait. If this unusual bacterium had caused the illness, it was unlikely to be an isolated occurrence. It would strike again.

In the labs Joy Wells was told about the Michigan illnesses, and at once she, too, suspected the outbreaks were related. "I was excited when I found O157," she remembers, "but when we had the second outbreak, I thought, 'This is it! Now we can prove it for sure.'"

They had the Oregon data, and the association with McDonald's had been strong. During the two weeks before the first sign of illness, the data demonstrated, twenty-one of twenty-five cases (84 percent) had eaten at one of the two McDonald's restaurants in town. Only thirteen of the forty-seven controls (28 percent) had eaten at the same restaurant. Three of the four ills who did not recall eating at that McDonald's remembered eating at another McDonald's within a week before the onset of the illness. And of those twenty-four who had eaten at McDonald's, twenty-one remembered eating the Big Mac.[11] Any of the well-known ingredients that almost any American could recite on request—"two all-beef patties, special sauce, lettuce, cheese, pickles, onions on a sesame seed bun"—might have been the culprit. "In the early days of an investigation," Wells says, "before you know anything, you just suspect everything."

In Traverse City, Bryce Turner had eaten a Big Mac. His brother and sister had eaten regular burgers. At first the CDC suspected the onions, with good reason. The onions were reconstituted from dry chopped onions and served without recooking. If they had been contaminated, microbes might have multiplied in the process. Remis had another idea. He thought it was the hamburger. But he remembers Dr. Mitchell Cohen, then assistant chief of the Enteric Diseases Branch, saying that it couldn't be. The meat was cooked. The McDonald's grills were kept at a temperature of 350 degrees, and there was no indication that the bacteria they were looking for were not susceptible to heat.

But Remis, a physics major in college, found that a thought kept nagging at him. McDonald's was careful with its meat patties. To make sure they were as fresh and safe as possible, McDonald's instructed its restaurant employees to keep the patties frozen in a cool cupboard near the grill; they were then tossed on the hot metallic surface with a glacial clunk. Remis wondered what a load of frozen meat patties would do to the grill temperature. He made some calculations, but there was only one thing to do to find out for sure. He and McGee returned to Traverse City with a primitive meat thermometer, timing their unannounced arrival to coincide with the busiest time of the week, Friday evening. The cooperative restaurant manager had a better thermometer that took instant readings from the grill surface. Remis's preliminary calculations had been correct. McGee took readings at sixteen reference points every minute and called them out to Remis. There it was. The proof. The cook was using part of the grill continuously—and it was cooling down. The burgers were not being thoroughly cooked. Remis felt vindicated.

But there was something else he had to do. He wanted to get a look at McDonald's meat-processing operation. The CDC decided to let both Riley and Remis make the trip. They met in Detroit and visited the Ohio plant together. It was a clean, efficient operation, Riley says, "so we really didn't think the contamination occurred at either the restaurant or the meat-processing plant."

Unless, thought Remis, the meat had arrived contaminated before being made into patties at the plant. The lot of meat consumed during the outbreak was long gone by this time, but the lot number had been recorded. While they were in the plant, Riley and Remis learned that the company had kept samples of the lot for quality control purposes. It was an incredible break. The samples were sent to the CDC labs.

Back in Atlanta tests had been run on the patient specimens from the Traverse City outbreak. The lab tested for the new *E. coli* serotype and found it—in about half of the samples. "'This has got to be the agent,'" Wells remembers saying. "We felt that very strongly, even though we didn't have the mechanism." Then the lab tested the sample from the burger lot and Wells's team found what they had been looking for. They isolated *E. coli* O157:H7. Wells had to be sure. "We did some plasmid profiles, and the strains from Michigan, Oregon, and the hamburger meat were all the same."

They had discovered a new pathogen.

To describe a pathogen as "new" can mean any one of several things. It can mean that it is simply "new to us," new to our experience in a particular place and time. "New" can also mean that a preexisting microorganism that has never caused disease before suddenly finds, because of some change in the environment, a friendly niche within which to expand. Or "new" for a pathogen can mean precisely that: a completely new organism created either naturally through bacterial mutation or conjugation, by which microbes exchange genetic information, or through science, which, in the past twenty-five years, has developed the ability to create new organisms using techniques of genetic manipulation or bioengineering—taking genetic material from one organism and transferring it to another. *E. coli* O157:H7 could have been new in any of these three ways, but in the summer of 1982 Riley, Remis, Wells, and the rest of the CDC team could be certain of only one thing: It was new to them, and new to the medical literature.

The public wasn't to know. The first they heard about the new pathogen came months later when Drs. Riley, Blake, Remis, and Cohen presented the CDC's findings at an October medical meeting in Miami. Although they referred to McDonald's only as "Restaurant A," Steve Sternberg of the *Miami Herald* discovered from someone in the audience what that referred to. On October 7, 1982, the *Herald* ran a story headlined "Undercooked Burgers Linked to Disease." Sternberg quoted Dr. Mitchell Cohen: "Contaminated meat carrying previously unknown bacteria apparently caused the intestinal ailment called hemorrhagic colitis."[12]

The CDC tried to lessen the blow to the fast-food giant. Cohen said he suspected that hamburgers sold by other restaurants carried the bacteria as well and that any undercooked hamburger meat, including that purchased at a grocery store and cooked at home, might cause illness if the bacteria was present.

McDonald's put the best spin it could on the leak, downplaying the seriousness of the disease, the numbers of ill people, and the conclusive microbial link. Its spokesman, Raymond Caruso, told the *Herald*, "Some months ago the health officials at the CDC made us aware of the possibility of a statistical association between a small number of diarrhea cases in two small towns and our restaurants."[13] He didn't mention the identification of the pathogen in the lot of hamburger meat. Behind the scenes they were planning what they would tell the rest of the world. They knew the story was out. The wires picked it up at once, and the other papers would have it the next day, but the stock market had reacted immediately. Trading of McDonald's Corporation stock on the New York Stock Exchange was halted at 2:45 P.M. after it fell one and a half points. The next morning the fast-food giant made the headlines in the *New York Times*: "McDonald's

Linked to 47 Stomach Disorder Cases."[14] Other New York papers were not so restrained: "A Big Mac Attack?: Hamburgers Linked to Intestinal Disease," the *Daily News* screamed.[15]

If stockholders were put off momentarily, they recovered. McDonald's did, too. "McDonald's Gets Break on Ailment," the *News* said, in an almost conciliatory headline the next day.[16] The stock had rebounded, climbing four and a quarter points in heavy trading. Nobody was going to mess with American burgers and get away with it. Investors leapt to the chain's defense with enthusiasm. A spot-check by the *News* "found that New Yorkers' appetites for McDonald's products seemed to be unabated."[15] But Cohen had been right. There was absolutely no reason to think that McDonald's was exceptional in any way. The CDC knew that O157 would strike again.

And because even small-town papers get the wire services, the people of Medford, Oregon, and Traverse City, Michigan, would finally discover what had hit them in the spring and early summer. "Illness victims ate at McDonald's," Allen Hallmark wrote in the *Medford Mail Tribune* on October 9. The local franchise owner called that information "ludicrous."[17]

As soon as O157 had been established as the culprit, the CDC looked around for other cases. Riley and Remis asked state and territorial epidemiologists to let the CDC know when they came across cases meeting the same criteria—crampy abdominal pain, grossly bloody diarrhea, absent or low-grade fever, and stool cultures negative for *Salmonella*, *Shigella*, *Campylobacter*, ova, and parasites. To get the word out further, they announced in various medical journals that they were looking for cases. "Then we started getting calls from various parts of the country," says Riley.

The CDC's *Morbidity and Mortality Weekly Report* soon reported that hemorrhagic colitis caused by *E. coli* O157 was occurring sporadically across the United States. "We learned 90 percent of what we know today from those two epidemics," Riley said in 1993. When the outbreaks occurred, he says, "both establishments were offering promotions. I don't think there was any obvious breakdown in the food handling, but it's possible that there were a lot of hamburgers being cooked in those periods, and it's possible some of them remained insufficiently cooked."

"Once we identified meat, cattle had to be involved," says Riley.

But they did not know for sure how the bacteria got into the meat. McDonald's tried to trace the beef back to its source, with no success. "It's not easy. It's a very complex network," says Riley. "The slaughterhouses, the distribution of meat items, it's practically impossible."

But the two outbreaks, the new pathogen, and the potential for further outbreaks of a serious hemorrhagic diarrheal disease set off a flurry of research in the medical and scientific communities. There was much to be learned, and everyone wanted to be the first to come up with the facts.

Northern Exposure

The CDC wasn't the only agency to notice the new *E. coli* serotype. In Canada at the National Enteric Reference Center at the Laboratory Center for Disease Control (LCDC), a group of researchers had long been accumulating an interesting collection of what they called verotoxin-producing *E. coli*. The term was created by Dr. J. Knowalchuk, a Canadian food microbiologist who had reported in 1977 that the toxin produced by certain strains of *E. coli* killed a line of African green monkey kidney cells called vero cells. Since something that kills cells is called a cytotoxin, he called it verotoxin, or VTEC. It was quite distinct from the other toxins known to be produced by *E. coli*, but at the time it was a toxin looking for a disease. The Canadians knew the illnesses had to be out there, and they were looking for and collecting examples of the toxin-producing *E. coli*. They had already spotted the serotype O157.

Hemolytic uremic syndrome was first identified in 1955 by a Swiss physician, Dr. C. Gasser. It was characterized, Gasser wrote in the medical journal *Schweiz med Wochenschr*, by severe anemia, a marked decrease in the number of platelets in the blood, and the failure of the renal system.[18] Patients—usually but not always children—became very ill indeed. The illness followed, more often than not, a respiratory infection or diarrhea, sometimes bloody, but the actual cause of the syndrome was not known. Other physicians began noticing the syndrome, and any number of different agents, including drugs, toxins, chemicals, and microbes, were considered as possible causes. It seemed likely by the late 1970s that an infectious agent was responsible because cases tended to cluster in communities and families; an assortment of viruses and microbes had been proposed, but none seemed likely enough even to be put to study. The disease was a mystery. Between the late 1950s and 1960s HUS seemed to be on the increase, but it was hard to be sure. Often when something new has been identified, it is difficult to tell whether an apparent increase is real or simply the result of a new awareness. Still, a clinic in the Netherlands, for instance, had reported only four cases between 1959 and 1964 while between 1965 and 1970 there were fifty. Something seemed to be happening.

In September 1980 Dr. Brian T. Steele, then a thirty-five-year-old pediatric nephrologist at the Hospital for Sick Children in Toronto, had been on duty when several children were admitted with similar symptoms. He diagnosed them correctly as having hemolytic uremic syndrome, but he was surprised to see so many. "We were used to seeing four or five kids a year come in with HUS," he says.[19] All of a sudden he had five at once. And they would keep coming in.

Talking to the mothers as he worked them up, Steele quickly discovered a common link. Eventually he would have thirteen patients with HUS, and most of them had attended a local fair. They had also drunk apple cider. Gradually a story emerged. A vendor at the market had offered samples of free, nonfermented apple juice manufactured the previous day from apples that had been stored at a local farm. All of the ill children had tried the juice at the market. All but one, that is. That child's father worked at the farm and had taken a bottle of the cider home.

Steele says,

> We asked the father what the apples had been like, and he said, "I'm ashamed to say, but they were absolutely disgusting." The fair took place in September, and the apples are still green on the trees then, so they had used windfalls from the previous year. I remember saying to the guy, "What do you mean by a bit disgusting? I mean, were there brown areas in these apples?" And he said, "Look, Doc, there were more brown areas in those apples than there were green areas. They were absolutely rotten."
>
> In fact, the more you hear about apple juice, the more disgusted you'll get. They can use pretty shoddy apples. The other interesting thing was, the weekend before they made the batch, they had slaugh-

tered a cow in the same barn. At the time we thought, "Well, that's interesting." What could the cow be carrying that would get into the apples and the apple juice? We had no idea what we were looking for.

Still, while the connection with the apple cider was clear—enough to be certain a pathogen was involved—they had no idea what had caused the HUS cases. The outbreak had caused a stir in the Canadian research community. At labs and facilities throughout the country studies were under way, variously looking for a fungal, chemical, viral, or bacterial cause for the illnesses. And more cases were being discovered. Two children had died of HUS in the past month in Alberta, and another was critically ill, but the cases didn't seem to be related to those in Toronto. Later, much later, the Alberta cases would begin to make sense. It is a cattle-raising area.

Then something serendipitous happened. Months later Steele was giving a presentation on the outbreak to a group of physicians, one of whom was a pathologist who had been abroad for a while and missed the widespread publicity the apple juice incident had garnered. He came forward and told Steele of an autopsy he had done on a child who died of HUS at about that time, and he remembered the father mentioning apple juice.

The child had not made it to the hospital. When his parents took him to the doctor because of his bloody diarrhea, they told the doctor that they had read in the paper and heard on television about the kids with kidney disease, and that their child had drunk the juice. The doctor responded with those immortal words, "You don't want to believe everything you read in the paper." The child was found dead in his bed the next morning. But even if he had been hospitalized, Steele was certain he could not have been saved. His organs had been destroyed. "That child," says Steele, "had the most florid hemolytic uremic syndrome you can imagine. His guts were infarcted—gangrenous. His kidneys were gangrenous as well; his brain was gangrenous. The thrombosis, or clots that form, can involve many organs. It was throughout his body, including his brain."

"The pathologist," remembers Steele, "had done a good job. He'd swabbed everything, and the swab had come back almost pure *E. coli.*" It had been identified as pathogenic and sent on to the labs in Ottawa, as all pathogenic strains were, but it had been misidentified. The *E. coli* group at the National Enteric Reference Center typed it as *E. coli* O113 verotoxin-producing. Research such as Knowalchuk's can lie buried in the medical literature, familiar only to a tiny group of highly specialized researchers. It may take years or even decades to trickle down to clinicians such as Steele. His reaction was, "What the hell is verotoxin?"

Dr. Mohamed A. Karmali, the chief microbiologist at the hospital, was very interested in Knowalchuk's VTEC-producing *E. coli* and what it might be doing. A few VTEC isolates had been identified from patients who'd had bloody diarrhea, and after the outbreak he asked Steele and other nephrologists to help him look for patients with this bacterium. In the next five patients they had no luck. Then they began looking for the toxin in the frozen fecal samples of the children who'd been ill with HUS. "Lo and behold," says Karmali, "we found it. The stools were actually loaded with the toxin."[20]

They had missed it before because everyone's stools are filled with *E. coli* and by the time the patients became ill the toxin-producing strains had done their work and weren't present in sufficient numbers to be obvious in a stool culture. When a culture is full of various *E. coli* growths, they may be difficult to isolate. But when Karmali knew what to look for, he began to find them by the painstaking method of picking out individual growths and culturing them. They found a number of verotoxin-producing *E. coli*, and in two of the fifteen HUS patients the *E. coli* was O157:H7. It was also found in two of the siblings of HUS patients who'd had diarrhea at around the same time. Shortly thereafter, the Canadians would announce the link between O157 and HUS.

The Canadians and the Americans had come at the culprit from different angles, but they were on the verge of a similar conclusion. The bits and pieces of the puzzle were coming together. The Canadians felt strongly that this was their verotoxin; the Americans that O157-related hemorrhagic colitis was their disease, the Americans even had a different name for the toxin, which they called Shiga-like toxin (SLT), because it seemed virtually identical to that produced by *Shigella dysenteria* type I. (U.S. researchers have since concluded that it is in fact identical to the *Shigella* toxin, and from here on I will refer to it as Shiga toxin to avoid the confusion of the dueling names.)

Pride, ambition, and chauvinism were playing a role, perhaps even contributing to the process. The researchers were nudging each other forward, whatever the motivation, by whatever means, enlarging in increments the pool of information about Shiga toxin, *E. coli* O157:H7, hemorrhagic colitis, and HUS. Gradually, in the awkward, haphazard way science is done, the story of a new pathogen was being written.

An Emerging Health Threat

At the CDC the Enteric Diseases Branch watched and waited. The new pathogen was out there; it was simply a question of where and when it would next strike. It happened on September 16, 1984, in Papillion, Nebraska. Patients at a nursing home began getting ill with severe, often bloody diarrhea. Because older patients are vulnerable, the staff contacted the Douglas County Health Department; Sarpy County, where the home was located, didn't have a health department of its own. The Douglas County Health Department realized it needed help and turned to the state. Gary Hosek, a registered sanitarian with a master's degree in public health, was asked to investigate. Tests for *Salmonella*, *Shigella*, *Campylobacter*, ova, and parasites were negative. The department preferred to handle the outbreak on its own, but when the number of patients reached twenty-two, with half of them hospitalized, it called the CDC.

Robert Tauxe was put in charge and began coordinating directly with the county health department. He had his suspicions as to what it might be. He arranged for the stool samples to be shipped to the CDC, and to make certain of the results, the samples were given to two different labs. Both labs isolated O157:H7 in at least one sample.

Tauxe felt that surge of excitement common to epidemiologists. This was the new bacterium they had been watching for. But in the time taken to process the cultures, two patients had died. That hadn't happened before. Even though the patients were more vulnerable because of their age and other health concerns, this looked very serious.

When you don't know what is causing an illness, you don't know whether it can spread. There is always the possibility that even a few related cases are signaling the start of an outbreak. The guilty food could be in the community—something widely distributed and widely available. Quick action could prevent a further outbreak of disease. Tauxe and Dr. Caroline Ryan, a new EIS officer, were assigned. They loaded up the Igloo coolers that went on every outbreak and set out at once.

On the way out Tauxe went over what they knew and did not know about the bacterium. (He still has his list of questions.) It had been found in meat once, but there was, as yet, no reason to think that meat alone posed a danger. With so little experience with the organism, even what they knew was suspect. What might be the case in one outbreak might not apply to the next. What they would have to find out was how widespread the outbreak was. Did it extend to the larger community? Was there a common exposure? What were the clinical signs? Just how sick were these people? How long had the patients continued to shed the bacteria? Was any treatment having an effect? What was killing the patients? And if the source were an animal product, could it be traced? It was pointless, even counterproductive, to speculate on the answers to these questions.

There was an eerie moment after they arrived, Ryan remembers. They had just driven up to the motel. Looking off in the distance, Ryan saw a water tower with the town's name and, just beneath it, the golden arches of McDonald's. It gave her a start. McDonald's would get a break: No one at the nursing home had eaten fast food. The illness was confined to the nursing home. The investigative challenge was working with elderly residents who couldn't remember what they had eaten. One lead: The nursing home got its eggs from a nonlicensed local farmer. But tests on the farm were negative. Finally Ryan had the idea of looking at special diets to see whether there was a relationship to those patients who had been sick. There was. Of the nineteen in whom the disease was confirmed, seventeen had been on the low-salt diet. And on September 13, when the residents on the regular diet were having ham, those on the low-salt diet had eaten hamburgers. That wasn't proof, but tracking the hamburger told them more. It had come from Great Plains Meat Distributors in Nebraska City, Nebraska. Running short on meat from its usual suppliers, Great Plains had purchased 500 pounds from Nebraska City Wholesale Meats, then sold 40 pounds of that supply to the nursing home. Of the 200 pounds the distributor sold to the public school system, 130 had been returned because of poor quality.[21]

Tauxe and Ryan had checked area hospitals for patients with similar symptoms, then set up an informal surveillance system at six hospitals. And they'd gone back through emergency room records. Several patients met the criteria they were looking for. One was a thirty-four-year-old mother from Auburn, Nebraska. Her family had cooked hamburgers around September 13, and she had subsequently become ill with cramping and loose stools, turning to bloody diarrhea, on the eighteenth. She was treated with antibiotics and after a few days seemed to improve. Then she developed HUS and died eleven days after first being taken ill. The hamburger she ate had been purchased in bulk from Nebraska City Wholesale Meats as well. Ryan wondered why other members of the family, who'd eaten hamburgers prepared from the same meat, hadn't become ill. The answer was simple: they liked their burgers thin, the victim liked hers thick. Her hamburger hadn't been cooked all the way through.[22]

In their final report to the Nebraska State Department of Heath, Ryan and Tauxe concluded that the outbreak was caused by *E. coli* O157:H7 and that, once again, the most likely source was hamburger meat. It could not be proven. Not surprisingly, tests were negative for the bacteria on meat samples. The bacteria were elusive. The evidence was epidemiological, and the Auburn woman was the vital link. In their final memo Tauxe and Ryan urged continued surveillance, reporting of suspect cases, and, if possible, identification of the herds of cattle from which the suspect meat came. They recommended that both the cattle and their feed be cultured. What the USDA actually did, if anything, will never be known because in the 1980s, under new paperwork reduction regulations, documents were no longer kept more than two years; the record was destroyed.

It was the early 1980s. The emergence of AIDS had changed the way medicine was reported. Medical coverage had gone from the science page to the front page. AIDS would also strike at our complacency as if specially ordered for the purpose. It would signal to us that our apparent mastery of disease was an illusion, and that factors outside our influence might be at work in the environment. It would call into question our basic assumptions about a basic human act: sex. The emerging foodborne pathogens would call into question another basic and essential human behavior: eating.

E. coli O157:H7, says Tauxe, had something else in common with AIDS: It "challenged the central dogma of modern medicine."[23] It masqueraded as other diseases, was often misdiagnosed, and produced little or no fever. While O157 and HUS had apparently been around for some time, no one had perceived the illnesses as infectious. Most agents that cause disease in humans also cause it in animals. This was not true of O157. And finally, says Tauxe, it upset the kingpin of the central dogma: Antibiotics fix infections.

"The only problem has been when bugs become insensitive to antibiotics, and yet when antibiotics were given for *E. coli* O157, it was very difficult to see any benefit, and some have concluded it may cause harm," he says. It wasn't that the bacteria were resistant; it was that antibiotics had no effect on the toxins the bacteria produced. And it was the toxin that was damaging cells and making people so ill. Indeed, even seemingly harmless antidiarrheal medicine appeared to make things worse. It seemed that diarrhea had a purpose: ridding the body of something harmful. Antidiarrheal medicine prevented that.

At nearly the same time as the Papillion outbreak, another was taking place in a day-care center in North Carolina. Thirty-six children out of 101 had diarrhea, and *E. coli* O157 was identified as the cause. Three children were hospitalized for HUS, but the disease was relatively mild. Hamburger was not implicated in this outbreak. No one knew where the first case had come from, but it seemed clear that it was transmitted from child to child. That worried the researchers. If this bacteria could be spread so easily by person-to-person contact, it meant that the number of organisms needed to make someone ill was very small indeed. The lower the infectious dose, the more contagious the disease. Nursery schools and day-care centers might be particularly vulnerable. In a changing culture in which day-care centers have become more and more important to the functioning of the economy, this had serious implications. There would be further day-care and kindergarten outbreaks. In one, the wall-to-wall carpet was implicated as the route of cross-contamination. The health implications of having so many children on a surface that could never be adequately cleaned had never been seriously questioned before.

There is a worldwide network among epidemiologists. Word travels fast even when material has not been published. There had been virtually nothing about the new pathogen in the national press, but even in the mid-1980s the CDC knew it was spreading. It was hearing of outbreaks in Canada, where the LCDC was keeping close tabs on the new pathogen. According to the *Canadian Disease Weekly Report*,[24] eighteen patients at Calgary General Hospital in Alberta had the now-familiar symptoms. All were positive for the pathogen. Then Dr. Anne Carter of the Bureau of Communicable Disease Epidemiology at the LCDC investigated a nursing home outbreak in which nineteen of fifty-five patients with *E. coli* O157 infection died; moreover, the death rate among those in whom the disease had progressed to HUS was a truly frightening 88 percent. Researchers noticed that patients taking antibiotics seemed to be more susceptible to person-to-person transfer in the second wave of the outbreak; they were also more likely to die. Carter believed that gastric activity had a role in the body's defenses against *E. coli* O157:H7 and that something that disturbed that activity, such as antibiotics, could make people more susceptible. Could antacids work in the same way? she wondered.

Looking back from what we know now, it's difficult to imagine that stage when what was known about the pathogen was so limited. But as outbreaks were investigated, each seemed to reveal something new. The CDC had attempted to relay some information to the public in *MMWR* reports, but the key was getting the press to pass it on. Journalists didn't seem very interested in writing about diarrhea, however, perhaps because it is so common and usually not terribly serious. But the outbreaks didn't stop.

In 1986 three older women from Walla Walla, Washington, a town in the eastern part of the state, were transferred to a Seattle hospital. They had thrombotic thrombocytopenic purpura (TTP), an adult version of HUS. An alert physician notified Stephen Ostroff, an EIS officer recently assigned to Washington State; after notifying the CDC, Ostroff set off to investigate.[25]

He would discover that a dozen people had reported to clinics and emergency rooms with bloody diarrhea in the weeks before, but the medical community in the town of twenty-five thousand had neither noticed the increase nor made the link between them. Clearly something was going on, and it looked like another outbreak from *E. coli* O157:H7. Another EIS officer, Dr. Patricia M. Griffin, was dispatched to help Ostroff with the investigation. The interviews implicated a favorite Walla Walla Mexican-style restaurant called Taco Time. It was a complicated investigation because so many foods contained similar ingredients: cheese, tomatoes, lettuce, and ground meat. When a nursing home nearby reported diarrhea cases and both the home and the restaurant were found to have used meat from the same source, the puzzle was solved.

One thing had been helpful since the beginning of this outbreak. The strain was unusual. Ostroff and Griffin would later discover that an HMO in Seattle had been doing a survey of the causes of diarrhea at around the same time, culturing every specimen, and had found a huge peak in the isolation of the same strain of *E. coli* O157:H7. There had also been children ill with HUS in Children's Hospital in Seattle. Ostroff and Griffin would realize that the contaminated meat had been distributed throughout the state and had actually caused a statewide outbreak that health officials had completely missed after the fact.

E. coli O157:H7 was one of the first things Griffin heard about when she joined the EIS. The Walla Walla outbreak was her first chance to see what the bug did in real life, and what she saw was frightening. When she and Ostroff discovered that hamburger was the likely cause, they notified the USDA to attempt a traceback. It was no easy task. Hamburger in the supermarket may appear to be meat from one source. It seldom, if ever, is. Most supermarkets buy coarse ground meat in ten-pound chubs and regrind it, sometimes adding scraps of their own. Distributed by a processor, the chub may represent cows from many farms and even different countries. It is prepared in forty-thousand-pound lots or greater. Three possible sources of the meat were implicated by the Walla Walla investigation. Two were large producers; the third was a small dairy processor in the southwestern part of the state.

Most people prefer not to think about what happens to dairy cows that no longer produce enough milk to please their owners. They usually became hamburger meat. Some slaughtering facilities specialize in these animals. After days of going through tedious records, Griffin came up with the most likely farms, and in a matter of days she and Ostroff would be slogging through mud swabbing cows with the help of EIS officer Larry Shipman, who was also a USDA veterinarian. They would find the organism in only about 1 percent of the 539 animals they tested. If hamburger were produced from a single animal, the chances of its being contaminated were apparently small. Mass-production and distribution had made a huge problem out of what might have been a serious but small one.

The investigation complete, Ostroff and Shipman would go on to other things. But Griffin found herself hooked on the new pathogen. It would become her bug. The growing number of outbreaks were of grave concern in the Enteric Diseases Branch. Griffin realized that her training had been in gastroenterology, but she'd never heard of the organism before coming to the CDC—nor, she discovered, had her former colleagues at the University of Pennsylvania and Harvard's Brigham and Women's Hospital. And yet *E. coli* O157:H7 was apparently an important cause of bloody diarrhea—something she had seen. How many cases had been misdiagnosed? The idea that no one in the field of gastroenterology seemed to be aware of this important pathogen was not just astonishing, it was intensely frustrating. The link to HUS that Mohamed Karmali had made in 1983 made sense. Autopsies of HUS victims showed the cellular destruction in the tissue that seemed to be caused by toxin. Few in the Enteric Diseases Branch had any doubt of the connection. Griffin felt strongly enough to start making calls around the country to tell pediatric nephrologists and to propose a study to look for the organism in the stools of HUS patients. She met resistance. On the whole, the doctors were skeptical and indifferent—understandably so, says Griffin charitably. The syndrome had been identified in 1955, and so many theories as to its cause had been advanced that this seemed like just one more. "None of them thought we knew the cause, and only a few thought it was worth going on another wild goose chase," says Griffin.[26]

Then Dr. Marguerite Neill, then an EIS officer in Washington State, and Dr. Phillip Tarr in Seattle did a study that showed that most cases of HUS were related to *E. coli* O157:H7. It was the evidence the CDC needed to make an even stronger case to physicians. Griffin had her mission.

Six years later, when the Jack-in-the-Box outbreak occurred, Griffin would be the expert the world turned to. Before long the network of people whose lives would be touched in some way by this pathogen would be calling her the "High Priestess of *E. coli*."

Cherchez la Vache

Between 1986 and 1992 the public remained virtually unaware of the new pathogen except when it struck in their own backyard, so to speak. And then the information about outbreaks or sporadic cases remained local. To make it worse, perhaps in a misguided attempt to simplify or describe in familiar terms for the public the serious involvement of the kidneys in HUS, the syndrome was at first described as a urinary tract infection in some news reports and even, on occasion, by the USDA, giving little hint of its horror.[27] Behind the scenes in research circles, however, there was much activity. When CDC investigators looked further for the pathogen in herds, they found it. The accumulating evidence pointed to a bovine host and the pathogen could be carried on products produced from cows—either meat or milk. It got into these products through contamination from the animal's fecal waste.

Despite its genuinely nasty traits, *E. coli* O157:H7 did have one desirable quality: It did not ferment sorbitol. That meant that a culture medium containing sorbitol could be prepared in which it would show up easily. Sorbitol-MacConkey quickly became the standard culture medium, but few labs were using it yet. When they did, they discovered that the pathogen was by no means rare as a cause of human illness. A study in Washington State showed that the rate of isolation was comparable to that of *Campylobacter*, *Salmonella*, and *Shigella*. The state would soon make O157 a reportable illness.

In the meantime, other researchers wanted to know where the new pathogen had come from. They were looking at its genetic makeup. What they found was odd. The one hundred strains of *E. coli* O157:H7 were quite different genetically from the other Shiga toxin–producing *E. coli*, yet they weren't very different from each other; they didn't show the wide diversity that would have developed over a long period of time from normal mutations and adaptations. What could be concluded from this was that they seemed likely to represent a single clone that had recently descended from a common ancestor and become widely disseminated geographically.

What had distributed it? Was the vast, worldwide trade in meat a possibility? There are reference centers where samples of bacteria are stored. An O157 that had been cultured from an Argentinian calf during the 1970s was found among them. Argentina was both an exporter of beef and a country with a high incidence of HUS. Could it have begun there? But now wherever scientists from North and South America, Europe, and Asia looked for it, they found it. There were places in

Canada where it was isolated more frequently from stools than *Shigella*. It had been isolated from Korean and Chilean children and from Thai adults. It had been found in England, the Netherlands, and Germany. It was new, it was different, it was virulent, and it was everywhere.

On the clinical side, studies were revealing that the bacteria didn't stay long in the human body. If you wanted to have a good chance of finding them, you had to culture patients within seven days of onset of illness—and before they had taken an antibiotic, as most strains were still not resistant. After that, it was shed, leaving the toxin behind to do its dreadful work.

Researchers concluded that when patients got HUS, it seemed likely that the toxins damaged the endothelial cells, causing damage to the intestines, which created the bleeding and also allowed the toxins to enter the bloodstream. Then arteries and capillaries became choked with the shredded particles of red blood cells and platelets destroyed by the toxins. In the kidneys the shredded material collected, disrupting their function. Body organs, because blood could no longer reach them through these clogged passageways, became starved for oxygen, swelled, and were damaged. Frustrated physicians had no treatment for this vicious, cascading pattern, only support—dialysis, respirators, IV fluids, transfusions—as patients' bodies fought the toxin, sometimes winning, but in 3–7 percent of cases failing before their eyes.

It was six years after the first O157 outbreak, and researchers were discovering more and more each year about the new pathogen. In 1988 an outbreak of *E. coli* O157:H7 infections in fifty junior high students was traced by the Minnesota team (there seemed to be something going on in the northern tier states) to precooked hamburger patties served in the school cafeteria. The patties appeared to be cooked enough to be served without further cooking, but they had not been cooked enough to destroy the pathogen.

With the encouragement of the Minnesota epidemiologists, the USDA at this point tried to raise the temperature to which the burgers had to be precooked; the agency ran into a balky industry. The meat industry feared new regulations and believed, probably correctly, that precooking a beef patty that was supposed to be recooked at home or school to the 160 degrees proposed by the USDA, and supported by the Minnesota epidemiologists, would leave it tasting like a hockey puck. It would also reduce the water content and thus the weight—the Holy Grail of profit-making in the food industry. The American Meat Institute resisted. Dr. George D. Wilson said the association saw no "grounds for associating a public health emergency with the production and consumption of precooked meat patties." The Western States Meat Association said that the proposed requirement of a minimum internal temperature for fully cooked patties would make it impossible to prepare a medium or medium-rare beef patty.[28] Little did they know that in a few years eating a rare beef patty would be considered the culinary equivalent of Russian roulette. The voice of industry is loud and powerful. The temperature to which patties had to be cooked before distribution wouldn't increase until 1993.

While hamburger appeared to be the chief villain for O157 infection, there were reports of outbreaks that seemed to have different causes. Several were linked to drinking unpasteurized milk, not unexpectedly, since cattle harbored the bacteria. People who had handled potatoes stored in peat (presumably contaminated with manure) were infected in England. In 1990 the largest outbreak to that date occurred in the small town of Cabool, Missouri, when the municipal water supply became contaminated after a hard freeze broke water pipes and allowed contamination. The outbreak, which took two months to control, made more than 243 individuals ill, hospitalized thirty-two, and killed four. Another deadly episode followed when two outbreaks in institutions for the mentally retarded in Utah in 1990 caused twenty illnesses that resulted in eight cases of HUS and four deaths.

Then, as if to confuse the issue, an outbreak in the late fall of 1991 in Massachusetts that resulted in four cases of HUS was found to be linked to drinking fresh-pressed apple cider. To members of the beef industry, who had been growing increasingly nervous, this was good news. They could point to apple cider and say, "See! It isn't just beef." But cows were probably ultimately responsible for the Massachusetts outbreak. The apples were likely to have become contaminated on the ground from cow manure—just as they apparently had in the 1980 outbreak that produced fifteen cases of HUS in Canada. Then an outbreak in Oregon from swimming in a fecally contaminated lake demonstrated that infection could be linked not just to a food but to an activity. Future investigations of outbreaks, noted Neill, would require creativity and insight. One wag had a word of advice: "Cherchez la vache." But it also seemed that the pathogen was now in the environment, not just in cows. It could pop up anywhere there was danger of contamination from fecal material, and that seemed to be almost everywhere.

In terms of outbreaks, 1992 proved to be a "relatively quiet" year for the pathogen, Dr. Marguerite Neill, now at the Brown University School of Medicine told a group of sanitarians in 1994. In fact, sporadic cases were causing heartbreak. Mary Heersink certainly would never forget 1992. Nor would Robert Galler and his wife Laurie of Long Island, New York.

Their daughter, Lois Joy, had become ill on Thursday, June 26, 1992. By the next day she had a stomachache. That was followed on Saturday by frequent and severe diarrhea, which by Sunday had become bloody. The family's pediatrician saw the child and sent her to the emergency room because she was dehydrated. At the hospital they rehydrated Lois, did a blood workup, and took X-rays of her stomach. Then they discharged her. They would take her back on Tuesday, "more and more lethargic with each passing hour," remembers Bob Galler.[29] One of the first questions the doctor asked was whether she had recently been to a fast-food restaurant or eaten hamburgers.

Then commenced, Galler remembers "the most emotional roller-coaster ride that any parent can experience." His daughter's kidneys stopped functioning on July 2, "never to function again. During her eighteen days in the hospital," he says, "she was given sixteen blood transfusions, fourteen with fluid, she had to be put on a respirator, she lost sight in her right eye, she suffered an infarction to her brain. We watched, simply helpless, as our daughter died before our very eyes."

On Wednesday, September 8, 1992, a similar ordeal began for Arthur O'Connell, a high school mathematics teacher from Kearny, New Jersey, a town about twenty minutes outside of New York City. Katie, his twenty-three-month-old toddler, had diarrhea.[30] By the next day she had dry heaves and was lethargic throughout the day. When her diarrhea became bloody, the family took Katie to the pediatrician, who immediately hospitalized her. The hospital began a series of tests; on Thursday Katie went into a seizure. Tests revealed HUS, and she was transferred to another hospital where she was put on peritoneal dialysis.

"On Tuesday Katie had the tubes removed from her nose. She

opened her eyes and talked to me," remembers O'Connell. "I asked her if she wanted anything, and she told me, 'Apple juice.' Those were the last words she ever said to me." Then, in the now-familiar pattern, crisis followed crisis until their daughter stopped breathing and was hooked to a respirator. She died on September 16, 1992. Scattered across the country—and around the world—were other deaths no less tragic and unnecessary.

In 1992 the food industry belatedly entered the picture by funding research on *E. coli* O157:H7 through the International Life Sciences Institute, which is supported by a wide range of corporate food giants. But lurking just over the horizon was the Jack-in-the-Box outbreak, which would finally galvanize the press, the public, the industry, and the regulatory agencies. Before the year was out Lauren Rudolph would go to a fast-food restaurant with her father, and the massive, widespread tragedy that would be called the western states outbreak would begin.

On January 12, 1993, a gastroenterologist at Children's Hospital and Medical Center in Seattle notified the Washington State Department of Health that physicians there were seeing many children with bloody diarrhea and some with HUS. Investigators quickly did a preliminary investigation and discovered almost at once that most of the cases, but none of the controls, had eaten at Jack-in-the-Box. Most had eaten a regular hamburger, which, being small, is a favorite with children.

The public first learned of the cases on January 17, when the *Seattle Times* and the *Seattle Post-Intelligencer* reported forty-five cases of foodborne illness in the western Washington area. The cases were linked to hamburgers. One day later the Washington Department of Health issued a public announcement urging people with bloody diarrhea to see a doctor. Hamburgers were recalled by the restaurant chain. The CDC would later estimate that the quick recall prevented eight hundred additional illnesses.

The beef industry was immediately on alert. Its PR people went into a huddle to plan what they should say and do. On January 19, Wendy Feik Pinkerton and C. J. Reynolds, spokespersons for the National Livestock and Meat Board, sent a memorandum to Washington State Beef Council executives telling them of the reports. "Here's what we know," they said.

> Of the reported cases, 13 children were infected and five of these are in intensive care. Of the reported cases, 37 people had been interviewed by the State Health Department; 27 had eaten at Jack in the Box restaurants in the Western Washington area and all 27 had eaten hamburgers. Jack in the Box has been the only restaurant chain implicated in this *E. coli* O157:H7 outbreak.

She also told the meat industry executives, although it made no sense at all and was clearly wishful thinking, that "Washington State epidemiologists said undercooking the burgers, and not the burgers themselves, appears to be the reason for the foodborne illnesses." Then she added: "While we do not anticipate widespread, national media attention, we have planned for it. Our goal, first and foremost, is to stay out of the media spotlight. The coverage, so far, has focused on cooking procedures at the fast food outlets, not beef industry issues. Let's try to keep it that way."[31] She had outlined the strategy that the industry would hold to. Hamburgers would be redefined.

Americans had been cooking and eating pink hamburgers for years; overnight, history would be rewritten. That would become "improper preparation." And the fault would rest with the consumer, who should have known better, although how the consumer would have discovered this mistake was mysterious and unclear. "There has been a subtle turning of this on to the consumer," says Steve Bjerklie, former editor of *Meat and Poultry* magazine, "and it's morally reprehensible."[32] Actually, it wasn't subtle at all.

It did not take Jack-in-the-Box's parent company, Foodmaker, long to trace the contaminated meat to a batch of seventy-seven thousand hamburger patties processed by Vons Company, a meat-processing plant in Los Angeles. Most had been shipped to Seattle, but ninety-two hundred were unaccounted for. They might have gone to San Diego, Los Angeles, Bakersfield, Las Vegas, or Hawaii; the company had no way of knowing. Two days later the *Seattle Post-Intelligencer* would report that while sixty-six people were already ill, the USDA had yet to begin tracking the source of the contaminated meat—for indeed, it was contaminated. The Vons spokesperson told the press that no changes had been made in the processing procedures. "The USDA maintains an office in our plant. There are not out-of-the-ordinary USDA tests going on."[33]

Both Phillip Tarr, a pediatric gastroenterologist, and Dale Hancock, a Washington State University epidemiologist and veterinarian, were already experts on O157. They told the paper they were worried about the lack of testing. "If there is a common source of infection," Tarr told the paper, "then it should be pursued. If there is a delay, one would be concerned about further spread of the infecting pathogen." Hancock was even blunter. "You'd think that an organization with a name like 'Food Safety and Inspection Service' would certainly monitor what happens at the plant."[34]

By the end of the month, three hundred were ill and it had become clear that the outbreak extended to four states. When the final count was complete, the case numbers would climb. Gaping holes in the public health system were revealed. Nevada, which actually had a reporting requirement for cases of O157, would not know it was in the midst of an outbreak until a parent of a child sick with HUS who had eaten at Jack-in-the-Box read the newspapers, made the connection and brought it to the attention of health officials.

The public was growing panicky. Hamburger sales were down. On January 21 in San Diego County, where Foodmaker has its corporate headquarters, the health department issued a calming release: "You can still eat at restaurants; you can still eat at home. Just be sure to have any meat cooked thoroughly." If the health department had made the link between Lauren Rudolph, who died from the pathogen in San Diego County three weeks before, and what was happening in Seattle, it didn't say so. The press would finally make the connection.

To the American public, the Jack-in-the-Box outbreak was about a new health threat from an ordinary, everyday food source, and it was about impressive numbers. (Although most reports stopped at three hundred and never updated their stories with the final figures.) The reaction, however, was not unlike the response after a hurricane or earthquake in some other part of the country: interest, concern, but ultimately the sense that "that was there and I am here." What the public did not understand was that the pathogen was everywhere and that everyone was a potential victim, for the routes of possible infection were now broad and varied; sporadic illness had become common across the United States. For those who were ill and those who lost children or suffered through weeks or months of hospitalization of their loved ones, it was about individuals, and it was pure hell.

Dorothy Dolan watched not one but both daughters, four-year-old Mary and three-year-old Aundrea, become terrifyingly ill after eating hamburgers from Jack-in-the-Box. Both children followed the familiar pattern of lethargy and fierce diarrhea, but Mary was the most severely ill, with HUS following its predictably horrifying course. Although the prognosis was poor, Mary, as well as Aundrea, would eventually recover. But after her long battle, Mary would need physical therapy, speech therapy, and occupational therapy. Six months later her blood pressure remained high and her speech was still slow. The Dolans would be left to wonder about the effect of the illness on her cognitive abilities, the risk of HIV from the transfusions, and her long-term kidney function. Even Aundrea, who fared better, would need routine testing.

Dorothy Dolan told her story to a Washington symposium in September 1993. It had been organized by Safe Tables Our Priority (STOP), a group of parents and friends of victims of *E. coli* O157. STOP would start a hotline for new victims, create a network, and act as a clearinghouse for information. With the help of the Safe Food Coalition, a group of organizations interested in food safety, it has become professional and organized, carrying its message to Washington.

"Not a day goes by," said Dolan, "that I don't hear about another child sick from *E. coli* and hemolytic uremic syndrome. I cry and feel sick to think another child and family has to go through what we had been through. We cannot continue to let our children suffer and die from something that can and should be prevented, from something that is stamped USDA-approved."[35]

The press, known for sensationalizing events, seemed strangely restrained to the point of disinformation. *Boston Globe* reporter Michael Rezendes on January 31, 1993, described Joseph and Dorothy Dolan as "the parents of two daughters laid low after eating cheeseburgers."

The Washington symposium was an emotional gathering. The Dolans could count themselves lucky. One child in the outbreak had been released from the hospital only to die in his car seat on the trip back to the hospital. Another family, Diana and Michael Nole, lost their only child, Michael James Nole. He was just over two when he died in January 1993 from a children's meal he'd eaten at Jack-in-the-

Box. Like Dorothy Dolan, Diana Nole was a nurse. "Some of the things my son went through were the most horrific things I have ever seen in my eight years working in the medical field and my most recent two years working in an emergency room."[36]

Ten-year-old Brianne Kiner survived after spending forty days in a coma; her colon, kidneys, lungs, heart, pancreas, and liver were damaged. After five months in the hospital and a miraculous recovery, she was still, a year after the episode, suffering seizures and had been rehospitalized a number of times. She was still receiving special nutrients through a feeding tube in her stomach because all of her large intestine, damaged by the bacteria's toxin, had been removed, and several times a day she inhaled medication to keep her lungs functioning as another machine applied gentle pats to help loosen the mucus in her lungs. With her pancreas only partly functional, she endured several needle sticks a day to monitor her blood sugar levels and receive insulin. Yet she was there at the hearing.

Her presence, her few words, and the testimony of others, would leave the audience damp-eyed and badly shaken. But they were mainly supporters and industry note-takers, along with a very few congressional aides. Despite the widespread attention the outbreak had garnered, few reporters or members of Congress bothered to attend. The public was fickle. The problem was behind them. But another mother at the hearing could have told them otherwise. Janis Sowerby of Saranac, Michigan, told the story of her son's illness following a camping trip with his father months after the Jack-in-the-Box outbreak. They had bought hamburger at a local store and prepared sloppy joes. His symptoms followed the same appalling, treacherous, downward spiral from diarrhea and cramps to bloody diarrhea, to HUS, to multiple complications and organ failure, to death.

STOP was disappointed by the small turnout for the symposium, but not discouraged. Its campaign would be waged at the USDA and with individual members of Congress, and much later, when President Clinton signed a bill to improve meat safety, he would credit the work of these parents, who had never given up the fight. But that would be three years away. In the meantime, the press and the public were beginning to ask what the USDA was going to do about meat safety.

What Is Safe Meat?

If a piece of raw meat harbors bacteria that can cause illness and death, should it be considered adulterated or contaminated? The answer to that question has preoccupied a frightened meat industry, caused deep divisions in the U.S. Department of Agriculture, possibly contributed to the short tenure of the last secretary of agriculture, and ultimately exposed the distasteful politics of food. In the meantime, 81 million Americans—or 260 million, according to which expert from the Centers for Disease Control and Prevention one listens to—continue to be made ill each year by foodborne bacteria, and the number is increasing.

Agriculture Secretary Michael Espy had no sooner been sworn in than he was faced with an epidemic that would eventually swell to 732 illnesses, 195 hospitalizations, and 55 cases of associated hemolytic uremic syndrome. There were four deaths. The meat responsible had been inspected and passed by his agency. It was a galvanizing moment.

The impact this pathogen was bound to have on the food industry has been obvious from the beginning, both to the industry and to the USDA. The pathogen is harbored by healthy cattle. Why it is so difficult to find in cattle (although Dale Hancock, a researcher with Washington State University, says he has been able to find it in every herd he's looked at, in about 1–3 percent of the animals) but relatively easy to find in human patients with diarrhea has to do with a number of factors. When other researchers began looking at cattle, they found any number of places along the path from field to fork where the bacteria could be amplified. An infected cow might shed the pathogen in the field and the next grazing cow pick it up. Even if only one or two cows in a herd were shedding the pathogen at the time of shipping, tired, frightened, and hungry cattle (they are given neither food nor drink during transport) attempting to keep their balance in a moving vehicle lose control of their bowels. In these tight quarters the fecal material gets on the hides of other animals. When they arrive, all the cows are filthy and distressed. Conditions before slaughter favor the pathogen as well. Routinely cattle, massed together in tight quarters, standing in the excrement of many animals, are starved a day or so before slaughter to empty their intestines. Research has found that this practice also produces more shedding of the bacteria. Given the rapid pace of meat processing today (as many as three hundred animals an hour, or five per minute), it's not surprising that O157 and other pathogens get onto meat during the slaughtering process through contamination with fecal material. Combining meat from many different sources in ground beef can spread any pathogens present throughout the product. Robert Tauxe has described the hamburger of today as "a mixture of one hundred different cattle from four different countries."[37]

Up until a few years ago slaughtered meat imported into the country had to be a certain size—large enough to identify what it was. That size restriction was done away with in 1993, and now, in effect, we import hamburger or coarse ground meat. The idea that the tons of ground meat and scraps imported into this country can be precisely identified and examined for the presence of pathogens is pure fantasy. Much of our imported meat comes from Australia, where outbreaks from O157 and other toxin-producing *E. coli* have occurred recently although there is no reason to believe that Australian meat is any more contaminated than domestic meat. In fact, we export about as much meat as we import in a mindless global transfer of not just meat but infectious agents.

While the pathogen may be only on the surface of steaks and roasts, where it will most likely be cooked away (if the meat is not prodded to tenderize it or pierced with a fork during marinating or cooking), grinding spreads the bug throughout ground beef. Ten years of research have revealed no guaranteed way to assure the consumer of an absolutely safe hamburger other than thorough cooking, although even industry representatives concede that more careful skinning and slaughtering could help reduce the microbial load of beef.

The microbe *E. coli* O157:H7, prior to 1993, seemed to put both federal agencies and the meat industry into a state of deep denial—at least when it came to taking action. While the USDA as early as 1988 suggested to consumers who were curious enough to call and ask for information that hamburgers should be cooked to 160 degrees, the FDA allowed restaurants to follow a 140-degree guideline until the Jack-in-the-Box outbreak, when it rushed to raise it to 168.3 degrees in the model food code, which states adhere to either voluntarily or, in some cases, not at all. (Temperature tests of hamburgers at one Jack-in-the-Box restaurant taken after the outbreak revealed that some burgers had cool and bloody areas; one spot in one burger had reached only 110 degrees.[38]) Prior to 1993 it is difficult to find evidence that the USDA did anything to attack the problem. Its own early tests had found the microbe to be only rarely present on beef, but according to a USDA source, the agency was not using the sorbitol-MacConkey culture medium, without which isolating the microbe is difficult.

Proposed regulations have met fierce industry resistance. Various sectors of the food industry fought attempts to raise the cooking temperature on precooked meat patties, despite an outbreak in a school cafeteria. The industry opposed warning labels on meat, including safe handling and cooking instructions, and even sought and obtained an injunction from Federal Judge James R. Nowlin, in Austin, Texas, that held up labels by five months. Two children died in Texas from HUS in the weeks following that decision.

Although the speed of meat-processing operations is implicated in the contamination of meat with fecal material, lines in the slaughterhouses accelerated even more during the deregulation frenzy in the 1970s and 1980s. Under experimental programs the number of inspectors was reduced. Unsanitary conditions were allowed to continue, and repeated violations failed to lead to USDA action. Some inspectors complained of being unable to do their jobs. A pilot program called the Streamlined Inspection System (SIS) brought forth horrifying descriptions of foul and unsanitary conditions from whistle-blowing meat inspectors. Line speeds often reached four hundred cattle per hour; inspectors were asked to check as many as thirty livers per minute and found the task impossible; grossly infected cattle with such conditions as urine-filled bellies, peritonitis, pneumonia, and measles (tapeworm) were slipping past inspectors into the food chain. Inspectors complained that they could no longer see the inside of the carcass and had lost the authority to trim to expose signs of disease. Cow heads, which are trimmed to provide meat for hamburger, were getting through with regurgitated food oozing out, and "fecal contamination and other filth are getting out of control." Plants that had once had clean floors were, under the new system, "full of guts, urine, feces and general muck, sometimes to the point of being so slippery it's dangerous to walk." "Much of what USDA calls wholesome today," said one affidavit, "would have been condemned in 1984, and it wouldn't have been a close call. Meat whose disease symptoms previously would have forced it to be condemned, or at most approved for dog food, now gets the USDA seal of approval for consumers."[39]

The SIS pilot program was eventually dropped, but the 3.5-percent rate of *E. coli* contamination in hamburger that industry itself found should have come as no surprise.[40] Many critics charged that USDA decision-making was doing more to please industry than to ensure food safety.

The USDA Response

When informed by the Washington State Department of Health of the connection between the Jack-in-the-Box outbreak and USDA-inspected meat, Jill Hollingsworth, assistant administrator of the Food Safety and Inspection system (FSIS), told the health department's Charles Bartleson, "We will take no action because this meat does not violate USDA standards." To which Bartleson replied, "I thought you guys were in the public health business."[41] He could be forgiven for his confusion.

Dr. Russell Cross was then the FSIS chief and much admired by industry. In a briefing memo to the new secretary, Cross apprised Espy of the department's position on *E. coli* O157:H7, echoing Hollingsworth's line. "Under current regulations," he said, "the presence of bacteria in raw meat, including *E. coli* O157:H7, although undesirable, is unavoidable, and not cause for condemnation of the product. Because warm-blooded animals naturally carry bacteria in their intestines, it is not uncommon to find bacteria on raw meat."[42]

Cross failed to mention to Espy that muscle meat is, in fact, sterile, and that fecal material transferred the bacteria to raw meat during the USDA-inspected slaughtering process. He would repeat this agency disclaimer that the meat was not contaminated several times over the next few weeks as public Senate hearings addressed the outbreak. It was frustrating, he said, that so little was known about the pathogen.

In fact, medical researchers and epidemiologists knew a great deal.

The information was as near as the nearest medical library in the great number of journal articles that had been published since 1983. But there seemed to have been a breakdown in communication between the USDA and the CDC—between the world of human health and the world of animal health. That sharp separation was no longer appropriate. The two were drawing ever closer.

The USDA position on microbial contamination was not new. It was based on a decision by the District of Columbia Court of Appeals in the case of *American Public Health Association* v. *Butz* (1974), which said that Congress had not intended that inspections would include "microscopic examinations." In 1987, when the pathogen was discovered in commodity meat, causing four deaths in two institutions in Utah, the USDA decided that the meat need not be sampled or recalled, because it was "safe" if "cooked properly," and that sampling or recalling it "would create more problem [*sic*] for the agency, especially if the press learns of the FSIS action on the product that has already been released by the state of Utah."[43] The meat from the contaminated lot was distributed by the state to other facilities.

The Jack-in-the-Box outbreak brought wide media attention. If Espy did not learn of the particular virulence and the source of *E. coli* O157:H7 from his undersecretary in charge of food safety, he heard of it elsewhere. He was concerned about the agency's reputation. Consumers would be unlikely to appreciate the subtlety of the argument that the meat wasn't, by definition, contaminated, when children were dying. In the following weeks and months the secretary would hire two hundred more inspectors, institute a policy of zero tolerance for fecal material on carcasses, and order surprise inspections of slaughterhouses and meatpacking facilities—although most reasonable members of the public could be forgiven for wondering why these steps had not been taken before. Only resistance from within the department kept Espy from doing more, aides said.

On March 2, thirty-five meat inspectors wrote Espy to say that they expected retaliation from plant managers for attempting to carry out zero tolerance. They explained that Cross had "almost immediately gutted" the standard by telling a plant official that the ruling applied only to "obvious" fecal contamination and inviting him to report any inspectors applying "knee-jerk" interpretations of zero tolerance. "We are now in a position where FSIS leaders may have invited industry to prepare a hit list of inspectors whose crime is following the agency's own orders," the inspectors wrote.[44]

"We are not talking about anything punitive," Cross would explain to the *Federal Times*. "I just wanted to know if there were any inconsistencies."[45]

A spokesperson for the Government Accountability Project, which represents whistle-blowers, said that Secretary Espy was being undermined "by a holdover FSIS bureaucracy. While Secretary Espy says the right thing in public, Dr. Cross undercuts him by doing the wrong thing in private."[46] Eventually Espy would decide that he was being ill served by Dr. Cross, and Cross and two others would resign.

Charges that Espy had acted unethically by accepting gifts of travel and entertainment from Tyson Foods, the giant Arkansas chicken producer, had begun to surface and had the secretary's staff on edge. The Tyson-Clinton connection, even as the Whitewater tale developed, made the story irresistible to journalists. Then from within the department came charges from Wilson S. Horne, former head of inspection operations, that Espy's chief of staff had suddenly stopped work on regulations to reduce pathogens in chicken. The charge was denied.

Espy loyalists, who had from the beginning found the agency's response to food safety weak, thought the attacks were originating with the beef industry. In fact, beef representatives have long felt that poultry has the regulatory advantage. Under USDA rules, poultry processors could add water weight, and did (8 percent); chlorine could be added to the water; irradiation had been approved for chicken; fecal contamination was allowed, as were pathogens; and in ground chicken products, lungs, sex glands, and skin could be a part of the mix. The beef industry wanted a level playing field.[47]

Battle lines were drawn. In a *Los Angeles Times* interview in July, Espy publicly said that he had the impression from officials in his department that they knew about the deaths in the 1993 outbreak but weren't concerned. "They considered the deaths acceptable," he told the *Times*.[48] Jill Hollingsworth, who'd been so quick to deny agency responsibility to Bartleson, told employees in a conference call that she couldn't imagine how the secretary "ever walked away with [that] impression."[49] An anonymous letter purporting to be from some USDA employees to the president and asking for his help came in

August. In it Espy was accused of being "desperate, malicious, and slanderous," and of using "food safety as his personal ticket to fame." The letter continued: "It is not acceptable for people to die from the food they eat . . . but it happens."[50] The writers said they feared retaliation if they revealed themselves. There was no way to tell whether the document was authentic.

Those in the department who knew what Espy was complaining about did their own leaking. Food safety advocates had long suggested that rapid microbial tests were needed to detect the presence of microbes and had pressed the USDA to develop one. In August, Dan Laster, director of the U.S. Meat Animal Research Center in Clay Center, Nebraska, wrote a memo detailing what he had previously revealed to a reporter, that several senior-level FSIS staff had stymied his efforts to develop the test. They had, he said, "tried very hard to prevent us from moving forward in any kind of timely manner to evaluate the Rapid Test under in-plant conditions."[51] The "stonewalling" was thought to have been an effort to embarrass Secretary Espy.

In fact, the attacks on Espy in the press were increasing. He knew time was short. In August he replaced Cross at the FSIS with Michael R. Taylor, who'd won praise from consumer advocates while at the Food and Drug Administration. Taylor stood at Espy's side as legislation proposing the new Pathogen Reduction Act was introduced to Congress. It would have given the USDA authority to recall products (now it can only withhold its stamp of inspection and encourage industry to recall) and impose fines, and it would have established microbial limits beyond which a product could be considered contaminated. It was a bold move by a wounded secretary. Heat was growing in the charges of misconduct. Days later he would announce his resignation and the Pathogen Reduction Act would, with advice from industry, consumers, and Congress, begin to evolve.

Taylor, too, planned to move quickly and decisively to do something about meat safety. His most dramatic gesture came on September 29, 1993, in an address to the American Meat Institute. When he got to the subject of *E. coli* O157:H7, the members of his audience, who'd been listening with half an ear, began frantically scribbling notes. The bacterium, he said, would from now on be considered an adulterant on meat. The shock waves could be felt through the audience. This was a historic change. Taylor went on. Because *E. coli*

O157:H7 was unusually virulent—some researchers had suggested that as few as ten microbes could cause illness—any amount of it would render the product unsafe, and the agency did have the authority to hold product that was clearly unsafe. Not only that, the agency would immediately begin testing for *E. coli* O157 at the retail and "grinder" level—although it would take only five thousand samples.

It was perhaps the single most important change in USDA history—the agency would do something it had never done, test for a microbe and hold the product if it was found—and yet the speech was not routinely released, as most official addresses are. Copies had to be requested. And the Associated Press delayed reporting the story for several days, for reasons that are not clear.

Taylor's speech marked an about-face that left the industry outraged. If *E. coli* O157:H7 could be considered an adulterant, so might *Salmonella* or *Listeria* or *Campylobacter* or any of the other pathogens that were getting into meat during slaughter and production. The industry protested vehemently in the days following, threatening to file a lawsuit, but Taylor was unmoved. He tossed the industry a bone: It could bypass the usual process for changing regulations and use organic acid sprays on the carcasses, although processors would still be required to trim. One study had shown acid sprays to be effective, and another that they were ineffective; spraying was cheaper, however, than the real solution of trimming away fecally contaminated meat. And besides, spraying added water weight—if "only one-half of one percent," in the words of an industry representative. But in a large-scale operation, a minute percentage can mean huge increased profits.

Taylor's gesture did not mollify the industry. On November 1, the American Meat Institute (AMI), the Food Marketing Institute (FMI), the Associated Grocers of America, and four other groups filed a lawsuit seeking a permanent injunction against the USDA to halt its testing of hamburger for O157. Their interesting rationale: Testing might lead consumers to think their meat was safe and to ignore the cooking and handling warnings that the industry had fought so hard to keep off meat. The industry took its suit to the same Judge Nowlin who had ruled in its favor before. The AMI-FMI lawsuit said the USDA failed to follow proper rule-making procedures in making the change. It used the agency's own inaction since the microbe first caused trouble in 1982 as evidence that *E. coli* O157:H7 neither represented adultration nor presented an emergency situation. The industry saw in the new definition a frightening potential for lawsuits. (One industry lawyer advised producers not to test their own meat, or to stop if they were already doing so, to avoid liability.)

The USDA was not surprised. It had anticipated something of the sort. But Taylor's staff had gambled that the weight of public opinion now favored food safety. And indeed, meat industry representatives were surprised when, meeting in Chicago to discuss foodborne illness three days after filing the suit, they were met by protesters from STOP bearing signs such as "Industry obstructs again" and "Testing = Accountability." The protests made some major newspapers.

The industry has argued for the past few years that USDA inspections are outdated. Inspectors, they argue, should abandon the traditional "organolyptic" methods, which use touch, sight, and smell, for "science-based inspection." Why, then, would the industry oppose the "science-based" microbial testing the USDA has begun?

The industry argues that testing at the retail level would not prevent the microbe from reaching the public—a negative test, as one industry scientist put it, "only tells you that you don't have it in that sample"—but might damage the reputation of the product. Grocers, for instance, resent risking their reputations when contamination probably occurred at the slaughterhouse level. There is some merit to that argument. But testing at an earlier point in the process can give an idea as to what and how many pathogens are present. It was that tack the USDA would pursue.

The "science-based" inspection the industry prefers is called the hazard analysis and critical control points system (HACCP, pronounced "hasip"), which establishes where dangers lurk in the process and monitors those points. HACCP could help; it has shown its effectiveness in other industries. In a sardine plant, for instance, the fish can be visually checked as they go into the can and are sealed, but the critical control point is the "retort," where the sealed can is subjected to intense steam heat for a certain length of time. The HACCP system requires that the time and temperatures be recorded and checked. But there is one essential difference between sardines and meat: Unlike ground beef, which generally carries the label of the store on the package rather than the producer or processor, accountability is built into the sardine can. And there is an easily identifiable critical control point. No pathogenic microbes (that we presently know about) can survive that process. But HACCP remains unproven with raw meat. The critical control points at which screening or testing might control this tenacious microbe are only now being identified, and there is no process that can render raw meat totally free of microbes. (Even irradiation leaves a few hearty microbes such as *Clostridium botulinum* and *Clostridium perfingens*.)

The USDA feared that Judge Nowlin, who had confessed to being a cattleman and wondered aloud to the lawyers whether that might constitute a conflict of interest, would rule from the bench on the AMI-FMI lawsuit. He did not. When he finally ruled on December 13, he declined to stop testing. The USDA had won a round. A furious meat industry promised to fight on, but in fact it would back off publicly. There wasn't a lot of good press to be gained by fighting food safety in the open. The industry would confine its considerable influence to getting the best deal it could from the upcoming HACCP regulations.

HACCP, in the version proposed by the AMI, would do more than systematize slaughter. It would have industry inspectors replace government inspectors, who would remain on the job but would shift from looking at product and process to looking at paper. It's the deregulatory aspects of HACCP that make it so appealing to the industry. The USDA began preparing new regulations that would include HACCP but also microbial testing. "We have a vision of HACCP that's going to be different from industry's vision," laughed Taylor.[52]

(The legislation would take time. Behind the scenes battle plans were drawn as industry, medical researchers, public health officials, and the consumer food safety organizations that now had a seat at the table vied to see who would have more influence. It would be 1996 before the legislation was finally introduced, and it would be a compromise version.)

E. coli O157 is an emergency situation, the USDA had argued in its brief response to the AMI-FMI lawsuit. The agency hoped that testing for the microbe would prompt the industry to take action. The aim, says Taylor, is to increase accountability. But in 1996, after two years of testing, the USDA had come up with only five positives—even as outbreaks and sporadic cases continued to occur. By contrast, David Acheson of Tufts University had devised a test to look not for the microbe but for the Shiga toxin produced by both O157 and some non-O157 *E. coli*. When he tested it on sample packages of raw hamburger, his baseline positive rate was 25 percent. He went back to these positives and confirmed them by finding the microbe. Any reasonable observer might wonder whether something wasn't wrong with the USDA's testing procedure.

Testing was not an entirely new thought. Florida had begun testing for the microbe fourteen months earlier than the USDA. A few days after the lawsuit was filed, Florida found it. Montfort, Inc., of Greely, Colorado, would recall nine thousand pounds of *E. coli* O157:H7–contaminated hamburger meat.

While thorough cooking is assumed to destroy the bacteria—a fact that has led the meat industry to call the consumer or food preparer the ultimate "critical control point"—the USDA now argues that many Americans are accustomed to eating their burgers rare or pink in the middle and are thus vulnerable. Not only that, cross-contamination in the kitchen can present a huge problem. In one case a teenager was preparing his own supper and began cooking a burger. He cooked it thoroughly. His brother came in while the burger was still cooking and tossed his own into the skillet, turning it and mashing it down with a spatula. The first boy used the same spatula to remove his burger. It was that boy who became ill.

The microbe has continued to demonstrate its adaptability. An outbreak of *E. coli* O157–associated illness in Washington State and California affecting eighteen people was associated with cured salami previously thought to be safe. When the manufacturer recalled ten thousand pounds, another segment of the food industry was jolted into the new O157 reality. The pathogen is more acid-resistant than most *E. coli*, and the traditional fermentation process hadn't stopped it. Since the "just cook it" message to consumers doesn't fit a product meant to be consumed as is, the cured-sausage industry is deeply troubled. But it should have anticipated the problem. Studies published in 1992 by Michael Doyle of the University of Georgia, a scientist who has researched *E. coli* O157:H7 for years, had indicated that the microbe survives both fermenting and drying. "Some folks in the industry were well aware of it," he says.[53]

But one food-safety advocate has advice for consumers, who may by now feel battered and vulnerable. Mary Heersink of STOP suggests putting a lump of any newly purchased ground beef and the receipt into a plastic bag and freezing it, thus creating what she calls a "frozen insurance policy and instant accountability" for that moment when the hospital asks whether you have any left. The thought may well have sent a chill through the industry.

Since the Jack-in-the-Box (western states) outbreak, there have been, according to STOP, which keeps an unofficial tally with their 800 number, more than two hundred outbreaks. Since the CDC estimates that there are 250–500 deaths a year from O157, and since ground beef is the primary vehicle, it is a real possibility that every day in this country someone dies from eating a hamburger, a thought that lends a new dimension to one of America's cultural icons.

The CDC had tried to trace the meat in the western states outbreak back to its source, but large-scale slaughtering and mass-distribution made doing so virtually impossible. The records simply stopped, and "no one could determine exactly what meat went into any lot," said Patricia Griffin. The agency did discover two points in the slaughter and deboning process that didn't look good. Some of the slaughterhouses had used "bed slaughter": laying the animal on a bed of rails inches from the floor, which "is often covered with feces and dirt; the animal is then skinned by hand." The other problem seemed to be the boning plant. "Meat is removed from the bones on a long table. A conveyer belt carries the meat to one end. The belt is sanitized once daily."[54] The boning belt was ideal for cross-contamination.

Griffin was able to establish a probable chain of events that resulted in the outbreak.

> A farm supplied cattle carrying *E. coli* O157:H7. A boning plant then supplied the contaminated meat to one hamburger-making plant. The hamburger-making plant mixed contaminated meat with other meat. Contaminated hamburgers were made in huge quantity on 2 successive days several weeks before the outbreak and were frozen for wide distribution. The final defense was lost when the hamburgers were insufficiently cooked in more than 90 Chain A restaurants in 4 states. As a result, a single strain of *E. coli* O157:H7 infected over 700 people.
>
> In conclusion, *E. coli* O157 colonization of cattle results in human illness and death. Current slaughtering methods can result in fecal contamination of meat. Central processing methods that mix contam-

> inated meat with other meat can result in many contaminated hamburgers. Fast-food chains can cause widespread outbreaks because they can have many restaurants using a uniform cooking method on meat from a common source.[55]

Griffin was optimistic. "The impact of this outbreak has been impressive. *E. coli* O157:H7 became a household word. Food safety became a hot topic. . . . The impact on the U.S. Department of Agriculture has been unprecedented. . . . Congress also awakened to food safety," she wrote. But it was 1994, and while Congress had been forcibly awakened to food safety, its actions after the November elections would be disappointing. And while it may have seemed to Griffin, from her perspective, that surely everyone knew about *E. coli* O157:H7 and the dangers of rare burgers or cross-contamination in the kitchen, she couldn't have been more wrong. The pathogen might have received massive publicity, but much of it had been inaccurate and misleading; it had minimal impact on a public conditioned to hyperbole.

In the summer of 1996 an expensive restaurant in Waitsfield, Vermont, would outline in its menu how it prepared its burgers. Rare was defined as pink and warm inside, medium was pink but hot. Well done, brown all the way through, came with a warning: "We are not responsible." It should have been the other way around. The message would be slow to penetrate the culture and then, at some levels, would be generally scoffed at. One more health scare. People were tired of hearing it. And most thought they were magically immune.

Vital questions remain unanswered: Where did the organism come from, and why was it in cattle? No one was sure, but clues as to why it was getting on meat pointed to a food safety strategy gone wrong. Slaughterhouses had assumed they could lower the level of contamination on meat if they emptied the intestines of animals by withdrawing food prior to killing them. But a number of studies would make this appear counterproductive.[56] And there were other changes that USDA researchers thought might be contributing.

Gregory Armstrong, Jill Hollingsworth, and Glenn Morris, Jr., of the FSIS identified these factors in an article that appeared in 1996 in *Epidemiologic Reviews*.[57] Over the past twenty years the beef cattle industry, they wrote, has become ever more concentrated at the feed lot level, decreasing from 121,000 feedlots in 1970 to 43,000 in 1988. The largest of these fatten as many as 16,000 animals at one time. The same trend can be seen in the dairy industry, which supplies spent animals for hamburger meat. While there were 600,000 dairy farms in 1955, by 1989 there were only 160,000, and the number of farms with milk cows decreased as well, from 2,800,000 to 205,000.

Changes in how animals are raised—from what they eat to how they eat—may also have contributed to the growing presence of this new pathogen in cattle. Computerized feeding has been linked to animals found to harbor *E. coli* O157:H7 and the feeding of antibiotics, a practice that began in the mid-1970s, parallels the emergence of this pathogen. Perhaps the spreading of manure slurry on fields may play a role, they suggested, as may, ironically, the success in reducing another disease in animals. Immunity to brucellosis, brought on by an animal's response to infection, can provide immunity to *E. coli* O157:H7, and there seems to be more O157 infection in areas, such as the Northwest, that have been successful in reducing the level of brucellosis in cows.

But changes in the way meat is processed are having a profound effect. The USDA researchers estimated that with the mass production now standard, one infected animal could contaminate 16 tons of hamburger meat. And the size of hamburger lots is, in turn, related to concentration and vertical integration in the meat industry. The four largest meat-packing firms increased their market share from 22 percent in 1977, to 32 percent in 1982, to 54 percent in 1987. Even more concentrated is the boxed beef industry. In 1987, 80 percent of the market share went to the four largest firms.

At the same time, consumers are changing their eating practices. While consumption of hamburgers has actually decreased since 1976, eating at fast-food restaurants increased 224 percent from 1967 to 1982. While these burgers are not intentionally undercooked (although it clearly happens), when they have a choice, either at home or in non-fast-food restaurants, 23–25 percent of Americans prefer their beef rare.

While no one change played the key role, it is likely that each change contributed. "In particular," say Armstrong, Hollingsworth, and Morris, "big may not always be better. Consolidation of the industry, widespread movement of cattle, increased use of large production lots for products such as hamburger, may all have played a role in the process—and may provide a setting in which other 'new' pathogens can rapidly move into human populations."[58]

At the molecular level the evidence seemed to show that O157:H7 had emerged recently when a strain of *E. coli* related to the enteropathogenic series picked up the ability to produce the toxin from *Shigella* by horizontal transfer—bacteria can share genetic information in DNA in a direct transfer process called conjugation. While chromosome transfer is relatively rare, plasmid transfer occurs frequently and rapidly among some bacteria.

Said Marguerite Neill,

> Judicious reflection on the meaning of this finding suggests a larger significance—that *E. coli* O157:H7 is a messenger, bringing an unwelcome message that in mankind's battle to conquer infectious diseases, the opposing army is being replenished with fresh replacements. . . . We have no cogent explanation for why this pathogen has appeared and we do not know whether we are fostering its dissemination. We have no detailed understanding of pathogenesis at the cellular level, leaving us without a scientific basis to design treatment strategies or prevent complications. We do not know what control measures work or where to apply them. We do not yet have a conceptual approach to *E. coli* O157:H7 which incorporates a comprehensive public health outlook with practical cost-effective control measures. It is sobering to note that all the features of the 1993 Washington State outbreak were already known by 1984.

That is, *E. coli* O157:H7 is a high-grade pathogen that needs only a few microbes to cause disease and is transmitted by a high-volume food item, she said, "whose preparation contains a compositional and thermal Achilles heel," and it is served to a "target audience [children] most at risk for complications of illness."[59]

10/New Pathogens and the Politics of Denial

> Our lives are inextricably interwoven with the lives of these creatures who we ignore until they cause us trouble.
>
> LYNN MARGULIS AND DORION SAGAN, *Microcosmos*, 1986

Two things are happening in the human contest against infectious disease. Some known diseases are striking humans more frequently, including many that were once considered under control, and the human population is being challenged not only by more pathogens but by a greater variety of them than ever before. Those were the observations of the Institute of Medicine's Committee on Emerging Microbial Threats to Health in 1992. New pathogens can arise, rarely, from a change in the genetic properties of a microbe, but much more often the sudden emergence of a disease-causing agent is the result of changes in the environment or in human ecology. If one thing is clear, it is that changes are coming at a faster rate than ever before in the history of the planet. There is no real way to measure, but it's a good guess that the number of changes in the past one hundred or even fifty years exceeds the total number of changes in the rest of human history—and no change, not even one of the smallest degree, is without consequences in the delicate equilibrium between humans and disease.

James Lovelock, a British scientist and the author of *The Gaia Hypothesis*, proposes that the earth is a self-regulating entity that can maintain the conditions necessary for its survival—for healing itself—by controlling the chemical and physical environment.[1] As the changes to the environment wrought by humans become more dramatic, and as outbreaks from emerging pathogens become more frequent, it seems clear that the two phenomena are connected. Has the corrective process begun? If Lovelock is correct, the push for a restored equilibrium would logically begin on the microbial level, for it is microbes that can adapt and change the most easily. Is the microbial world eking out revenge for our hubris and our exploitation of the planet, or are we simply bystanders as the earth attempts to right itself and restore a healthy balance? What Lovelock and others remind us is that the result of the earth's self-regulating may not be to our advantage, nor is the process likely to be pleasant.

That worry hovered just over the horizon in 1996. The year began with "mad cows," and by year's end a host of challenges to the food supply had caught the public off guard. No one at the CDC, the WHO, the GAO, or the National Academy of Sciences was surprised. All had predicted an increase in foodborne disease, but few outside those circles had taken the predictions seriously.

Among consumers, even those who waited patiently for the seasons had their strawberry days spoiled by another intruder. *Cyclospora* was the uninvited guest at graduations and weddings during the months of May and June in North America. It was the parasite that turned good memories into very bad ones for people from Texas to Cape Cod. As sanitarians tramped through vegetable markets looking for suspect vehicles, as the story made the front page of the *New York Times*, as earnest epidemiologists ran through their endless questionnaires, the simplest pleasures of summer were tainted. A veterinarian friend who also runs a bed-and-breakfast in Vermont told me that his wife made him scrub every single berry.

"It ruined them, of course."

Were we looking at the future? Was every food pleasure to be spoiled?

Cyclospora: The Berry Bad Bug

The story began in Texas in early May for the victims, but some weeks later for public health officials. It can take that long from ingestion of the parasitic *Cyclospora* oocyst (spore) to symptoms (the protozoal stage), to culturing, to the realization by the health department that something is amiss. When the parasite finds a comfortable berth in the human intestines, it begins to multiply, causing physical distress. The symptoms appear three to seven days after ingestion and, if untreated, can continue for a long time with periods of improvement as the parasite goes through its life cycle. It produces watery diarrhea, loss of appetite, bloating, increased gas, cramps, nausea, vomiting, extreme fatigue, muscle aches, low-grade fever, and, because it blocks food absorption, severe weight loss. Like most other foodborne diseases, it is spread by infected fecal matter that contaminates food or water. Person-to-person transmission is unlikely. And puzzlingly, not everyone who is exposed to *Cyclospora* becomes infected.

The first known case of infection with *Cyclospora* was diagnosed in 1977 in Papua, New Guinea. Before 1996 only three outbreaks had been reported in the United States, one at a country club in New York. Previously, the infection had occurred mainly in people returning from international travel, but the travel of the victims in the country club outbreak had been limited to the golf course; in fact, those who drank water from portable coolers on the course were more likely to become ill. The CDC team suspected that the cooler water had been contaminated by dirty hands or by contaminated backflow from a hose used to fill the coolers. But where had the pathogen originated? How had it entered the country? That they couldn't answer. Nevertheless, the investigators knew they were looking at an emerging pathogen that was bound to cause more trouble. They warned physicians in their report of the outbreak to consider *Cyclospora* if they were called to treat serious and protracted diarrheal disease. It was good advice.

The CDC was alerted to the 1996 outbreak when it received a call from a Canadian physician in Calgary. He had patients with persistent diarrhea who tested positive for the parasite, and they had been abroad—to Texas. The CDC arranged for the Texas Department of

Health to get in on a conference call. Dr. Kate Hendricks, director of the Infectious Disease Epidemiology and Surveillance Division, is a physician who trained at the University of Illinois, earned her M.P.H. at Tulane in the School of Tropical Medicine, and had served as an EIS officer before accepting the Texas position. At once she began organizing an investigation based on what the Canadian physician had been able to tell them about his patients' trip.

The business gathering had been held at a Houston athletic club that had an attached restaurant and meeting facility. Hendricks's team contacted that city's medical officials, and a quick survey found that of the twenty-six people attending, sixteen had become ill. Nine of the sixteen experienced diarrhea for two weeks or longer. In eight of these people, *Cyclospora* was identified. The health department quickly issued a press release on *Cyclospora* and secured a "function sheet" from the manager of the restaurant to try to figure out what people had eaten. Unfortunately, the listing wasn't clear enough for the department's purposes. It supplied only general descriptions of the food served, such as "salad bar." The investigators would have to work on-site.

Hendricks and Jeffrey Taylor, an epidemiologist on Hendricks's staff, left on May 31 for Houston, where they were met by Houston health officials. Together they checked out the club's water, the sources for the ice, the plumbing, and the employees; everything seemed in order. At least one employee was ill, but he seemed to have gotten the illness at the same time as the guests—perhaps from eating the same food. Then began the difficult task of trying to determine what people had eaten three weeks before.

Their first discovery was that there had been substitutions on the menu. They put together a questionnaire based on what they learned and that very night returned to the health department and began calling cases. On the first call Hendricks realized that their questionnaire had to be revised. The investigators hadn't known until then that the same group of people had eaten together the next day and that they had had dessert both at lunch that day and the night before. The case remembered quite accurately that he had insisted that everyone have the strawberries at lunch because they had been so good the night before served in wineglasses with a Romanoff sauce.

In the subsequent investigation, twelve of the thirteen cases said they had eaten strawberries, while only one of the well controls had done so. Implicating the strawberries seemed to be supported by the facts.

Other outbreaks were occurring around the city. Most were at business gatherings or at expensive restaurants, and again the analysis of food histories showed an association with strawberries. At one, strawberries had been used as garnishes on the plates. Ten out of ten cases remembered eating them, while only one of the controls had. That seemed to clinch it. The source of the strawberries appeared to be California. Cases in Toronto seemed to be linked to strawberries as well. The press reported the results of the investigation, the story traveled across the country, and strawberry consumption dropped, to the dismay of the strawberry growers and the California Strawberry Commission, which immediately started testing for the microbe but could not find it.

Cases were appearing elsewhere in the country. Florida and New York had outbreaks. A bridal luncheon in Osterville on Cape Cod had a disastrous aftermath. A week later eighteen of the twenty-two who had attended—including the bride honeymooning in Spain—became frighteningly ill. "No one had a mild case of this," the bride told the *Boston Globe*. "Every person I know has spent a few nights sleeping on the bathroom floor." The *Globe* reported that most lost ten to fifteen pounds and that one woman lost thirty. Another required hospitalization. One woman remembered eating a fruit salad containing strawberries, raspberries, blackberries, and blueberries.[2]

The CDC would eventually report cases in nine states. Twenty-one were ill in Toronto, and on June 21 the Canadian Press Agency reported the first death, a fifty-one-year-old man. But if Houston and Toronto thought the link was to strawberries, the CDC was being cagey. It wasn't sure. Epidemiology elsewhere implicated fresh fruit in general—a mixture of berries and other fruits, especially raspberries. Then the investigation of an outbreak in South Carolina—sixty-four people attended a luncheon and thirty-seven developed *Cyclospora* infection—made things clearer. The infections seemed to be linked to raspberries, strawberries, and potato salad. But on the same day another luncheon had been held in an adjacent room. Strawberries had been served from the same source, and no one had become sick. Well, one person had. But she had eaten raspberries at the same establishment that evening. The CDC would later announce that it suspected raspberries imported from Guatemala and was sending someone to take a look at the situation there.

When Hendricks heard this news, she was troubled. Could there have been something she missed? Her identification of strawberries had thrown an industry into shock at the height of the season. The decision to announce a suspected vehicle is always made cautiously, balancing the economic consequences against the health of the public. She had been cautious, but she had thought she was right. Now she went back over what she had done. She tried to reinterview the cases, but most were tired of talking about it and uncommunicative. It had been so long ago now. She did manage to talk to the reservations manager and the sous-chef at one of the restaurants where people had become ill, and new facts emerged—details they had forgotten to mention. She discovered that the restaurant had a special policy for guests it considered VIPs: Their servings were garnished not just with strawberries but also with raspberries. No one could remember whether this group had been considered VIP. Then Hendricks managed to talk to a sous-chef she hadn't questioned before. He remembered mixing the berries on the plates.

Epidemiology is science, but it is also an art, one that involves sorting out the "noise" that accompanies any investigation. Hendricks was chagrined to discover that she'd apparently been wrong. This was one investigation she wouldn't be able to publish. It was also an investigation that other public health officials fear has set them back in urging that epidemiology be accepted as an effective tool for determining the vehicle in the absence of finding the pathogen in a food.

Hendricks clings to the hope that she might, after all, have been right, but when the Guatemalan raspberry season ended, the outbreaks ended. It seemed safe to go back to fresh fruit—much of the public had never stopped—but some of the joy had been diminished. No longer could a pile of sweet, succulent berries be delightedly consumed with a pile of soft whipped cream without that nasty little thought of wicked *Cyclospora* creeping in to spoil the pleasure.

The Smallest Troublemakers: Viruses

Cyclospora was a new worry—it had changed, momentarily at least, what some people were eating. But it was hardly the only emerging foodborne pathogen out there. In the ten-day period December 4–14, 1994, more than three hundred students had reported to the health service at Harvard University with acute gastroenteritis. People tend to call such sudden, mass outbreaks of vomiting and diarrhea "flu," but "stomach flu" does not exist. Influenza is a respiratory disease in which the virus is spread by air or close personal contact. The viruses that cause diarrheal disease, like bacteria and parasites, are either in contaminated food and water or transmitted by the fecal-oral route.

Most of those affected at Harvard were freshmen who shared a dining facility. A team of local and CDC epidemiologists quickly investigated. The specimens were negative for all the usual bacterial pathogens, but the investigators did discover something called small round-structured virus (SRSV), which is the name that Norwalk virus is better known as outside the United States. Food histories linked the illnesses to eating salad. The infection was traced back to one individual—a salad chef who had reported illness on December 4. The DNA sequence of the case strain and the salad chef's strain matched. One infected food handler had made three hundred people acutely ill.

Viral agents are often named for the place where an outbreak is first identified. Norwalk virus was one of those. The first outbreak occurred in Norwalk, Ohio, in 1969, and it was three years before the agent was identified by electron microscopy. The viruses are still difficult to detect because they are generally shed for only two or three days. While the illness the microbes cause is relatively mild (but unpleasant), these tiny creatures are thought to be the most frequent nonbacterial cause of gastrointestinal disease and thus have an important adverse economic impact. They can also be transmitted with amazing ease. Minnesota epidemiologists investigated two outbreaks of Norwalk or Norwalk-like gastroenteritis in which an employee who vomited at work subsequently contaminated seventy-six liters of butter-cream frosting in one outbreak and several hundred hamburger buns and oatmeal cookies in another. The attack rate in this outbreak was impressive: Of those who ate the ten thousand contaminated frosted bakery products, 60 percent had become ill; 30 percent of those who ate the contaminated unfrosted baked goods became ill. Moreover, investigators found that the virus continued to be transmitted at the bakery for more than a week, contaminating baked goods each and every day.

Once again the need—even the cost-effectiveness—of food service workers being given paid sick leave becomes obvious. The Minnesota outbreak points out something else as well. Government regulators are often mocked for paying such close attention to seemingly mundane things as working toilets and the availability of handwashing facilities with soap and hot water. But such facilities are vital to safe food preparation.

Viruses can be a handy scapegoat when looking for an explanation for foodborne disease, but they do not begin to cause the number of outbreaks that bacteria and parasites are responsible for. Nevertheless, their importance is beginning to be recognized, because they may be responsible for illnesses that have been assumed to be chronic conditions. And yet an air of mystery surrounds viruses. Where do they come from? How do they get into the food supply? Is there anything we can do about them? The answers—other than practicing basic hygiene—are often unclear.

Sometimes the agent is never found at all, even when the study implicates it clearly. That was the case with an outbreak in Brainard, Minnesota, in the 1980s. Many of the patients had episodes that lasted for as long as a year and involved from ten to twenty bowel movements a day—effectively causing them to be homebound. Some cases were identified as functional or irritable bowel syndrome. The investigation by the Minnesota team revealed a clear link with the consumption of raw milk. But the etiological agent was never identified.

When the article appeared in *JAMA*, Dr. Martin Blaser's accompanying editorial explained that this agent probably had never been recognized before in humans. If we needed reminding after AIDS, Lyme disease, and Legionnaires' disease, this was one more sign that new microbial pathogens were out there. Blaser urged physicians to consider the possibility when confronted with what seemed to be irritable bowel syndrome, "functional" bowel disorders, or nonspecific colitis that didn't respond to antibiotics, that a foodborne infectious agent might be responsible. Ask about eating patterns, he recommended. Raw milk was always suspect, but plenty of ordinary, supposedly safe foods had also been implicated in outbreaks.

The implications were startling. For many physicians, irritable bowel syndrome is a "rule-out" diagnosis when they can't or won't take the time to find out what is actually going on. That such cases might be caused by an infectious viral agent that could apparently be transmitted by food was news indeed.

Cryptic *Cryptosporidium*

Other emerging pathogens are showing up in places they haven't been seen before. *Cryptosporidium* is one of them. It has caused widespread outbreaks, including the one in Milwaukee that infected more than 400,000 people and caused 104 deaths. The immune-compromised, which now includes 25 percent of the population (young children, the elderly, AIDS patients, transplant recipients, and those undergoing cancer therapy) are especially vulnerable. But virtually all of the outbreaks have been traced to water, and to get rid of the parasite requires special filters rather than simple chlorination. Around the country water control boards are having to catch up to a new reality. Recently an outbreak of "crypto," as it is called, has been linked to a new vehicle.

In October 1993 the principal of an elementary school in central Maine telephoned the office of the state epidemiologist, Dr. Kathleen Gensheimer, to report that many of his students—too many—were out sick. A quick check revealed that another elementary school and the local high school were reporting the same thing. Altogether 230 people were ill.[3] Both principals noticed that those ill were in the classes that had attended an agriculture fair eight days before. Organized and run by high school students, the fair had included agricultural demonstrations, a petting zoo, a hayride, a cider-pressing demonstration, and light refreshments. Two days after the outbreak was reported, the stools of three of the ill children revealed *Cryptosporidium*.

At the time Gensheimer was fortunate to have attracted to her department a young physician, Dr. Peter Mallard, who was undergoing the CDC's EIS training. Together they conducted an investigation to identify the source. The students who attended the fair were asked to complete a written questionnaire, with their parents' help, about what they had done at the fair. Drinking cider was reported by 54 percent of those who were ill, as compared to only 2 percent of those who were not. In addition, only those who had attended the fair in the afternoon were sick; there were no cases from the morning attendees.

As accurate as epidemiology can be—and in this case there appeared to be no doubt that the apple cider served in the afternoon was the culprit—it is a wonderful bonus to be able to confirm the findings by discovering the pathogen in the food product as well. More often than not, the food is gone. This time, however, a little questioning revealed that there was in fact some cider left. It's well known among country folk—even among young country folk—that unpasteurized cider goes hard when stored. That is, it develops significant alcohol content. Mallard was to discover that one enterprising student had taken a jug home with just this transition in mind. *Cryptosporidium* oocysts were found in a sample.

Bit by bit Mallard and Gensheimer were able to piece together what had happened. The five bushels of apples used during the morning pressing had been bought from a commercial grower, but the high school students had gathered more apples from old trees on the edge of a pasture where cows had recently grazed to use in the afternoon. They collected them by shaking the tree and then picking up the fallen apples off the ground. The apples were stored in clean boxes and sprayed with municipal water before the fair. Oocysts of *Cryptosporidium* were found in a calf that had been grazing in the pasture where the apple trees stood. And they were found on the press. Not washing the apples thoroughly enough before pressing probably allowed the oocysts to spread throughout the juice.

This was one outbreak that made the papers—too many children got sick for it to be ignored. Gensheimer says that she worried about hurting the apple cider industry, which is important in Maine, but the investigators couldn't ignore what they had discovered. And *Cryptosporidium* could lead to a very serious infection indeed. The organism is resistant to common disinfectants. Pasteurization is the only solution. Gensheimer and her team didn't want to advise that—at least not yet. After an outbreak of *E. coli* O157 from apple cider in

Massachusetts a few years earlier revealed that the organism had a high resistance to acids that other *E. coli* didn't have, Maine had mandated washing and brushing apples used for making cider. The state's cider producers were formulating voluntary sanitary guidelines, and all seemed well with commercial ciders. But clouds were gathering over the traditional industry, and one of the grand pleasures of fall suddenly became a lot less fun.

Listeria: The Ultimate Opportunist

Some emerging enteric pathogens are actually just newly identified as the cause of foodborne diseases. Scientists know that *Listeria monocytogenes* has infected animals since 1911 and that the first case of human infection was detected in 1929. It was thought at first to be a zoonotic disease, or one that animals gave to humans. That was reasonable. Some of the cases were in veterinarians—it can infect the skin—who had delivered lambs from *Listeria*-infected dams. Some researchers suspected that it was also foodborne, but it was only after an outbreak in 1981, when forty-one cases of listeriosis caused eighteen deaths, that *Listeria* was conclusively linked to a food vehicle. The implicated food was coleslaw.

Like most foodborne pathogens, *Listeria* is carried by fecal material. The cabbage from which the coleslaw was made had been grown in a field fertilized with manure from *Listeria*-infected sheep. Then it had been held in cold storage at a temperature that inhibited the growth of other organisms but that *Listeria* doesn't mind at all. In fact, under these conditions it could reproduce slowly but steadily; *Listeria* has been found to grow at temperatures as low as 4 degrees Celsius.

In 1983 an outbreak in Boston caused 49 cases of listeriosis and 14 deaths. It was linked to pasteurized milk. The most widespread outbreak was the one most publicized by the media. In 1985, 142 cases of listeriosis in the Los Angeles area, resulting in 46 deaths, were traced to *L. monocytogenes* on soft, Mexican-style cheese. A few years later an outbreak was linked to eating microwaved turkey hot dogs; once again, the microwave wasn't heating thoroughly and evenly. Outbreaks have been connected to soft cheese from Switzerland (the case-fatality rate was 28 percent), to salami in Philadelphia, ice cream, and pâté. When babies in a Costa Rican hospital came down with listeriosis, it was traced to a common supply of mineral oil that had become contaminated with the pathogen.

The growing number of cases and the realization that *Listeria* is widely present in the environment threw food scientists into a tailspin. It could exist and reproduce at every stage of the food chain: farm, slaughterhouse, processing plant, retail store, and even in the household refrigerator. They thought in the 1970s and early 1980s that they had most things under control in processed foods. *Listeria* undid all that. It was back to the lab to look for new interventions in processing to solve the problem. But first scientists needed to know where the problem was. When the American Meat Institute went looking for it in meat-processing plants, its investigators found *Listeria* in one-third of cultures from floors and drains and in more than one-fifth of cultures from cleaning aids, wash areas, sausage peelers, and food contact surfaces. It was endemic in something called the biofilm, a coating that built up in the equipment in which certain pathogens that were resistant to normal cleaning could hide out.[4] Microbes can attach themselves to even seemingly smooth surfaces, such as stainless steel, by growing tiny filaments and producing a gooey substance that resists sprays of water, even mild rubbing and weak detergent solutions. "Eventually unrelated families of microbes move in. The resulting cosmopolitan community forms biofilms that further protect its inhabitants," Edmund A. Zottola of the University of Minnesota told *Science News* in September 1996.[5] Really intense scrubbing with an abrasive is necessary to remove them. And biofilms didn't just happen in the food-processing environment—they can build up in the home, on knives, can openers, sinks, drains, even around counters.

Part of the trouble with *Listeria* is that both food processing and eating have changed. Who, thirty-five years ago, would have considered buying cold, vacuum-packed precooked chicken ready to eat? There were deli counters here and there, but few supermarkets purchased from distant sources coleslaw and potato salad already prepared in mass quantities.

My grandmother, remembering Upton Sinclair's revelations about meat and meat processing, had warned about hot dogs for years—I was not allowed to eat them—but now it was turkey franks that were implicated, and they were being marketed as a particularly healthy food.

Michael Doyle at the University of Georgia, one of the country's premier food microbiologists, discovered that *L. monocytogenes* was especially well adapted to processed poultry products; even low concentrations of the bacteria could reproduce under refrigeration to become a health hazard.[6] Researchers had good news though. They found the point in the hot dog process where the product was most likely to be contaminated. It was suspected that the pathogen got onto the turkey franks as the turkey skin was being removed. The machinery was contributing to contamination. The industry could focus on fixing that particular problem.

But while industry fretted and looked for answers to control this emerging pathogen, most people had never heard of it. The worry and concern took place behind the closed doors of industry and the regulatory agencies. Consumers remained ignorant of the risks from some of their favorite foods.

The public was not precisely kept in the dark about all this. Both the USDA and the CDC distributed information sheets—if you knew enough to ask for them—on which information was carefully phrased not to alarm but to caution. The problem was not given any widespread publicity, however. And yet, in 1986 alone, the CDC's Listeria Study Group estimated that listeriosis was responsible for 1,700 illnesses, 450 deaths in adults, and an additional 100 deaths in infants.

Pregnant women were especially at risk. Of the 1,700 cases, 27 percent occurred in pregnant women, and 22 percent of those cases occurred at or around the time of childbirth and resulted in stillbirths or neonatal death. These figures were determined not by passive surveillance and reporting—methods known to underestimate cases—but by conducting a nationwide study for which numbers were obtained from hospitals and other acute-care facilities. No other foodborne disease, the study group concluded, including botulism, had a higher fatality rate, and even this study, the researchers knew, was likely to have underrepresented the illness.[7] A French study found that 1.6 percent of pregnancies ending in premature labor or spontaneous abortion were linked to infection with *Listeria monocytogenes*.[8] And *Listeria* is costly when both illness and product recalls are factored in: an estimated total of $209–233 million a year in the United States.

There were constant surprises as researchers began to learn more about listeriosis. The sporadic cases were especially puzzling. Where were they coming from? Then a study linked them to the consumption of chicken—probably undercooked—and still another linked the infection to ready-cooked chicken. Changing eating habits again, and changes in packages and marketing. For every food company more processing carried the seductive allure of value-added. Not only was there potentially more profitability in such products, but busy people liked them.

More and more people were buying plastic bags of fresh, precut, and prewashed vegetables and salads. Since *Listeria monocytogenes* could be found in water, soil, and manure, it was potentially easy for those prepared vegetable products to become contaminated. They were washed—sometimes triple-washed—but in what, and where? There were no regulations. The bags of salad were kept chilled, and some were packaged in gases that reduced the oxygen content, which controlled some of the bacteria. But *Listeria* was happy both with and without oxygen. Karen Sizmur and C. W. Walker of the Clinical Microbiology and Public Health Laboratory at Addenbrooke's Hospital in Cambridge, England, examined sixty samples of bagged salads and found that four of them (6 percent) contained *L. monocytogenes*. What was worse was what happened to these products in the refrigerator. When researchers left samples at 4 degrees Celsius for four days, the number of *Listeria* increased roughly twofold. Other reports from Britain suggested that because *Listeria* was happy at refrigerator temperatures and capable of growing with or without oxygen, new techniques such as "sous-vide" (in which perishable products are cooked after being enclosed in vacuum packing), and the vacuum packaging of cooked foods could present some new risks. The food industry, in an effort to make these products more appealing and to reduce water loss, had lowered cooking temperatures as well. The combination was sure to have microbial consequences.[9] The World Health Organization found that when these packaged foods with extended shelf life were contaminated, they posed a major potential risk.

It might not be possible to eliminate *Listeria* at the farm level, the WHO group thought, but they did believe the number of *Listeria* excreted in animals would be reduced by using feed with lower levels of contamination. Using such feed could decrease levels and rates of contamination further along the food chain. And reducing *Listeria* should be based on keeping animals in clean conditions with clean food and water—basic principles of good farm management practice. It was common sense.

The next best step to cut down on the presence of *Listeria* in finished products, the WHO group suggested, would be to decrease the amount of fecal contamination and cross-contamination of the final product by equipment and other raw products. But an additional problem was that *Listeria* could be spread in the air—especially, and paradoxically, during the cleaning of equipment with high-pressure hoses. Modern cleaning methods were apparently spreading the pathogen further. The food industry was perplexed.

The presence of *Salmonella* in the egg products used in cake mixes, egg noodles, and liquid eggs had been resolved by pasteurization in the early 1970s, but now egg processors had a new worry. While no *Listeria* outbreaks had yet been linked to pasteurized eggs, studies suggested that, especially in salted products, the danger was there. They revealed that the HTST (high-temperature short-time) pasteurization conditions for liquid eggs were not enough to ensure a *Listeria*-free product. Salt could deter many pathogens, but *L. monocytogenes* was more tolerant. It could endure wider ranges of acid. It was versatile and dangerous. "Safe" pasteurized eggs were a food vehicle waiting for an outbreak.

The food industry seemed to be taking two steps forward and one backward. Just as everything seemed to be under control and food processors were basking in their successes, there were new processes, new packaging, and new pathogens to worry about—unfriendly microbes that seemed perfectly capable of slipping in between the cracks of food technology, or even adapting to safe-food technologies. Within a decade, *Listeria monocytogenes* had become one of the most serious challenges to the industry. It didn't seem to be causing mass outbreaks; many people seemed to be able to tolerate it. But when it did cause illness, its fatality rate was frightening.

The CDC knew about *Listeria,* the food industry knew about *Listeria,* the government regulatory agencies knew about *Listeria.* But it virtually ended there. Warnings had been made to the public, but they had not been widely disseminated. The CDC had strongly suggested in the *MMWR* that pregnant women avoid certain foods. The agency depends on the media, however, to get its messages out, and somehow that one got lost.

On the corkboard in the hall of the Foodborne and Diarrheal Diseases Division of the CDC is a poster of a beautiful woman. She is obviously even magnificently, pregnant, and yet her pose, in a flimsy, snugly fitting T-shirt through which her nipples clearly show, is quite sexy in a charming, outdoorsy way. The poster, which most effectively draws the eye, is a warning to pregnant women—and to their mates, who would presumably relay the information—to avoid eating certain foods such as soft cheeses and processed meat products during pregnancy. The poster is not a product of a U.S. regulatory agency. It is Australian.

In July 1995 Julia Langdon wrote an op-ed piece for the *London Telegraph* that told how her life and that of her newborn daughter had been saved in 1987 because doctors quickly and accurately diagnosed the listeriosis that had caused her to deliver the child early.[10] The two were treated promptly with antibiotics, and after three weeks they were discharged from the hospital. She counts herself very lucky. In pregnant women listeriosis can and does often cause meningitis or bacteremia and then miscarriage because the bacteria first enters the bloodstream and then the uterus. Infants who survive may have brain damage or cerebral palsy.

Langdon wrote the article because the British press had just revealed that when her listeriosis occurred the British government in fact had already known about the dangers for two years and yet until 1989 failed to alert the public. During that time at least twenty-six babies were known to have died. It is thought that this number, because the disease was not then widely understood or recognized, represents substantial underreporting. The placenta and other material from miscarriages is seldom tested for *Listeria;* listeriosis could have been responsible for many unexplained spontaneous abortions, researchers suspect. Langdon's first real information about this new and strange disease came from the public health official who visited her after her release from the hospital. He asked whether she had eaten dairy products, whether the cheeses were hard or soft, and where she had bought them.

"I had eaten a great deal of cheese, hard and soft. Like most pregnant women, I had tried to eat a wholesome diet to the baby's advantage. The doctor from the laboratory shook his head; I shouldn't have touched soft cheeses, chicken I hadn't cooked myself, or meat products such as pâté." She was stunned. She had never heard of this disease or the bacteria that caused it; certainly she had never been warned about avoiding any particular foods. She should also have been warned about deli foods, as they have been shown to be sources of *Listeria*, but she was not. Her (former) doctor, when she asked about *Listeria*, seemed never to have heard of it and thought she had made some mistake. When Langdon questioned Virginia Bottomley, the former British health secretary, in a radio interview in 1995 as to why the government had not warned the public, Bottomley replied that the government was anxious to avoid causing undue panic: "You have to be careful to inform rather than alarm," she said.[11] Apparently they had done neither.

In the United States, information about *Listeria* is just as scarce. Dr. Kim Cook at the CDC had to warn his pregnant wife, also a physician, to avoid certain foods. She had never heard of the danger before. An informal survey of new mothers that I conducted revealed that none had been warned about the dangers from the pathogen in specific foods.

But why is there so much *Listeria?* Certainly some of the increase could be related to mounting awareness among health officials. In addition, researchers have found that there is a seasonal pattern in both animal and human listeriosis. Contaminated silage, which farm animals eat in late winter and early spring when there is little grazing, seems to be contributing. Once again contaminated animal feed is implicated.

The spreading of manure on croplands was also implicated. Call it the manure cycle. Wherever one looked at the growing problem of pathogens in animal foods, manure seemed to be a factor. It was too nitrogen-rich, it was being recycled too casually, and there was just too much of it. The environment can handle a certain level of contamination, but we seemed to have reached some kind of limit. All around the world humans are eating more animal protein than ever before in human history. We have developed a taste for it, and few want to give it up. But we are paying for the pleasure in ways we never anticipated.

The Bug That Wouldn't Go Away

By the summer of 1996 consumers everywhere were being shaken by the new reality of foodborne disease. Just as the *Cyclospora* outbreaks seemed to draw to a close, the public learned through the press that Japan was in the midst of a rapidly expanding outbreak of infections from *E. coli* O157:H7. It began in May with isolated cases of HUS, then picked up speed in June, apparently from school lunches. The epidemic grew rapidly from eight hundred, to one thousand, then three thousand victims. While some areas were especially hard hit—in Sakai one out of eight elementary school children had become ill—cases were appearing all over the country. By July 25, eight thousand illnesses and ten deaths had occurred, and the fanatically hygienic Japanese were understandably in a state of panic. The reactions of ordinary Japanese, though surprising—families and even towns with many victims were literally shunned—should be put in context. The Japanese approach to disease is very different from the contemporary Western approach. Rigid—if irrational—quarantine rules are still in place for those with diseases labeled infectious, even though good hygiene and sanitation can prevent the spread, even among intimate family members.

The experience of friends of mine who lived in Tokyo illustrates the Japanese approach. The husband, after returning to Tokyo from a trip elsewhere, came down with acute diarrhea. By the third day he was in real pain, and someone in his office recommended an excellent Japanese physician. Unfortunately, his stools tested positive for *Shigella*, and the Japanese response was immediate. He was sent at once to a quarantine facility with nothing more than a pair of pajamas and a robe, where he was put in isolation. A troop of sanitation workers garbed in white biosecurity costumes then descended on my friend's home and completely disinfected it. She managed to keep them out of her kitchen with their chemical sprays because she told them her husband never went into it, information that they accepted as plausible. The husband managed to escape his imprisonment only when he convinced his doctors that he would be leaving the country on business. They insisted he go straight to the airport.

In the 1996 mass outbreak, Japanese health officials implicated first one, then another food as the source of the infection. They found

E. coli O157:H7 in eel, radish sprouts, and beef liver but seemed unable—or reluctant—to identify a primary source. Part of that reluctance was a fear of insulting, and reviving the prejudice against, the traditional meat workers, the Burakumin people, who had in the past suffered widespread discrimination. That sensitivity led Japanese newspapers to avoid references to slaughterhouses or the fact that the pathogen was associated with meat processing in the United States. And that confused the issue. Eating habits were transformed because virtually everything, including the air, was suspect in the resulting information vacuum. The Japanese custom of consuming some animal protein foods raw began to look very risky. People were cooking everything. Sushi restaurants, afraid of going out of business, added fried tofu, a dish usually served in the autumn, to their menus. The entire pattern of eating was disrupted by a pathogen that seemed to be everywhere. Actually, the Japanese had been changing their eating patterns for some time. One thing they had certainly increased was their consumption of beef and such American foods as hamburgers.

Although nothing appeared in U.S. newspapers, behind the scenes the American beef industry was very concerned. While *E. coli* O157:H7 may appear in many different vehicles from cross-contamination, the ultimate source is the cow. Japan had resisted for many years importing significant quantities of American beef, but those days were long gone. Now it imported a great deal of beef from the United States—nearly 5 percent of total U.S. production; indeed, in 1996, from January to May, Japan had been the destination for 58 percent of total U.S. beef exports. When one case was linked to beef liver, much of which was imported, orders for U.S. beef livers plummeted at once, even though U.S. beef had not been specifically implicated.

The Japanese outbreak could not have come at a worse time for beef producers; sales to Japan had dropped in the second quarter because consumers there were worried about mad cow disease and beef in general. Confidence was only just returning when President Clinton signed the new meat regulations, which had alerted the world to the fact that U.S. meat was not as safe as it was touted to be. Then on August 26, *Food Chemical News* reported that Japanese authorities had indeed found *E. coli* O157:H7 on imported U.S. beef intestines to be used in soups. That was to be expected, replied the U.S. embassy, since *E. coli* O157:H7 originates in the intestinal tract of cattle. Nevertheless, it was the last thing the American meat industry wanted to hear. When the STOP spokesperson Mary Heersink was asked on a Japanese television program later that fall, "Is America exporting its own *E. coli* problems?" her extended remarks were reprinted and distributed in the National Meat Association's newsletter.

By the first of August Japan's health and welfare minister, Naoto Kan, had decided to designate *E. coli* O157:H7 as an infectious disease. But some of the draconian measures of Japan's Infectious Disease Prevention Law would not be enforced. There would be compulsory stool tests of suspected patients, homes would be disinfected, and doctors would be required to report infections, but patients would not be isolated, and their homes would not be cordoned off. Good hygiene, good hand-washing, and not sharing a bath with a patient would probably prevent secondary illnesses, officials said. But their cautious approach to the epidemic brought criticism. Authorities were blamed for acting too slowly and for failing to develop an effective strategy for dealing with the outbreak.

As always, politics, cultural traditions, and bureaucratic inertia play what seem almost preordained roles when a society is confronted with a serious and unexpected disease threat to social order. In such a crisis the weaknesses of our systems are revealed, and attention to the real cause of the problem may be lost in the tumult. In the western states outbreak the food vehicle was quickly identified and recalled because Washington State was fully informed and ready to act decisively. But the outbreak had, apparently, begun in California—where the systems that might have prevented the outbreak failed.

U.S. officials were worried by the Japanese outbreak. Because so many different foods had been implicated, one USDA official mused, O157 just might have infiltrated the Japanese food supply. Dr. Patricia Griffin, the CDC's *E. coli* expert who spent three unhappy weeks in Japan, wanting to help but finding her help rejected, thought the radish sprouts might be the cause but suspected that it would never be proven. Rumor put the blame on a beef-rearing facility upriver of the sprout grower. There had been recent flooding that could have spread pathogens. It might have been the answer, but facts evaporated as the trail grew cold.

The concern, both at the CDC and the World Health Organization, was that the incidence of infection with the pathogen was apparently spreading worldwide. There was reason to worry. Outbreaks in Alabama and New Hampshire were occurring at about the same time, although they were only being reported in the local papers. In Germany concerns were growing as the number of HUS patients was increasing following infection with enterohemorrhagic *E. coli*, the term used to distinguish the illness from the many Shiga toxin–producing *E. coli*, of which O157 and 150 other serotypes had been identified. What bothered epidemiologists in Germany was that in earlier outbreaks it had been possible to identify one suspected food and the illnesses had been restricted to one area. Now forty-seven cases were reported in Bavaria, and no one food could be identified. All of the victims, except one older man, were small children. Seven died, and others suffered severe kidney and brain damage. The immunologist Dr. Helge Karcher at the University of Wurzburg predicted an enormous increase in cases—a prediction that was bolstered by the increase in the detection of the bacteria in cattle in southern Germany from 5 percent in 1990 to 30 percent in 1995.[12]

No sooner was the Japanese outbreak under control than the West Coast of the United States was hit with the shattering Odwalla apple juice outbreak from *E. coli* O157:H7. The forty-five confirmed illnesses and the death of a child who had consumed the juice caused the public to reevaluate a food it had considered a safe and healthy choice. Not long afterward the regulatory agencies and epidemiologists gathered in Washington to decide whether pasteurization should now be required of all juices. They also wondered if perhaps they should require that only apples picked from the tree go into juice rather than those that had fallen to the ground—a move that had cultural and historical implications that many had long forgotten.

Once the whole purpose of making cider was to use up "windfall" apples, those that had fallen to the ground and were bruised. Because bruising accelerates rot, these apples could not be successfully stored, and the traditional way to put them to good use was to make cider. That explains why fresh cider is a seasonal product—or should be. But now windfalls were presumably becoming contaminated with *E. coli* O157:H7 on the ground. Previously it had been thought that cattle grazing beneath the apple trees was the cause, but that seldom occurred, cattle and dairy men said, because eating too many apples made cows sick. Thus, they were kept out of orchards if at all possible. Some epidemiologists suspected the cause was actually the proliferation (in the absence of regular hunting) of white-tailed deer, which sometimes grazed in the evenings with cattle, where they could have picked up the pathogen. Some had tested positive for O157. Deer could also leap nimbly over the fences that kept the cattle out of the orchards.

There was another possible factor. Once again a sweeter, less acid fruit had been selected for by growers over the years as they aimed to please consumer preferences. Did less-acid fruit decrease the acidity of the juice, thus allowing the pathogen to survive? A final factor was the origin of the apples. If apple production was seasonal but Odwalla juice was produced year-round, where were the apples coming from? How long had they been stored, and what effect might the length of storage have on the growth of bacteria? Or was the juice frozen, and if so, did that have an effect? All of these were questions that epidemiologists, regulators, and food scientists were newly considering.

The newspapers were still filled with news of the Odwalla outbreak when it began to seem that CDC and WHO fears of a worldwide epidemic were close to the mark. An outbreak of O157-related illness began in Scotland that in its death rate would surpass the Japanese outbreak. By December there were four hundred suspected cases, of which 216 had been confirmed, and a horrifying mortality rate: Eleven deaths, all of them older adults, had been reported. In mid-December four children remained in serious condition on dialysis. By January the number of deaths had climbed to sixteen, and it was still climbing. Eighteen would eventually die. The cause seemed to be vacuum-packed precooked meat, and the Scottish officials were coming in for strong criticism for not tracing the source sooner. Accused of protecting business interests, they were considered by the public responsible for the many illnesses and deaths that might have been prevented had they acted sooner to stop the distribution of contaminated meat from the suspected source. From the ordinary to the exotic, the simple act of eating seemed riskier than ever.

The Other *E. Coli*

Shiga toxin, the toxin that makes *E. coli* O157:H7 so effective as a pathogen, was identified in 1977 (see chapter 9). Further research identified two types of the toxin; the most virulent strains of *E. coli* O157:H7 have both. Canadian scientists, among them Hermy Lior and Dr. Mohamed Karmali, have collected over one hundred toxin-producing *E. coli*. With the realization in 1982 during the McDonald's outbreak of how virulent the toxins were in humans, scientists had wondered and waited for the other Shiga toxin–producing *E. coli* serotypes to show up in outbreaks.

In May 1994 the CDC was asked by the Lewis and Clark County Department of Health and Environmental Sciences in Montana to help with an outbreak of bloody diarrhea and severe abdominal cramps. Four people had the symptoms, but their stools were negative not only for the normal pathogens but also for *E. coli* O157:H7. When isolates were examined by the CDC labs, the cause was found to be a rare serotype, *E. coli* O104:H21. It was what they had been waiting for. It produced the Shiga toxin.

The EIS officers Dr. Cynthia Whitman and Dr. Nicholas Banatvala, as well as Erica Dugar, a veterinary student, joined the county and the Montana Department of Health in conducting an investigation. The team issued a press release describing the illness and asking that people with the symptoms contact the state health department. New cases turned up. Eventually ten confirmed and nine suspect cases were identified. Each of the nineteen was asked about the food and drink they had consumed in the seven days before their illness, and any food common to more than 50 percent of the ill persons was included in the matched case-control study. All of the patients had drunk milk. So had 92 percent of the controls. But when specific brands of milk were asked about, one stood out. Eighty-seven percent of the ills who drank milk reported drinking Brand A at home; that was true of only 51 percent of the controls. The team investigated the dairy. After examining the plant's complex piping, they concluded that the milk had become contaminated after pasteurization, possibly at one of several places, by water that was positive for fecal coliform bacteria. It wasn't the first outbreak from the other Shiga toxin–producing *E. coli*, but it was the first in the United States.

Dr. Karl Bettelheim of the Victorian Infectious Diseases Laboratory in Australia, was born in Austria. His family emigrated first to Shanghai, where he spent his childhood, then to England, where he attended university. His first work involved *E. coli*, and he has been working on the pathogen more or less ever since. He points out that as early as 1980, non-O157:H7 strains of *E. coli* that produced the Shiga toxin were associated with cases of diarrhea in New Zealand and in the late 1980s he reported cases of HUS due to O111 strains in Australia. An outbreak in Italy in 1992 had been linked to O111:H-, as had periodic cases as far back as 1988 not only in Italy but in France and Germany. *E. coli* O157:H7 seemed to be causing the most trouble, but clearly it was not alone.

Not long after the Montana outbreak, the Adelaide Women's and Children's Hospital in South Australia reported that they were treating three children with HUS. Public health officials looked for more cases and found them. By January 1995 nine children with HUS had been identified. By February 3 there were eighteen cases of HUS in children, all requiring dialysis, and two cases of thrombotic thrombocytopenic purpura (TTP) in adults. One of the children died. And the cause of the outbreak was not *E. coli* O157:H7 but *E. coli* O111.

All but two of the patients had eaten mettwurst, a type of sausage. Of those two, one was an infant whose mother had eaten mettwurst, and the other lived in a household in which mettwurst had been purchased. Thirteen identified the same company as the manufacturer of the mettwurst they had consumed. The laboratories were able to isolate the same toxin-producing *E. coli* in the mettwurst, and two of five samples taken at random from shops tested positive for the Shiga-toxin gene. The product was recalled at once, and many more cases were avoided.

In the April–June 1996 issue of *Emerging Infectious Diseases*, Bettelheim and P. N. Goldwater published a letter entitled "An Outbreak of Hemolytic Uremic Syndrome Due to *Escherichia Coli* O157:H-: Or Was It?" They raised an annoying question. Testing for O157 was fairly easy because a special medium could be used that would screen out many other *E. coli*. Once labs learned of the medium and doctors learned of the importance of asking labs to look for the pathogen, testing had become simple.

But in the mettwurst outbreak, Goldwater and Bettelheim pointed out, the pattern of isolations had been confusing. More than one strain of *E. coli* had been isolated. The laboratory had been conducting an ongoing research project, and using sophisticated molecular biologic techniques, it had identified not only O157:H7, which had been easy to find, but another strain, O111, in nineteen patients and in the mettwurst.

If the lab had simply used the standard means of looking for *E. coli* O157:H7, it would have found it in two of the patients and assumed that was the cause of the outbreak. In fact, the outbreak seemed to be caused by a number of different *E. coli* serotypes of which O111:H- and O157:H- were the most prominent—which was not really surprising since mettwurst is made from meats from various sources, many of which apparently were contaminated. By correctly identifying the real cause, the outbreak had quickly been brought under control. But it raised the question of how many outbreaks might have been attributed to the wrong pathogen because the more sophisticated techniques weren't routinely used. While *E. coli* O157:H7 is widespread throughout the world, it might well be that the other serotypes are causing infection but are being missed.

Patricia Griffin at the CDC and the members of STOP had worked hard to make O157:H7 a reportable disease, and she and Joy Wells at the CDC enterics lab had gone to great trouble to educate labs across the country to use the Sorbitol-MacConkey agar. Now Goldwater and Bettelheim were suggesting that reliance on that medium might actually be concealing the true cause of some outbreaks. Sorbitol-MacConkey had seemed such a good solution; it worked so well. Now just as things were getting better—just as more labs were testing for O157 on a regular basis—it seemed that a new way of testing for other *E. coli* strains might be needed.

Another problem loomed just over the horizon. There were also reports of O157 strains that did ferment sorbitol that the Sorbitol-MacConkey medium missed. So far they had been found only in Europe; they had not been reported in the United States or elsewhere. But was it just a matter of time? It had been hard enough to get labs to test for O157. The thought of having to ask them to do more was depressing to Griffin. And yet she honestly felt that in addition to the twenty thousand illnesses O157 was thought to cause, an additional ten thousand to twenty thousand cases of food poisoning might be caused by Shiga toxin–producing strains other than O157.

Others agreed. Several researchers at the Fairfax Hospital in Falls Church, Virginia, had decided to look for Shiga toxin–producing *E. coli*. From 270 stool samples, they found 11 that were positive for the toxin, but the surprise was that only 6 were O157:H7. The other 5, almost half the total, were non-O157 types, including 0103 and 088.[13]

From the very beginning, Lior and Karmali had worried that *E. coli* O157:H7 might represent only the tip of a pathogenic iceberg. The toxin-producing *E. coli* were out there, looking for a niche, and more and more frequently they were finding it. Infections from all the Shiga toxin–producing *E. coli* had been identified in Argentina, Australia, Belgium, Canada, Chile, the Czech Republic, Denmark, France, Germany, Hungary, Italy, Japan, Lithuania, Mexico, the Netherlands, Nigeria, Norway, Poland, South Africa, the United Kingdom, and the United States. But perhaps most worrying of all, Germany had reported an outbreak from another bacterium, *Citrobacter freundii*, that was found to be producing the Shiga toxin as well. It looked as if either the ability to produce this toxin was more prevalent than anyone had thought or microbes were sharing, through conjugation, the ability to produce it among themselves. Clearly the viral interloper that produced the toxin was spreading, taking up lodging in suitable bacteria where it might either remain or even be merely a temporary resident. This perapatetic behavior would present all sorts of new detection problems. Shiga toxin might be anywhere.

Researchers everywhere were looking for solutions. Dr. David Acheson, Dr. Gerald Keusch and Dr. Arthur Donohoe-Rolfe at Tufts New England Medical Center thought the answer was to focus on testing for the toxin rather than the strains of bacteria. Acheson saw the toxin as the common denominator. Its presence could turn ordinary *E. coli*—or another bacterium—into something much more dangerous. In fact, anyone secreting the toxin, whatever the strain, was in danger of getting HUS, he thought. Acheson and Keusch's test has now been approved by the FDA.

It was this test that he and his colleagues applied to a random selection of hamburger from his local market—25 percent of which turned out to be positive for Shiga toxin, which further testing revealed to be from a number of serotypes, none of them O157. The discovery might well explain some of the serious sporadic cases of diarrheal disease that have never been identified. It might even explain the cause of illnesses for which appendixes were removed and found not to be infected. It might be the explanation for a lot of things.

Dr. Griffin is interested in these non-O157 Shiga toxin–producing bugs but isn't sure they are all capable of causing disease. She suspects that the ability to cause disease in humans has to do with other qualities of the bacteria the toxin-producing phage has managed to infiltrate. In the case of O157, the bacteria have the unique ability to adhere or attach themselves to human cells and it is this combination that makes them so virulent. She remains convinced that of all the *E. coli*, O157 is the most serious. Testing with the Sorbitol-MacConkey medium might be missing something, but that may be the necessary price for more frequent testing and awareness of what is certainly an important pathogen.

What labs are testing for, what methods are being used, what compromises infectious disease experts feel they have to make—they seem small, perhaps inconsequential matters. They are not. Once again chaos intervenes. The seemingly small choice of what medium to use or what pathogens to look for can mean the difference between spotting or missing what may be a widespread outbreak of disease. And the quick identification of an outbreak can mean—as it did in Australia—controlling an epidemic and avoiding more disease and possibly death.

11/The Madness Behind Mad Cows

> You may drive out Nature with a pitchfork, yet she still will hurry back.
>
> HORACE,
> *Epistles, Book I*

In the early spring of 1995 Anthony and Michelle Bowen were living not far from the city center in Manchester, England. Their house was on the edge of a housing estate, and if there was no countryside around, there were fields that had not yet been developed, giving at least the feeling of openness. Both Bowens had good jobs. Anthony did construction work, laying pipe along the roads for the gas company, and Michelle served drinks at the bingo hall in the evenings. That way they could divide caring for their two daughters, eight-year-old Jacqueline and four-year-old Natalie. Life was going pretty well for the Bowens except for one thing. Michelle seemed increasingly absent-minded.[1]

"At first it was very slight, like the normal things we all do. Like losing the keys and thinking, 'Oh, where have I put them,'" remembers Anthony. But then she began to be depressed as well. At about the same time she discovered she was pregnant, and they both thought that might be the explanation. If there was something called postpartum depression, perhaps this was the opposite—pre-partum depression. Besides, her uncle, of whom she was very fond, had just been diagnosed with terminal cancer. That was enough to make anyone depressed.

"We put it down to that," says Anthony.

But things got worse. Michelle's energy and enthusiasm for life seemed to wane. The normal chores needed to keep the household running weren't getting done, Anthony remembers. The dishes would pile up in the sink. The dirty clothes collected in mounds. He would step in and get things straightened up, but in a few days disorder returned.

"Finally she stopped cooking tea," says Anthony, and he knew they needed help. He took her to the doctor's, and her physician diagnosed her as having a serious depression. "He gave her tablets, but they didn't seem to make her any better," Anthony remembers. In fact, things became even worse. Michelle was getting progressively more confused, and Anthony progressively more concerned. "She would forget to lock the door when she went out and would make up excuses, saying the key wouldn't work. But then I'd put my key into it and it would work. I'd find the door open and I'd say, 'Why is the door open?' and she'd say, 'It blew open.' She was still driving then, and she'd forget how to disengage the lock on the steering wheel."

Anthony was beginning to be very worried now. And Michelle's condition made things difficult at home. More and more of the responsibility of the house and children fell to him. It was hard. He didn't seem to be able to talk to her anymore, or to get her to make sense. This wasn't the Michelle he knew.

But things would get worse still. Michelle began stumbling. "All of a sudden her legs would go. She would drop things. And she was becoming neurotic. She'd be aggressive. If you corrected her about something she'd said or done, she would get angry and say that what she'd said was right."

Her doctor referred her to a psychiatrist. The diagnosis was the same. He said she was depressed and gave her a mild medication—concerned that anything stronger would affect the pregnancy. She had trouble walking and holding things. She had to give up driving. Finally a neurologist was called on to the case, and Michelle was hospitalized at the Manchester Royal Infirmary. Her condition deteriorated rapidly from then on. She stopped being able to feed herself and was put on a drip feed. She suffered countless complications, says Anthony, including infections that couldn't be cured. "It was as if her body couldn't defend itself anymore."

Then, when she was six months pregnant, she began to have convulsions. The doctors felt that any medication they could give her to stop the convulsions might harm the baby, and so eleven weeks before her due date the baby was delivered by cesarean section. The baby boy was small, only three pounds, eleven ounces, but healthy. He was named Anthony James, after his father.

"She didn't have long after that," Anthony says.

Michelle was almost completely irrational now, pulling at her cesarean stitches until the wound opened. And clearly she was in pain. That was the last time Anthony was able to communicate with her. Michelle's doctors would put her on heavy medication, she would sink into a coma, and a few weeks later she would die.

Not long after she was hospitalized, Anthony remembers, Michelle "had a visit from Edinburgh." Professor Robert Will, a consultant neurologist at Edinburgh's Western General Hospital, is an expert in an unusual disease. The government had set up a surveillance unit in Edinburgh to watch for it around the country. Dr. Will was put in charge, and he and his staff investigated cases that had the symptoms they were looking for. Anthony was told that the experts could not be sure, but they thought Michelle was likely to be suffering from something very rare called Creutzfeldt-Jakob disease (CJD). They had a lot of questions. Among them, had Michelle had any particular contact with beef? She had. Ten years before she had worked for a butcher. And Anthony had once worked at an abattoir.

The British Reaction

That was in late November 1995. In March 1996 the Spongiform Encephalopathy Advisory Committee advised the British government that it had become aware of ten cases of a variant of CJD for which "the most likely explanation" was consumption of or contact with beef infected with bovine spongiform encephalopathy (BSE), popularly known as "mad cow" disease, an epizootic, or animal, epidemic that in ten years had spread in England from a few cases to more than 150,000. The next day Health Minister Stephen Dorrell announced the findings to a stunned Parliament.

With those words the beef industry worldwide felt the ground shake beneath it. This was not a typical foodborne pathogen, but an agent that produced an inevitably fatal disease of the nervous system; an agent that researchers estimated might take as long as ten to twenty years from the time of infection to manifest itself. Nor was it an agent with which the industry was unfamiliar. Agricultural and medical researchers had quietly watched and worried for years for fear BSE might affect humans, but now those worries had reached the general public around the world. Mad cow disease, whatever it was—and few knew much about it—was suddenly part of the language.

The link between BSE and CJD had far-reaching implications precisely because so little was known about the disease. Although there were credible theories, no researchers were certain precisely what the infectious agent was, and what they did know—that it could withstand intense heat, radiation, and most disinfectants—was more troubling still. The infectious agent passed through the filters that detected bacteria, so it had to be viruslike or even smaller than a virus, for it could not be seen even under powerful electron microscopes. But if it was a virus, it was slow-acting and like nothing seen before. It did not seem to leave the telltale signs in the body that can reveal viral infection.

In fact, no one could even be absolutely positive that the ten new cases of CJD unequivocally demonstrated an association between the bovine and human forms of the disease. The meat industry eagerly latched on to the fact that it might take years to prove a connection to the satisfaction of all. But as scientific research accumulates, each study adding to our knowledge and understanding, there are points at which reasonable people can draw reasonable conclusions. Given what the researchers now knew, the possibility of a link could no longer be dismissed. In fact, all the available evidence pointed to that conclusion.

To the scientific community, accustomed to studied restraint in public announcements, Dorrell's choice of words indicated real alarm on the part of the researchers. To even suggest such an association, with the possible repercussions for the food supply and the beef industry well understood, there had to be very good evidence indeed. Had the species barrier, the firewall that we've always hoped keeps animal disease well separated from humans, been breached? And there were other frightening possibilities. Had mad cow disease spread? Were the herds of other countries infected? The dangers of BSE, as the epidemic grew, had been to animals and to the cattle industry. Now there were clear implications for the human food supply. Was anyone who had ever eaten beef in England at risk for the terrible death Michelle Bowen and nine others had undergone?

The reaction was instantaneous. The announcement ran on the front pages of newspapers and was the lead on television news around the world; consumers in Britain reacted predictably. It was thought that the agent was less likely to be in the muscle meat and more likely to be in burgers, which contained "who knew what," but few were willing to take a chance. Beef sales plummeted. About one-third of the schools in Britain said they would remove beef from their menus out of concern for children. (Of the ten CJD cases, two were assumed, because of the long incubation period, to have contracted the disease as youngsters of ten or eleven.) McDonald's in England said it would use only imported beef from Argentina, and the fast-food giant that had made its worldwide reputation from beef burgers posted advertisements touting "veggie" burgers in its windows in response to widespread consumer anxiety. Other burger chains soon did the same. There were accusations from the beef industry that they were acting too quickly, but the fast-food empires had their own reputations to protect. Consumer confidence in their products had to be maintained. British Airways added its name to the list of boycotters of British beef. And around the world people who had visited Britain in the preceding ten years were frantically trying to remember exactly what they had eaten on their trips. Out in Ohio somewhere Harry would poke Muriel and say, "I had the roast beef, but you, you insisted on the hamburger." To which she would reply, "Yeah, but you had all those meat pies." And highly suggestible people everywhere found themselves paying closer attention to the number of times they couldn't find their keys or remember their second cousin's married name.

France and Belgium immediately suspended the importation of British beef and cattle. A day later the European Union would place a total ban on the export of British beef and beef products; the EU was also concerned about imports of animal products from countries that had themselves imported British beef, on the grounds that infected imported material could be reprocessed—in sausage or other products—and then exported. There was urgency to the fears. An amazed tourist watched as emergency vehicles with lights and horns blaring rushed up to a fast-food restaurant in Amsterdam to impound cases of hamburgers. Not everyone was unhappy. Argentina, a major beef producer, announced cheerfully that its cows were BSE-free.

It was just one more food scare, and yet it was much, much more. It seemed to crystallize in the minds of consumers the growing distrust of all food—the niggling but persistent worries about chemicals, antibiotics, pesticide residues, bacteria, hormones, and genetic engineering that were harder and harder, now maybe impossible, to forget. For the British, something even more important was at stake. National identity is intertwined with food until we find it difficult to separate the two. We dine with cultural expectations. People with a common ethnic heritage in a land not their own come together to share memories, dances, stories, and customs, but they inevitably cement that bond with familiar food. Even when an ethnic group has been assimilated by a dominant culture, its food customs survive long after other habits have been laid aside. Daily diets from Boston to Bangkok may have been handed over to the fast-food franchises and global technology, but special dishes will be brought out on special occasions because we need that reminder of who we once were. A traveler who wants to know a country well samples its food. A guest in a house shows her lack of enmity by sharing food. A politician shows his affinity for a group by consuming its food. British identity is wrapped up in beef with its inevitable associations with strength, vigor, and prosperity. A country that has an abundant supply of quality animal protein has wealth and stability and solid resources. The Dickensian table groaned with the haunch of seared beef. Britain was its Sunday roasts and its steak and kidney pie. The British had survived the crumbling of the empire, emerged from the Blitz unvanquished, endured the humiliating behavior of the young royals, but now the very food that had sustained them through it all was thrown into question.

Members of Prime Minister John Major's government, having announced the possible connection between BSE and CJD, then did an about-face. Hoping to have it both ways, they rushed to defend British beef to the European Union, saying the reactions were excessive and driven by a hysterical press. Their protestations fell on the ears of government officials tuned only to the concerns of consumers on the

Continent and their own meat industries. The relationship of the British government with the EU, of which Britain seemed a reluctant member, was touchy in any case. The EU's peremptory ban bolstered the country's "Euro-skeptics," who had warned of the dangers of letting a distant centralized bureaucracy dictate British national policy. The government's message was wildly mixed. Safeguards were in place to make sure infected meat stayed out of the food supply, the public was told. And yet the government agreed, at first, to slaughter some 3.4 million animals to keep the spread in check. The ten people who had recently died of CJD, government officials said, had presumably contracted the disease years before. There was little or no danger now. But even as they spoke they were being contradicted by their own scientists. One member of the Spongiform Encephalopathy Advisory Committee said he would not allow his granddaughter to eat beef; another said he himself would not eat meat from an animal he knew to be infected. The microbiologist Richard Lacy, an expert on BSE, said, "I simply cannot understand why anyone eats beef at all." Many agreed.

British farmers were understandably devastated at the prospect of killing so much of their stock, built up carefully from bloodlines they were proud of. Moreover, they were facing prices on the market of less than half of what they'd been getting a few weeks before. Most held their animals back from slaughter—waiting for the government to work matters out with Europe, waiting for it all to go away, waiting for a miracle that was unlikely to occur.

Unfortunately for the industry, it was crystal clear that no matter what reassurances they were given, consumers no longer believed the government. It had told them too many times that there was no danger from BSE. Even if they decided to continue eating beef, and many did, it was simply because they chose to take the risk, not because they believed there wasn't one. A young man would walk into a McDonald's and boldly ask for a "Creutzfeldt-Jakob Burger." An elderly couple would decide that they were too old to change—and besides, with the disease's long incubation period, what did it matter to them? The press, sensing a mortal wound to John Major's Conservative government, stayed on the story.

The aftershocks would eventually extend into myriad other industries that used animal protein or gelatin (a product derived from hides, hooves, and spinal columns), from pet food to candy, from pharmaceuticals to cosmetics. Every use of beef and beef by-products originating from British beef—and there were many more than the average consumer had ever realized—was now thrown into question. Animal protein went into yogurt, jelly beans—even into a favorite of Anglophiles, Altoids peppermints. Medical appliances that stayed in the body after surgery were sometimes made or coated with animal protein. Sutures were suspect, and some vaccines. It seemed to be everywhere. Was it possible that these seemingly benign and ubiquitous substances could be capable of causing a fatal neurological disease? The likelihood was low, but then the experts had been wrong about the potential for a BSE-CJD connection. Who wanted to take chances?

In government offices worldwide, in the departments of agriculture and food and drug regulation, in the infectious disease agencies, in every department that dealt with food, there was widespread alarm and an immediate reaction. For all the assurances given to the public that the products on their grocery shelves were safe for consumption, government officials knew that the possible threat to the world's food supply was real and frightening. In the United States, experts from the USDA, the CDC, and the FDA were dispatched at once to England and to Brussels to get the facts firsthand and to sort the hard science from what was appearing in the popular press. The World Health Organization organized a consultation in Geneva on April 2–3 at which a group of international experts reviewed the public health issues related to BSE and the new variant of CJD and within weeks made recommendations. Hasty meetings were held to exchange the gathered information, then to formulate the agency position. Even the U.S. Defense Department got involved. The reason why was not at first clear.

The public affairs officials in these agencies worked overtime. At the height of the concern the press contacts in these departments distributed their home telephone numbers to journalists—a rare concession. And all the while the Internet buzzed as veterinarians, neurologists, infectious disease experts, agricultural officials, journalists, food safety advisers and activists, and concerned individuals searched for and exchanged information on mad cow disease. The first order of business for the beef and diary industries was to replace the noxious term "mad cow" disease, which conjured up an entirely unacceptable image in the public's mind, with a term more difficult to pronounce and spell—and therefore less likely to be used—bovine spongiform encephalopathy. They were almost entirely unsuccessful in this effort.

It was a disaster—there could be no other word for it as the repercussions became more fully understood. Implicated was the very agricultural revolution that for the past fifty years had promised to feed with ease a growing and increasingly hungry world population. Something had gone terribly wrong with one of the systems we absolutely relied on to nourish us. As one vegetarian group suggested, it was the revenge of those poor beasts we had herded and milked and slaughtered and skinned and worn and ground up and consumed. "The Revenge of the Cows," they called it. In the political fallout fingers were pointed in almost every direction but the right one. It was not simply Britain's Conservative government that was culpable, although it bore its share of blame for inaction and deregulation of the pertinent industries; nor did the whole blame lie with the agriculture industry, the rendering industry, the food industry, or the consumer's taste for cheap animal protein. Something far more basic and pervasive was wrong. It would be no one's and yet everyone's fault. Call it an attitude. The story had begun not with Michelle's illness but more than a decade before.

The First Mad Cows

On a fine spring day in April 1985, Colin Whitaker, a forty-three-year-old veterinarian from Kent, was called to tend to a sick cow on a local farm. "She was behaving oddly," he remembers. "She was being aggressive to other cows and seemed nervous and apprehensive."[2] That had struck the cow's owner as very strange, because Jonquil, as she was called, had always been a friendly, good-natured animal. Whitaker looked her over thoroughly, but except for cystic ovaries, which can lead to aggressive behavior and which he treated, he could find nothing wrong. The ovarian infection improved, but the cow continued to deteriorate.

The next time he looked at her, he says, "she seemed to get worse, unsteady on her feet, and very unhappy." He considered a number of neurological explanations for the unsteadiness. "I thought it might be a brain abscess or a tumor, or a magnesium deficiency farmers know as 'staggers.'" Eventually the animal had to be destroyed. Its carcass was sold for pet food.

Whitaker assumed it was a "one-off case" but didn't really know since there was no postmortem. Lab work was expensive, and few farmers wanted to assume the cost. He went back to his work at Sambrook and Partners, the veterinary clinic he'd been with since graduating from the Royal Veterinary College in London nearly twenty years before, and didn't give the animal much thought until six months later, in January 1986, when he got another call. It was the same farmer. Another cow in his herd had the symptoms. Whitaker went through the same routine: examining the animal and finding nothing wrong, yet watching her steadily deteriorate. Again he thought about the possibility of a brain tumor or "staggers." When a call came about a third cow, he sensed it was no longer coincidental. He knew he had to investigate further.

The symptoms were puzzling. Some aspects looked like one disease, some looked like another, and still other symptoms fit nothing he was familiar with. The first step was a thorough laboratory analysis to look for abnormalities in the blood; evidence of some sort of infection or foreign material. He called in a specialist, who found nothing unusual. He then compared symptoms to other diseases and found that some aspects of sixteen diseases seemed similar without being a perfect fit. He copied them down in his notebook and described the seventeenth himself: "New scrapie-like syndrome." Looking back, it would be the first mention of the possibility of a frightening connection: that scrapie, a common disease in sheep, might have jumped the species barrier to infect cattle.

These would not be isolated cases; Whitaker began hearing of others on nearby farms. He sent the brains of two cows to the Wye Investigative Center (one of the many regional labs in England that have now been closed by government cutbacks). After a preliminary investigation, the heads were sent to the Central Veterinary Laboratory at Weybridge in Surrey for further testing. They went to the pathologist Gerald Wells, now the head of neuropathology at the lab and an old friend of Whitaker's. He examined the brains. He found that they looked oddly like sponges and very much like those of sheep infected with scrapie. Because of the vacuolar lesions, or holes, he found there, he named the disease bovine spongiform encephalopathy. He also identified fibrils—threadlike particles—very similar to those associated with scrapie.[3] Despite trying to keep an open mind—there are other diseases that cause vacuolar lesions—he was immediately concerned that it must have been transmitted from scrapie-infected sheep.

"Our first inclination," Wells says,

> was to question how, after so many years of cattle not, apparently, having scrapie, could they have acquired the disease? Obviously, initially we thought of this as a sporadic occurrence. Colin Whitaker was able to give us information that there had been, if I recollect correctly, at least five other cases with a similar clinical picture that had occurred on that farm or nearby prior to the two in which we confirmed the disease. That was interesting in itself, and it was obviously followed up by an investigation.[4]

By January 1987 the Weybridge lab had received three more cow brains, with the same symptoms, that upon analysis revealed the same spongelike characteristics. "As soon as we had these cases we thought about the potential for an epidemic," says Wells. By May 1987 there were a total of ten cases from widely differing parts of the country. "At which point we realized we did indeed have a national epidemic," says Wells.

The epidemiologist John Wilesmith undertook the task of uncovering the cause. In June 1987 the Weybridge laboratory would put out a nationwide request to veterinary surgeons to look for more cases. Wells would publish a short communication in the *Veterinary Record*, but it would not appear until October 1987—and by that time the cases were proliferating.

The symptoms of scrapie-infected sheep are similar to those Whitaker saw in the cows. Their behavior changes, they tend to rub up against fences, as if scratching themselves, until their fleece is rubbed away—hence, "scrapie," for scraping. They lose their appetites as well as their sense of balance and direction, and they eventually waste and die. Scrapie has been endemic in sheep in England for more than two hundred years, and France has a problem with it as well.

Quick and radical action can eliminate scrapie. When the disease appeared in Australia and New Zealand, the affected herds were immediately destroyed, and both countries now appear scrapie-free.

In 1947 the disease was first detected in the United States, introduced, it is thought, by Suffolk sheep imported from Britain. Two more outbreaks occurred in the United States in the early 1950s, transmitted by sheep imported from Britain by way of Canada. The number of outbreaks was low until the late 1970s, when it began to increase. Between 1984 and 1987 there were an average of thirty-seven outbreaks a year. Cases continued to climb. In 1989 scrapie was diagnosed in fifty-two flocks from twenty states. Morbidity has increased as well: As many as 30 percent of the animals in some flocks succumb to the disease where once the figure was around 5 percent. The disease has spread from flock to flock (by 1990 only eleven states had not had cases) with the transfer of young sheep who are infected but show no signs of the disease. At first entire flocks were slaughtered when the disease was detected, but farmers complained that when they'd built up a flock for years that could be totally destroyed if they happened to bring in a new infected animal, it was "overkill" in every sense of the word. In 1983, when enthusiasm for deregulation and for letting industry set its own standards was at an all-time high, the program was changed. Only dams were required to be removed from the flock, an approach that was not as effective in containing the disease.[5]

Scrapie in England had also increased in even greater numbers in recent years; sheep rearing had increased at the same time, so that, with the concentration of animals, the level of infection was high and growing. In England another important factor, which hadn't been clearly understood, would eventually explain the persistence of the disease.

When John Wilesmith and his colleagues at Weybridge looked for a link between sheep and the sick cows in Great Britain, they hunted for epidemiological associations. Since they had no idea what they were looking for as a cause and the possibilities were almost endless, a broad questionnaire was developed for all farmers with at least one confirmed case. It asked about the age, sex, breed, origin, and date of clinical onset for the disease. This information was then related to herd type and size, the presence of sheep on the farm, feeding practices, and details on the use of medications, pesticides, and herbicides.

They then created a computer model to look at when the animals had gotten sick, how long the disease had taken to run its course, and other relevant factors.

They decided, on the basis of what they learned, that BSE was indeed a new disease. It had not been introduced into England by imported animals or animal products but was a common-source outbreak: The cases had appeared at about the same time from a source apparently common to all. Each animal, in effect, was an index case. The disease was widespread but had a higher incidence rate in some areas. It was more prevalent in dairy cows than in beef cattle, and in large herds rather than in small ones, and only a few animals, all of them adults, were affected in a single herd. They determined, although there would be disagreements about this later, that there was no particular association with breed, gender, or season; that there was no association with the use of vaccines, pharmaceutical products, or agricultural chemicals; and that the animals did not contract the disease from direct contact with sheep. But there was one association they could not eliminate. All the diseased animals for which a feeding history was available had been fed commercially prepared feed containing meat and bone meal.[6]

The Offal Truth

If there is a prevailing mind-set in modern agricultural production, it is one that seems, on its face, worthy enough: Waste nothing. Everything that can possibly be recycled in agriculture is. "We use everything but the squeal, the cluck, and the moo," Dr. Raymond L. Burns, coordinator of the alternative uses program for the Kansas Department of Agriculture in Topeka, told the *New York Times*.[7] This practice is, of course, driven more by cold, hard economics than environmentalism. Farmers have always tried to avoid waste, which is by its very definition loss, but recently the practice has been pushed to extremes—the triumph of efficiency over reason, you might call it. Sheep that died of scrapie were not buried or incinerated, but recycled into food for animals.

Protein food has the same effect on animals that it does on humans. It causes them to grow faster or, if they are milk cows, to produce more. Often animals are fed vegetables high in protein, such as soybeans, which have the same effect. But where the supply of vegetable protein is low and the availability of animal protein is high, the latter is used. Cost is generally the deciding factor in choosing a protein source. Thus, the dead, diseased sheep had been sent to the rendering plant where, along with other animal waste, they were reduced to meal. Then the material was mixed with other ingredients we think of as feed, but also, depending upon which was available and cheap, with such materials as citrus pulp, chicken feathers and litter, and sawdust, before being fed to swine, chickens, pets, zoo animals, and, of course, cattle. Food animals were being used as garbage dumps for agribusiness waste.

While among the nonfarming population the reaction to learning this is instinctive revulsion—on an Oprah Winfrey show where the practice was revealed, the audience gave an unrehearsed collective gasp of horror at the thought of making cows, natural herbivores, eat dead cows and sheep—few farmers or veterinarians, if they actually knew what was in feed (and often it was concealed under the words "protein content") openly questioned the wisdom of this practice.[8]

Colin Whitaker is one who does.

> It's asking for trouble, feeding herbivores back to herbivores. It's absolute nonsense. It's done for cost reasons because it's a cheap form of protein. In those days [the 1970s] it was a trade secret between the feed companies. We live in a world run by accountants with no knowledge of the real world. Of course, it's all easy in hindsight. There were mistakes made all along the way.

But what exactly had gone wrong with feed? Was it just the inclusion of infected material, or was it something else? The good epidemiologist, as Robert Tauxe at the CDC often says, goes beyond that first "Eureka" when a clear association is pinpointed to ask, "Why?" The best way to discover the answer is usually to ask, "What has changed?" In this case, as often happens, it was a number of things. Using computer modeling, researchers determined that most of the diseased animals had been exposed as young adults in 1981 or 1982. Other researchers looked intently at the rendering industry, which transforms the leftover animal bits into animal protein. Three things had changed in the British rendering industry at about the same time that the sick cows were assumed to have been exposed. The use of strong solvents to extract the tallow and fat had been abandoned. Coincidentally the plants had also abandoned the batch method of rendering, which might have limited the spread of an infected lot, for a continuous system, which would allow contaminated material to spread throughout. (This is precisely the same sort of change that now allows an infected cow to potentially contaminate not fifteen pounds of hamburger but a good proportion of an entire day's production.)

Renderers haven't been worried about diseases being recycled because most infectious agents are susceptible to heat. The rendering process should have destroyed whatever had made the sheep ill. It didn't. The heat had been reduced considerably in the late 1970s and early 1980s to enhance taste, to increase the nutrient value of the material, and to lower energy costs. In the United Kingdom, it was a bare 100 degrees Celsius. In France and Italy, temperatures ranged between 130 and 140 degrees Celsius, which was still not enough. "A committee warned the British Government in 1980 that the low temperatures could be dangerous," but the warnings were ignored, *The Economist* reported.[9]

The rendering method was the American Carver-Greenfield process (used in several U.S. processing facilities as well). Any or all of these changes might have allowed an infectious agent to survive the rendering process. The researchers also noticed that these changes occurred at about the same time as an increase in sheep rearing, as well as a jump in scrapie cases. Later tests would demonstrate that the feed from the diseased animals could cause transmissible spongiform encephalopathy (TSE) when fed or injected into the brains of mice. The species barrier, which supposedly prevents a disease from going from one animal species to another, was not impenetrable.

As more sheep were infected, more were going to the renderers to be recycled back into feed. This coincidence, the researchers concluded, was the most likely explanation for the exposure to the "cattle-adapted scrapie-like agent."[10] And one reason scrapie itself had been spreading so rapidly was the recycling of scrapie-infected animals in sheep feed. (This recycling probably explained the increase in scrapie in the United States as well.)

Then, to make matters worse, the dead, BSE-infected cattle had themselves been sent to the rendering plants and were apparently recycling or "amplifying" the disease even as the contamination of feed with diseased sheep continued. Animals were getting a megadose of infected material. It was also clear why the larger herds suffered more cases of the disease. On smaller farms the chances of getting an infected batch of feed were less. With the huge purchases of feed made by the large operations, all the cows might be exposed time and time again to infected feed. BSE in a cow can be produced with an oral dose of about a teaspoonful of highly infective cattle feed concentrate.[11]

The implications for feeding cattle worldwide were obvious. Agricultural agencies around the globe suddenly took a look at their systems and considered regulations that might keep the disease out of their country's herds. But at the same time some industries were loath to change the process. Meat producers still wanted the cheapest animal protein, and the rendering industry understandably wanted to stay in business. And there was another concern. If animals weren't rendered, what precisely would be done with them? Would they be buried, only to contaminate the ground water? How would they be gotten rid of? The consumption and demand for animal foods had increased greatly over the past hundred years worldwide, and this demand was putting stress on agriculture at every level. Dealing with the by-products of increased animal production was not just a British problem. It was a problem virtually everywhere animals were raised intensively. The United States was no different, except that it raised fewer sheep.

In 1988, neither Canada nor the United States banned the inclusion of diseased sheep or cows in animal feed. But in both countries a ban on the importation of British cattle was put in place. Many animals had been imported to each country for breeding purposes, and both countries began tracking them down to see if they were infected. And everywhere, as soon as the nature of the disease was understood, veterinarians, farmers, and slaughterhouse inspectors began looking for sick cows, hoping not to find them. But their primary concern was the health of their herds and their agricultural industries.

Whatever horrors were now in store for us—and no one could be certain what the extent of the human outbreak might be—we had clearly brought it on ourselves in the relentless search for cheaper and cheaper ways to produce more and more food.

The Risk to Humans

From the very first some were concerned that the disease might be transferred to humans through consumption of beef. In 1988 the British government asked Professor Sir Richard Southwood to investigate and report on BSE and its implications for the human food chain. As an interim recommendation, the Southwood Committee advised the immediate slaughter of infected animals and the listing of BSE as a notifiable disease—meaning that diseased animals were to be reported (although there were no penalties for not doing so). In addition, the committee recommended the discarding of milk from infected animals. The Ministry of Agriculture, Fisheries, and Food (MAFF) responded by banning at once the use of scrapie-infected sheep offal in cattle feed, and a month later the destruction of BSE-infected animals began. Farmers were told they would be compensated for their destroyed animals at 50 percent of the actual value—not as much as they had hoped, but something. In December 1988 milk from cows suspected of having BSE was banned for human consumption.

The public was concerned but waited for the committee report. In early 1989 the Southwood Committee issued its recommendations. To everyone's great relief, the committee declared that it was "most unlikely that BSE will have any implications for human health."[12] But at the same time doctors thought they were seeing an increase in the number of patients with a disease that looked a lot like a human version of BSE. Called Creutzfeldt-Jakob disease (CJD), it previously had generally occurred in older people. It caused neurological symptoms similar to those in animals: Victims had increasing trouble thinking, speaking, seeing, and finally moving. CJD left the brain with similar holes and deposits, and it was invariably fatal. But it could be diagnosed with certainty only after death, when the brain could be examined. The choreographer George Balanchine died of the disease in 1983.

CJD is one of a number of what are called transmissible spongiform encephalopathies (TSEs); it is similar to a disease called kuru once found in indigenous peoples in the highlands of Papua–New Guinea, where it was called "laughing death." Dr. D. Carleton Gajdusek, then a young researcher, noticed back in the 1950s that it was more prevalent in women and children than in adult men. Eventually he would make an association between the disease and the unique funeral practices of the Papua–New Guineans, including the rubbing of pieces of the deceased's tissue on the body and even occasionally removing and eating parts of the deceased, including the brain. It was the eating of what we wanted to remember carried to extremes. The choice bits of meat went to the men; the offal and brain, the most infected tissues apparently, were left to the women and children. When the practice was stopped, the disease gradually declined and almost disappeared. Gajdusek would later win the Nobel Prize for demonstrating that kuru and Creutzfeldt-Jakob disease were transmissible to animals.

Although CJD is not particularly contagious under ordinary circumstances, it is highly transmissible in certain very specific ways. In the 1980s cases began to appear in young people. It was determined almost at once that they had been infected by receiving human growth hormone made from the pooled pituitary glands of diseased cadavers. Today the growth hormone is synthetic and safe, but in some countries, including the United States, the cadaver-derived bovine growth hormone was given until 1985; with the disease's long incubation period, cases of hormone-induced CJD continue to crop up. CJD has also been contracted by people receiving cornea transplants from infected cadavers, and it has been transferred in the operating room—once by instruments that had gone through the normal disinfecting procedure but nevertheless remained contaminated by the infectious agent. Following a brain probe on a suspected CJD patient, two young epileptic patients were examined with the same probe and both developed CJD after a short incubation and died. Medical personnel have also apparently contracted the disease from contact with infected material, and there are now guidelines for surgeons and pathologists that have led to a reluctance to perform certain procedures.

For the scientists, the medical researchers, the agricultural officials, and the epidemiologists, the most intriguing question of all was, "What was the agent?"

The Presence of Prions

The existence of a disease agent can be demonstrated without its being visible. That happened with viruses. Their presence was detected long before they were seen. In one experiment concentrated material from a diseased plant that had been filtered for bacteria was nevertheless able to transmit disease to a healthy plant on which it was rubbed. Scientists knew there was something there, some infectious agent they couldn't see, and eventually electron microscopes would reveal the tiny, fuzzy shapes of the virus that had caused the disease.

This new agent was equally illusive, but it could not be seen with the most powerful microscopes, and it apparently left none of the customary signs or evidence of a virus. While some clung to the theory that it was a virus with unusual characteristics, Dr. Stanley B. Prusiner at the University of California at San Francisco had another idea. He proposed in 1982 in an article in *Science* that the infectious agent that caused scrapie was something he called a prion, a protein particle that could replicate itself, startlingly, with no DNA or RNA—something no other living particle could do. Apparently it performed this trick not by reproducing itself but by sidling up to other protein particles in normal brain tissue and inducing them to change. If Prusiner's theory was correct, it revealed an entirely new form of contagion. It was a radical proposition, and one that until fairly recently was met with intense skepticism. What was known about the prion's ability to transmit disease came as the result of tests in which mice, cattle, and other animals were fed or injected (in the brain) with material from TSE-infected animals. Some parts of the animal were clearly more infectious than others.

But even as scientists and government officials in Great Britain assured the public that the species barrier between cows and humans was more difficult to cross, they sent one more mixed message to consumers by banning certain organ meats from cattle for human consumption because they were seen as the most likely source of the infectious BSE agent. Called offal, these meats included the brain, spinal cord, spleen, thymus, tonsils, and intestines. While these organs are not commonly eaten in either the United States or Great Britain (although they could be in ground and processed products), the thymus and sometimes the pancreas of veal are called sweetbreads and are considered delicacies to many—especially the French.

Growing up I remember well the occasional creamed sweetbreads on rice that my mother would get excited about. They were, to my mind, pale, rather lumpy, odd-textured, and fairly tasteless bits of something that definitely was animal and yet not quite meat. My father was a great consumer of brain, tripe, and other oddments that seemed highly questionable, and now, with the news of the BSE-CJD link, would be unthinkable. While they weren't particular favorites of mine, the elimination of these parts of the animal represents one more incremental loss in the food supply.

The British government also announced that these bits of cows and steers would no longer be going into hamburger or meat pies either. Most consumers were more than a bit surprised to learn that they had been. In fact, offal didn't need to be sold in recognizable particles to enter the food chain. In an effort to get every tiny scrap of meat from slaughtered animals, carcasses were subjected to a mechanical deboning process that, using super-strong jets of water or a vacuum process, flushed the last bits of meat off the bone, out of which a kind of meat slurry was created that could be tinted the appropriate color and used in various products such as luncheon meats and hot dogs. (The British would be horrified to discover that the mechanically recovered meat was used in baby foods until 1995.) It was quite possible that the end product contained spinal column and other highly infective parts of the animal. Despite these revelations, Minister for Agriculture John Gummer would display his own confidence in British beef by bringing his four-year-old daughter to a press conference and having her, on cue, eat a hamburger, an act that would be looked back upon with scorn and ridicule.

The possibility that this British problem might spread was generating concern among government agricultural officials in the United States and elsewhere. In June 1989 an international workshop on BSE was held at the National Institutes of Health. Presenting papers were scientists from the United Kingdom, France, Italy, Germany, and the United States. They acknowledged the worry among public health officials, veterinary regulatory agencies, cattlemen, milk producers, and the pharmaceutical industry in their countries. The outbreak was apparently restricted to cows in the United Kingdom, but what seriously concerned them was both the widespread incidence of scrapie throughout the world and the virtually universal practice of using rendered dead animals as a protein source in animal—especially cattle—feeds. That practice might lead to more outbreaks elsewhere—or even to the establishment of cattle as carriers of BSE around the globe.

One of the first questions at the conference was how serious the possibility might be that BSE could be transported to humans in the form of CJD. It was probably greater than the risk of humans getting scrapie, Dr. Richard H. Kimberlin of Edinburgh said, because material from cattle was widely used not only as food but in such a vast range of other products. There was a good possibility, he pointed out, that since the disease seemed to have jumped the species barrier from sheep to cows, the cows might have selected strains that were easier to pass to man. The risk might be small, he said, but it couldn't and shouldn't be ignored.

Almost everyone at the gathering assumed that the United States was free of BSE. That assumption was seriously questioned by Dr. Richard F. Marsh from the School of Veterinary Medicine at the University of Wisconsin at Madison.[12] He believed that there might already be a scrapielike disease in cattle in the United States.

Marsh was no stranger to the scientists looking at BSE in England. He had long been interested in transmissible diseases of the nervous system, and his studies had been cited in the article written by John Wilesmith and his colleagues for the *Veterinary Record* on their first investigation of BSE. Much of Marsh's work had been with transmissible encephalopathy in mink—which seemed to be related to feeding mink scrapie-infected sheep. His team had looked for a virus but failed to find it.

It was an unusual outbreak of mink encephalopathy in Stetsonville, Wisconsin, in 1985 that caused Marsh to question whether the United States was actually free of BSE. It was unusual because the mink farmer had used no commercially available animal by-product mixtures in his feed, Marsh said, "but instead slaughtered all animals going into the mink diet," most of which (more than 95 percent) included "'downer' dairy cows, a few horses, but never sheep." To test the hypothesis that "downer cows" had transmitted the infection, Marsh inoculated two six-week-old Holstein bull calves with mink brain from the farm. The bulls developed neurologic disease eighteen and nineteen months after inoculation. Both brains had spongiform degeneration at necropsy, and both were transmissible back to mink either by injection in the brain or orally.[13]

This could mean, Marsh says, that we had our own form of BSE in some cattle, and that it had been amplified by ongoing ruminant to ruminant feeding. In fact, it might be one cause of "the downer cow syndrome," a general description that refers to anything from a broken leg to "milk fever" to a condition in which the animal simply loses the ability to get up and walk—without the apprehension and aggressive behavior of BSE-infected cattle in Britain. In the United States these animals have all too often ended up in hamburger. Downer cows, for instance, went into the hamburger implicated in the Jack-in-the-Box outbreak, investigators discovered.[14] It was Dr. Marsh's contention that a U.S. strain of BSE was likely to look different both in its clinical presentation and in the brain from that in the United Kingdom because sheep in the United States had a different strain of scrapie. Cows would not turn into the worried, nervous, clumsy wrecks they become in England but would look more like typical downer cows. They would simply become recumbent.

Because Marsh's work is widely respected, his view cannot be ignored entirely—he has received a hearing—but because his theory is so horrifying, it rarely gets more than a token mention. He is generally drowned out by the loud voices of reassurance that BSE does not exist in the United States. But he is not alone. C. Joseph Gibbs, a researcher with the National Institutes of Health, has said, "I'm convinced that BSE has occurred here."[15] But because the general view was that Marsh and Gibbs were mistaken, no measures were taken to restrict the diseased cows and diseased sheep that were being recycled into feed for U.S. herds. Marsh has estimated that thirty-five thousand downers go to renderers every year in Wisconsin alone, with most carrying no diagnosis as to the cause of their illness. "An increasing number of these downers are fed back to lactating cattle in the form of meat- and bone-meal," wrote Joel McNair, the state editor of *Agri-View*, in September 1993.[16]

If we have the disease, we are doing our best to amplify it.

The Mad Cat

It was a cat that sparked a scare in 1990 that was to undermine all the reassurances British consumers were getting about the lack of danger to humans from the new disease. Max, an English cat, died of a brain disease that looked remarkably like BSE; since pet food still contained the diseased sheep material, this probably meant that the agent could indeed cross the species barrier.

There were also rumors that farmers, unhappy at receiving only 50 percent of their diseased animals' value, were, rather than taking a loss, rushing cows with symptoms to slaughter, where they entered the food chain. And, of course, cows without symptoms might nevertheless be infected as well. Beef cattle were slaughtered long before they would show signs of the disease but after they may have become infected. In retrospect, the government's penuriousness would look like a contributing factor to the cases that appeared in younger animals that should not have been exposed. As beef sales plummeted, the government in 1991 organized the National CJD Surveillance Unit in Edinburgh to monitor Creutzfeldt-Jakob cases and banned the use of offal in fertilizer. The number of vegetarians in England grew. Gradually, after a period of low sales, consumers—or some of them at least—returned to their beef-eating habits. And yet sales of beef never quite came back to their previous levels in England. Some people were just permanently put off.

In the next few years each period of renewed confidence in British beef, nurtured by the government, would be broken by another worrisome announcement in the medical and scientific journals—which would then be repeated in the popular press. Two farmers were reported to have contracted and died of CJD. Some scientists responded that the deaths were only what was to be expected if the normal worldwide rate of CJD was one in a million. As they pointed out, the only profession showing a number of CJD cases beyond the average were clergyman, a fact for which the surveillance team could find no explanation. (Of course, all those Sunday roast beef dinners with parishioners come to mind.)

Although journal articles were accumulating rapidly, American consumers had heard virtually nothing by 1991 about the problems with BSE in Britain, but in the December 6 *Federal Register,* where U.S. regulatory documents appear, the USDA posted new rules to apply to the importation of animal products and by-products from countries where BSE existed. These now included France, Great Britain, Northern Ireland, the Republic of Ireland, Oman, and Switzerland. It was reckoned by that time that more than twenty-three thousand cattle on more than ten thousand farms in Great Britain had died or been destroyed. The importing rules were strict, and some industry requests for exceptions were denied because "BSE is not known to occur in the United States and its introduction would be a major economic disaster for our animal industries."[17]

The USDA did not, for instance, allow gelatin into the country for use in pharmacological products for animals—such as the covering of capsules. Gelatin is made from hides and bones that have had the fat and marrow removed with hot water. The spinal column also goes into the mix. The material is then subjected to three to five days in an acid bath of 4–6 percent hydrochloric acid, followed by fifty to sixty days in a lime pit. The U.S. companies that imported gelatin objected to the restriction because they were sure this process would destroy the infectious agent. The USDA didn't agree. It had seen no studies that showed that to be the case.

The FDA did, however, allow gelatin and other beef by-products from BSE-infected countries into the country if they were solely for human use. This included extracts and products from "ruminant organs" such as collagen products, amniotic liquids or extracts, placental liquids or extracts, serum albumia, and serocolostrum, all of which could be imported for use in cosmetics.[18] After all, there was no real evidence at the time that BSE was transmittable to humans. Just those odd cases in farmers.

The USDA also kept trying to find the 499 animals that had been imported for breeding purposes between 1981 and 1988 (when importation was banned). By March 1996, 35 were still unaccounted for. The agency calculated that only 9 of these would still be alive, based on their age, but tracing them was still important. In Canada in 1993 an animal had exhibited signs of BSE. When after death it was diagnosed with the disease, the Canadians slaughtered 400 animals that had been imported since 1985.[19]

The news of Michelle Bowen and several other cases of CJD among young people in the United Kingdom raised the issue with the British public again in the fall of 1995. And two more farmers had died. Again there was widespread alarm. Sir Bernard Tomlinson, a former government health adviser, announced in an interview in December 1995 that he would not eat beef liver, beef pies, or beefburgers. Tim Lang, professor of food policy at Thames Valley University, said he would not allow small children to eat beef in any form.

Schools began removing beef from their menus, and again sales dropped significantly. But the government was still reassuring. Minister of Agriculture Douglas Hogg announced that he was still eating beef and happy for his children to do so. And even the experts disagreed on whether there was a possibility of transmission to humans. Paul Brown, medical director of the Laboratory of Central Nervous Studies at the National Institutes of Health in Bethesda, Maryland, blamed the media and a "flair for publicity" on the part of some scientists for the continuing alarm among British consumers. He urged continuing surveillance to accumulate more epidemiological information. The title of his article said it all: "The Jury Is Still Out."[20] At the other end of the spectrum, Professor Richard Lacey, a well-known microbiologist and expert on BSE, was making dire predictions—that there could be an epidemic of CJD that might affect annually as few as five hundred people or, at worst, five hundred thousand people.

It was only in January 1996 that American consumers got wind, through the *New York Times* writer John Darnton, of what had been turning British stomachs, but it seemed to be some sort of odd English problem—irrelevant to the United States. The announcement on March 20 by British Health Secretary Stephen Dorrell on the floor of Parliament that the most likely cause of the ten cases of CJD was exposure to bovine spongiform encephalopathy changed all that. There was no concealing the significance of what Dorrell said. The British media, already primed by month after month of mad cow–related stories, responded instantaneously. British television that evening reported the story even as newspapers were making room on their front pages for what they said might well be the public health calamity of the century. For ten years British consumers had heard government assurances that there was no risk from eating beef; now the *Daily Express*'s edition of March 21 echoed their mood perfectly: "Can We Still Trust Them?"

What made the situation worse was that the science to back up the announcement wasn't available. The National CJD Surveillance Unit, which had notified the Spongiform Encephalopathy Advisory Committee a few days earlier of the link, planned to publish its findings, but the article wasn't to appear in *The Lancet*, the prestigious British medical journal, for another two weeks. And with the public announcement by the health minister, the members of the unit were suddenly very hard to reach. According to *Nature* magazine, the Edinburgh CJD group had been due to present an update on the epidemiology of CJD in the United Kingdom to an international conference on spongiform encephalopathies in Paris but were recalled by the British government the night before Dorrell's announcement. The French researcher Olivier Robain, a prion specialist, openly questioned whether this move represented censorship. In the days between the announcement and the publication of the article, without access to the data, scientists everywhere would be in the dark as well. Around the world, anyone with an interest was turning to both official and unofficial sources on the World Wide Web to glean what they could.

If the missing information fed public fears, the normally cautious scientists, in the absence of the facts, could react only with skepticism. "There is no proof of a link," was the general response. And indeed there wasn't—because there probably never would be in the sense of being able to say, "This hamburger from this BSE-infected animal caused this case of CJD." To put the matter in perspective, while virtually all of the scientific community now accepts the more than thirty-year-old claim that cigarette smoking causes cancer, the precise mechanism by which it seems to do so has only just been identified. The compelling evidence has been the accumulation year after year of epidemiological data. And the CJD surveillance team had epidemiological, clinical, and pathological evidence. When *The Lancet* of April 6, 1996, finally arrived, the collective grab of scientists and journalists could be heard around the world. They flipped quickly to the section discussing the possible link with BSE and read in the dry, emotionless tones of scientific speech: "The first aim of the CJD Surveillance Unit has been to identify any changes in CJD that might be attributable to the transmission of BSE to the human population. Although the small number of cases in this report cannot be regarded as proof, the observation of a potentially new form of CJD in the UK is consistent with such a link."[21]

Variant—CJD

What was unique about these cases (for reasons of confidentiality none was identified, but some of their families, such as Michelle

Bowen's, revealed their names) was their youthfulness—the average age was twenty-seven. CJD normally occurs in individuals over fifty. Their disease course was also unusually long—they survived from six to twenty-two months, as compared to sporadic CJD cases who lived only two and a half to six and a half months. The pattern of the disease was unique as well. It began with behavioral or psychiatric changes. In fact, none of these cases would have elicited a diagnosis of CJD based on the standard clinical presentation or symptoms. Their brains on autopsy revealed the expected spongy holes, but the holes were in a different place in the brain and looked different from the holes in the brains of other CJD patients. In fact, they looked like the flower-shaped lesions typically seen in scrapie. They also looked like the lesions found in the brains of individuals who died of kuru. What made these ten cases unique as well, the *Lancet* article said, was that none had a history of exposure to CJD through hormones, surgery, or transplants, and none had had a blood transfusion. (This last item was a bit of a surprise. It had been widely assumed that blood did not transmit the infectious agent. If it did, new routes of infection were possible, and the safety of the blood supply could be called into question, a freshly horrifying thought.)

But what of the farmers with CJD who'd caused such concern only a few months before? They were dismissed in a sentence: "None of the four farmers showed the neuropathological features described here." If farmers had higher than normal levels of CJD—and the article said not only that they did but that this pattern was also seen in countries that weren't infected with BSE—then the disease had to be related to something else. The first thought was of Richard Marsh and his hypothesis of a different sort of BSE in U.S. downer cows. Could there be a downer cow CJD as well? Such talk was still wild speculation.

Government officials moaned about the sensational press, but most newspapers in fact took a cautious tone now that the worst fears seemed realized. The *Daily Telegraph* said that "eating beef must rank as [a] minor hazard" in life's risks. The *Evening Standard* called for "a calm and rational assessment of risks." In the United States some newspapers, like the *New York Times*, covered the issue in depth, but other papers quickly quoted the USDA assurances that there was no BSE in the country, that the country had an adequate surveillance program, and that the CDC had ascertained that the rate of CJD was right where it was expected to be: at about one case per million. Beef consumption hardly wavered in the United States, and meat industry polls found that most people were still confident that U.S. beef was safe.

The assurances went mostly unquestioned, but there were areas that needed further investigation. True, the USDA had examined more than twenty-six hundred brains of cattle and reported no BSE. Some had been sent from labs that checked them for rabies when the cows exhibited the classic neurological symptoms. Some had come from inspectors who spotted cows acting oddly. These assurances contained an obvious flaw. Why would these brains be positive for BSE when we generally slaughter our animals long before they would be showing any signs of the disease? Beef steers are usually slaughtered at twelve to eighteen months, in both the United States and Great Britain. Dairy cows, which have been most affected by BSE because they live longer and receive more feed containing animal protein, are slaughtered between five and eight years in Great Britain. In the U.S. throwaway culture, however, they are generally kept for only two to five years, during their peak production, after which time they go into hamburger. Marsh said it was "absolutely" unlikely that inspectors would be seeing cows with the classic BSE signs if the United States did have it, since it takes five to eight years to develop. The BSE-infected animals, if we had them, would be preclinical but possibly still capable of passing the disease on. And they were going into America's favorite food.

Moreover, both farmers and inspectors had reasons to avoid identifying an animal with suspected BSE. For the farmer it could mean massive trouble for not only his herd but his industry. And the inspector at a slaughter facility would have to justify to an angry processor his subjective decision to take an animal out of the line. The pressure on inspectors not to slow up the process can be extreme. Not only that, there were real questions about the USDA examination of the brains it had seen. Some critics complained that the samples were ill prepared and unreadable.[22] Marsh and others wondered whether the USDA was looking for the right patterns. Back in 1978, before BSE was discovered, cattle had been inoculated by a USDA researcher in Mission, Texas, with scrapie-infected material from sheep to see whether the disease was transmissible. The animals developed neurological symptoms not like those of scrapie, but similar to those seen in "downer cows." While the brains didn't look precisely like the scrapie brains the researchers were accustomed to seeing, on review they decided that the infected cattle brains represented a variation. A more recent analysis of their brains has revealed the prion protein (PrP) that distinguishes BSE. The USDA should have been looking for those patterns, not the British patterns, said Marsh, because our BSE, coming as it probably did from a different strain of scrapie, would look different.

The USDA is perfectly aware of the shortcomings of its surveillance. In the wake of Dorrell's March 1996 announcement of a possible CJD-BSE link, even as the agency gave blanket assurances to the public, it quietly revised its procedures. It began a program to reeducate its inspectors and veterinarians, using videos prepared in Great Britain to identify possibly affected animals. It had begun—and would now increase—the spot-checking of downer cows at slaughter facilities, depending no longer on the heads voluntarily sent in for testing. And it had already started, in 1994, to take a new look at some of the brains it had rejected for BSE to test them instead for the protein. So far it has found nothing, the agency's BSE expert says.

At the Food and Drug Administration one official admitted that the people who really knew about BSE could fit into a tiny office. The agency didn't plan an emergency rule to restrict gelatin in pharmaceuticals and cosmetics because it would have to demonstrate that gelatin represented an immediate threat to human health; no one at the FDA thought it did. Unlike the USDA, the FDA thought the lime process that gelatin went through was enough to render it safe. The agency's spokespeople weren't sure whether the United States imported animal by-products or feed, although the USDA's *Agricultural Fact Book* said that it did. The question of feed or meat and bone meal importation was important. We might not have imported it from the United Kingdom or Switzerland or France, where BSE had been identified, but what if we imported it from South Africa, to which the United Kingdom had exported cattle? The worldwide trade in agricultural products was a tangled web of relationships. Who knew, for instance, the origin of the meat contained in salami imported by the United States from Europe?

The big question was what the FDA would do about the continuing

U.S. practice of feeding ruminant feed to ruminants. The agency had considered halting the practice in 1991, but the industry protested and said it would voluntarily ban the material. But when the FDA's Center for Veterinary Medicine looked into rendering in 1993 to see whether sheep were being eliminated, they found that more than half of the renderers they checked who processed adult sheep were still "selling rendered protein by-products to cattle feed producers."[23] The voluntary ban wasn't working. The same practices that had produced the BSE crisis in the United Kingdom were continuing in the United States. The matter was so serious that the Foundation on Economic Trends brought a petition asking that the FDA and the USDA order a permanent halt to all feeding of ruminant animal protein to ruminants, especially cows and sheep. The FDA proposed a rule banning specified offal from adult sheep and goats from animal feed, but the agency failed to follow through, deciding instead to stick with the voluntary ban—one that it had already determined was ineffective.[24]

In the spring of 1996 the situation was very different. The World Health Organization was making the same recommendation that ruminant-to-ruminant feeding be stopped, and the FDA had obtained the support of the beef and dairy industries, which were concerned enough about their own products to want real protection. Yet to the public the cattle industry maintained a steadfast denial. Picking up on the hottest story around, the television host Oprah Winfrey did a program on April 16, 1996, on "Dangerous Foods." The American public was astonished to hear what the animals they ate were being fed. Oprah gave Gary Weber, the representative from the National Cattlemen's Beef Association (NCBA), little time to respond to the devastating accusations of Howard Lyman, a former cattleman and now a vegetarian and a representative of the Humane Society's "Eating with Conscience" program. Not only was Weber unhappy about his lack of response time, but cattle futures took a nosedive after the program aired. The NCBA was furious and complained loudly. A week later Oprah would give the meat industry what it wanted—another program, virtually all its own.

Without Lyman or a USDA representative present, Gary Weber got no argument when he told the audience that there was no BSE in the country as well as something that clearly was not true—that U.S.

cows were not fed animal protein. Animal protein didn't even cause the disease, he said. Ignoring the reams of scientific material to the contrary, he said that BSE was a disease unique to cattle in Great Britain and that the beef industry didn't know why it happened there. His theory was that cattle got the disease and then spread it. It was a startling position, contrary to all the available scientific data on BSE, but it went unchallenged.

For the beef industry BSE presented a new dilemma. There was no possibility of passing on the responsibility to consumers, as there had been with *Salmonella* and *E. coli* O157. Consumers couldn't just be told to "cook it" since the infectious agent survived extreme temperatures—possibly up to 360 degrees Celsius of dry heat. (Steam heat is more effective at a lower temperature.) Presumably U.S. beef was safe, but the downer cow issue left questions that needed to be answered. The only recourse for the industries involved—from rendering to cosmetics—was to attempt at once to maintain public confidence in the BSE-free status of the United States by emphasizing the doubts and downplaying the evidence. While in trade publications the information about the possible connection between BSE and CJD was usually straightforward, public communications emphasized, not the several compelling connections between the animal and human disease that the U.K. researchers had outlined in their stunning announcement and subsequent published article, but the lack of scientific proof. Typical was the Issue Brief of the American Veterinary Medical Association that announced, "There is no scientific evidence which indicates that BSE can be transmitted from infected cattle to humans through contact orconsumption of beef or dairy products." And further, "There has not [their emphasis] been any scientific evidence to support a link between the spongiform encephalopathies found in animals with those found in humans." [25] *Beef Today* ran an article entitled "BSE-human link alleged," which emphasized that CJD occurred in countries where BSE had not been found and also at a "consistent rate among vegetarian groups and meat eaters alike," an observation that was attributed to Gary Weber, whose source for that unlikely information was not revealed.[26] But the cosmetics industry, which had a different problem—many of their products contained bovine-derived ingredients—reassured cosmetic users that "the industry has not, and will not, use bovine ingredients from countries known to have a bovine spongiform encephalopathy (BSE, a.k.a. mad cow disease) problem."[27] It was urgent, however, to ban ruminant-to-ruminant feed. Who knew what infectious agents it might be transmitting? But the idea of a ban was in trouble throughout North America. Dr. Graham Clark of Agriculture Canada predicted that such a major change would cause widespread disruption in a complex system of interdependence that had been built up over a period of years. The waste from one industry was recycled for another. If dead animals were refused by the renderers because carcasses could no longer go into feed, something would have to be done with them. What would that be? They couldn't just be sent to the local transfer station. And what about the masses of meat and bone meal now produced? Countries that had discontinued the practice of feeding ruminant feed to ruminants had solved the problem of what to do with bone meal by exporting it. Now it seemed unlikely that there would be any place to export it to. Several countries now proposed banning the use of bone meal in agriculture. Bone meal would soon be accumulating worldwide, creating a huge problem. The public was once again being forced to face the consequences of the worldwide increase in the consumption of animal foods. There was a price to pay both literally and figuratively. One thing that had not been widely considered was whether, in fact, the poultry and swine getting this feed had developed their own form of TSEs. Eighteen zoo animals had. In the United Kingdom both nyala and kudu had developed prion disease, presumably from contaminated feed given to them before the ban. Paul Brown, writing in the *British Medical Journal* not long after the announcement of the possible link, suggested as much—after offering something of a mea culpa for his article "The Jury Is Still Out," which had appeared six months earlier and in which he expressed his doubts about the connection between BSE and CJD. "It now appears I was wrong." Still, the link between BSE and the human disease was only presumptive.

> How ironic, for example, if 11 million British cattle should be slaughtered in a preemptive strike to eliminate the risk of zoonotic Creutzfeldt-Jakob disease, only to find belatedly that the true villains were pigs or chickens which were also fed contaminated nutritional

supplements but were brought to market at such a young age that the disease had not had time to become manifest.[28]

Even researchers in the field winced at Brown's suggestion. "We don't need to bring any other animals into this," one said.

In England, they were, in fact, doing poultry and swine studies. The preliminary information indicated that when pigs were bombarded with the infectious agent, eight of ten came down with the disease. The chickens seemed healthy.

And, of course, pet owners were yet to be heard from. The fact that so much of this rendered animal protein, another 34 percent, was already going into pet food, which had apparently already resulted in eighty cases of feline encephalopathy in England, had not yet registered with the public. (Dogs seemed curiously immune.) In fact, no one at all seemed to be looking at whether the United States had feline encephalopathy. The American Veterinary Medical Association had no information on the disease, and the USDA didn't concern itself with pet diseases. To make matters potentially worse, dead pets were being recycled into animal protein in the United States, which had the potential to amplify the disease in animal feed.[29] Studies had shown that when the infectious agent was passed through animals again it became more virulent.

But these details didn't worry the American consumer particularly. The USDA had said it wasn't here, and the government still had its credibility. There was one group of Americans who had, however, been exposed to British beef, to the chagrin of the military. It explained why the Defense Department became interested when the BSE-CJD connection had first been announced. The defense commissary system has 311 stores and serves 11 million patrons by providing all the food items available in supermarkets to army military dependents and retirees at 5 percent above cost. In the European division there had been a problem getting U.S. boxed beef to some areas. While vacuum-packed, chilled "boxed beef" has a shelf life of sixty days, it was being delayed en route from Germany by overzealous customs officials. The substitute was British beef. Military dependents in Europe south of the Alps had been consuming the now-suspect meat. "We have the dubious distinction of creating the largest single population exposed to the BSE risk," admitted the distressed commissary spokesperson.[30]

But Colonel Colin G. Meyer, the director of food safety, says that these military men and their families were exposed to British beef after the ban on specified offal in Great Britain. In fact, American soldiers in Europe from 1985 to 1990 had been exposed to British beef that was more likely to be infectious. It didn't escape the notice of epidemiologists that either of these groups would be an ideal population in which to track the incidence of the "new variant" CJD (V-CJD) as it related to the general population.

The Scare That Wouldn't Go Away

It would have been helpful to the beef industry if the story could have just faded into the background, as so many "food scares" do. This one refused to do that. Within a very few weeks the number of cases of the new CJD variant had grown to fifteen. One was a girl of fifteen, who was now the youngest victim. Her doctor, Professor Peter Behan at the Southern General Hospital in Glasgow, Scotland, bluntly told the *Telegraph* that he believed her disease had been brought about by eating hamburgers. "Her parents have told me she has a predilection for hamburgers."

With fifteen confirmed cases and many more suspected in England, the wild projections of Dr. Lacey didn't seem quite as outrageous. Researchers were quietly saying that the number of suspected cases—as many as thirty by late 1996, some said—was showing a small but significant increase each month, the beginning, perhaps, of the epi-curve—the predictable bell-shape of all epidemics. Cases were being reported elsewhere in Europe as well. French officials announced in May 1996 that they were slaughtering a herd of 130 cattle in Cherbourg. It would be the fourth herd destroyed in France because of BSE since the beginning of the year. "Domestic beef is safe," said the French government, but French consumers weren't buying it—at least not many of them.

And as countries around the world looked at what they were feeding animals, they received other rude shocks. Two maternity hospitals in Zurich, Switzerland, were found to have been disposing of human placentas in a novel way. For years they had been sold to a supplier of offal and making their way from there into animal feed. Cows were not only eating cows, they were eating humans. And then the humans were eating cows. According to *The Lancet*, "horrified authorities intervened immediately."[31]

Over the next months scientific confirmation of the link and other disquieting information would build quickly. More than twenty-three thousand cattle born after the British feed ban had come down with BSE. That was a puzzle that needed solving. The collective judgment of researchers was that stored feed accounted for some of the cases and that infected material had probably gone into the feed chain despite the ban. But there were still three thousand animals whose illness couldn't be explained away. Preliminary data from a British study seemed to indicate that maternal transmission was possible. If so, it would take a longer, harder effort to contain the outbreak. And maternal transmission had implications for blood and milk that required still more research, although the prevailing idea was that blood and milk did not carry the infection. The preliminary results certainly threw into question the practice of breeding from the offspring of BSE-infected cattle.

In the United Kingdom strict feeding, reporting, and slaughtering practices were having an impact. While there had been one thousand cases a week of the disease in animals in 1993, the numbers were down in 1995 to three hundred cases a week. EU members were generally unsympathetic to Britain's protests at having to cull so many animals in an attempt to end the epidemic because they felt that Britain had created the problem in the first place. But in July 1996 experts concluded that European cows were more widely infected with BSE than had been suspected. Only about fifty cases of the disease had been reported in the region; Portugal had just reported thirty, but that was assumed to be an underestimate. Large amounts of potentially contaminated feed had been imported before the ban, and France, for one, was furious to find that from 1989 to 1992 it had received thousands of tons of potentially BSE-contaminated feed exported after being banned from consumption in Britain.[32] This fact coupled with the potentially infected breeding stock the Europeans had imported meant that there had to be a great many more BSE-infected cows—perhaps as many as two thousand—that had not been reported. One possible explanation for the low count: In some areas farmers were not fully compensated for BSE cows and therefore were not likely to report them.

European veterinarians meeting in Paris in June 1996 acknowledged that BSE was now a "European problem," not just a British one. They called on the EU to "design, apply, and enforce" a comprehensive food hygiene program "from the stable to the table," the British journal *Nature* reported.[33] Belatedly the realization was beginning to dawn that the responsibility for safe food could not simply be left to the consumer, or even to the processor, but that it began with the animal or the crop.

Although the United Kingdom was negotiating to reduce the number of cattle it would cull—something on the order of 400,000 rather than several million—Switzerland, with the second-highest incidence of BSE in Europe, announced in September 1996 that it would subsidize the slaughter of 230,000 cows born before the 1990 ban against feeding calves with animal protein. The move was meant to restore consumer confidence in an industry badly damaged by the disease. The domestic agriculture market had fallen by one-third since January 1996 because of falling beef prices and the expense of stockpiling quantities of meat. The EU had banned the importation of Swiss beef. Economic pressures, however, triumphed over common sense. The Swiss planned to include the remains of cows born before 1990 in feed destined for pigs. But with consumer awareness heightened, it was difficult to see how that would work. Two major supermarket chains, Migros and Coop, announced that, in an attempt to restore consumer confidence, they would no longer sell meat raised on feed containing animal protein.[34] Falling beef prices throughout Europe were having an effect. European consumers were beginning to shape European agricultural practices.

Further proof of the BSE-CJD link came in October of 1996. The "new variant" Creutzfeldt-Jakob disease (V-CJD) was revealed by John Collinge and his team of researchers to be a particular strain that was different from the other types of CJD but resembled those of BSE transmitted to mice, domestic cats, and macaque monkeys. This finding was "consistent with BSE being the source of this new disease."[35]

In the United States by year's end, the USDA still had found no evidence of BSE in any of the cow brains it examined and it had begun looking, Dr. Linda Detweiler said, for the patterns Marsh said might be there. "We're not sitting quietly, by any means. We're looking, among other things, for better ways to enhance our surveillance." Among those efforts was improving, with the agency's limited resources, the downer cow sample. But still the critical factor would be to find cows old enough to have manifested symptoms and signs of the disease. In the United States there were few of those around.

Was the Prion the Cause?

While the prion theory has its supporters as the cause of BSE and CJD, it is not the only theory around, and it does have problems. It is difficult to see, for instance, how different strains could develop in a substance without DNA. There are other less popular theories as to what might cause the disease, and some fit better with what is known about BSE. One theory comes from Dr. Frank Bastian, director of neuropathology at the University of South Alabama, who believes that a more conventional infectious agent is involved. His research indicates that the cause is likely to be something called a spiroplasma, an organism about the size of a virus. These microbes, first identified only in 1976, are similar to mycoplasmas in that they do not have a cell wall. (They may have some other type of protective coating.) They are fastidious—difficult to isolate—and can tolerate a wide range of temperatures. Present in the hemolymph of almost all insects, they are resistant to antibiotics.

Bastian's early experiments at the NIH demonstrated that spiroplasma were able to produce neurological symptoms in laboratory animals very similar to scrapie symptoms—rats even developed rubbed spots on their backs, just as sheep did. He believes that the fibril protein structures found in BSE- and CJD-infected brain material represent the remains of the spiroplasma. While much of the funding for neuropathological research in CJD has gone to Prusiner and the prion theory, Bastian has continued his investigations and recently, in studies that in the spring of 1996 were as yet unpublished, found genetic material unique to spiroplasma in CJD brains, providing apparently solid evidence for the theory.

Spiroplasma infection would also explain some reports that cannot be accommodated within the prion theory. In 1996 a letter to *The*

Lancet reported that investigators had produced scrapie in mice with material from ground-up mites from Iceland. Bastian is virtually certain there were spiroplasma in the hay mites. If the infectious agent was present in an insect, it would also explain why Irish calves that had not eaten infected feed came down with the disease when their mothers were well. It would explain the failure of attempts to eradicate scrapie by disposing of infected flocks, since new sheep put onto the same fields and into the same barns also came down with the disease. Spiroplasma as the source of infection can accommodate most of the other facts of the disease as well. It is likely that the organism would become more virulent as diseased animals were recycled through feed. The only question might be whether it could survive the rendering process. Bastian says that while sufficient heat has been shown to destroy most of the infectious agent, a subpopulation of highly resistant material inevitably remains. He links this to recent reports of thermophilic bacteria—they survive, and seem to thrive, in temperatures as high as 140 degrees Celsius. The spiroplasma may, he thinks, be able to take resistant forms that protect them from adverse conditions.

Among the other theories about the cause of BSE is the idea that the infection followed a massive campaign to eradicate the warble fly in the United Kingdom in the 1980s and early 1990s with intensive application of organophosphate insecticide. The chief proponent of this theory, a self-trained scientist named Mark Purdey, proposes that exposing the bovine embryo to these pesticides was the primary trigger for the epidemic. While his explanation suggests a direct toxic effect, it could, in fact, fit with Bastian's theory if the application of organophosphates selected for a resistant insect carrier of the spiroplasma by killing off the competition.[36] It might also explain why a higher number of ordinary CJD victims are farmers.

Bastian believes the prion theory is a dead end. One major problem is that the prion particle, which Bastian refers to as "something from Mars," would frustrate any attempt to stop the spread of disease. There are no means of combating the process by which the prion is thought to be infective. When BSE became a human health issue, British researchers looking for a solution began to take more interest in Bastian's theory. "If it is a spiroplasma," Bastian says, "the infection is potentially treatable, and it offers the possibility of testing animals and people before they show signs of the disease."[37]

The Regulators Regulate

The best course of action in the absence of a cure or a preclinical test is to prevent further transmission—especially in countries, such as the United States, that appear to be free of the disease. The WHO-supported effort to ban the feeding of animal proteins to ruminant animals in the United States got a push in March 1996 when the news of "mad cows" belatedly penetrated the odd insulating curtain that keeps much of the world's news from the American public. But the FDA accepted the rendering industry's call for a voluntary ban. Months later, it wasn't working.

"Dairy industry officials say they've seen almost no change in dairy cow rations since the British 'mad cow' scare of late March and early April," wrote Joel McNair in *Agri-View*. "Renderers report that any initial losses in ruminant protein sales due to the publicity have since been recaptured. Indeed, meat-and-bone meal sales volumes appear to have risen in recent weeks as dairy farmers cope with rising soy meal prices."[38]

Dr. Don Franko, director of scientific services at the National Renderers' Association, agreed. He saw no signs of a decrease in the feeding of rendered animal protein. While the National Cattlemen's Beef Association said that the feeding had decreased, Franko believes that "what the NCBA said was different from what they did."[39]

What was causing renderers and dairy farmers to ignore the voluntary ban was both economics and a lack of pressure from the consumer. Substituting the safe soy meal for meat and bone meal added about a nickel per day per cow, said dairy experts. For a 100-cow herd, that could come to about $1,500 a year.

In early January 1997 the FDA made a long-expected announcement on the use of ruminant-to-ruminant feeding. It proposed banning the use of any animal protein material coming from sheep, cattle, goats, deer, elk, and mink in feed for ruminants. The material could still be fed to swine and poultry. The public would have forty-five days to comment, but the agency expected that few changes would be made to this rule, which would go into effect later in 1997.

Despite the cases of BSE in cats in England and the incidents of scrapie that had the potential to produce cases in the United States, there would be no ban on ruminant animal protein going into pet foods. The chief concern, said the FDA's Dr. Steve Sundlof, was to avoid infecting the cattle population, which could then spread the infection further through the use of rendered animal protein. It might be possible for a cat in the United States to come down with the disease, said Sundlof. And dead pets were being recycled into food for other animals, he confirmed. But he felt that the possibility that this kind of recycling could become an important route of infection for cattle was slight. (Of course it does mean that pets may be eating pets, an unpleasant thought.)

To prevent the prohibited material from being fed to ruminants, the FDA proposed requiring that it be labeled. Farmers who continued to use it would be subject, if detected, to warning letters, fines, and possible imprisonment if they did not stop. FDA officials said they had the means to enforce the proposed rules and the authority to inspect feed records on farms.

A far greater challenge would be determining whether renderers were following the rules, since there would be virtually no way of detecting the origin of animal protein in the finished product. The FDA would rely on inspecting the written procedures and records of the plants to see that systems for separating animal species for rendering were being followed. The EU was finding this a problem. Despite a ban on feeding mammalian protein to ruminants, *Nature* reported in March 1997 that inspectors had found that bone meal often contained as much as 5 percent. The most likely explanation, they said, was cross contamination. "Most feed mills use the same equipment to produce all feed, so meal destined for pig and poultry feed can find its way into cattle fed." The solution: either separate the mills or ban mammalian protein from the feed of all farm animals as the U.K. had done a year ago.[40] The effort to stop the feeding of ruminant protein to ruminants in the U.S. (they could still consume animal protein from hogs and poultry) would very much depend on how much funding the agency received for enforcement, but in the wake of the British crisis, it was an important step.

The Aftermath

BSE is the Chernobyl of food safety. Just as the world's worst nuclear accident transformed public thinking about the wisdom of producing electricity by a means with the potential to be so damaging for so long a time, BSE is the warning shot across the bow of intensive farming practices, the worldwide distribution of agricultural products, and the demand for cheap food. It underscores the dangers inherent in creating a division between animal and human medical science and making the erroneous assumption that they are not directly related. It also underlines the inherent flaw in entrusting the safety of food to a government agency that is at the same time mandated to protect the agricultural industry.

With BSE-CJD and the new evidence of possible maternal transmission of the disease in cattle, as well as the growing awareness of the potential for blood products to carry the infective agent, it again looked likely, as it had for other foodborne pathogens, that while interventions at critical points (such as feed) could play a vital role, no one factor from seed (animal or vegetable) to slaughter could be singled out as the critical moment when food safety could be guaranteed. No single technology could wipe a contaminated product clean. Ensuring food safety, and, indeed the safety of products used on and in the body in other ways, was a matter of synergy and interconnectedness in ways that had never been considered. The process was a continuum, and a misstep anywhere along the line could, and had, spelled disaster.

Part IV/Our Just Deserts

12/Caught with Our Defenses Down

The destiny of nations depends on how they nourish themselves.

JEAN ANTHELME BRILLAT-SAVARIN,
La Physiologie du gout, 1825

The indications of a food chain gone haywire are everywhere. The outbreaks themselves are the outward and visible signs of systems gone awry. There are new pathogens, and there are old pathogens with new tricks, or in new places, or in new numbers. There are clear indications that changes in human behavior from the farm to the fork have enabled pathogens to reach the human population in new ways. We have created the problem, and yet there is as yet no real acknowledgment of that among regulators and the agribusiness community. We have started trying to correct the flaws in the systems, but our attempts are narrowly focused.

The blame for the lack of progress can be divided between the players—the industry, the regulators, the politicians, the scientists, and the media—and the systems each works in. Each has failed to contend effectively with the problem of emerging foodborne disease because of specific and inherent weaknesses within each system. The situation is made worse by an institutionalized disinclination to work together toward a common goal, although attempts are being made to correct this. At the other end of the spectrum, the public health systems we depend upon to respond to impending health crises have been weakened and undermined through lack of attention.

For the public, the explanation as to what precisely has gone wrong with food, and why, is increasingly difficult to ferret out. A press accustomed to reducing complex matters to a sentence or two, with a related tendency to focus on events rather than patterns, parcels out clues in a piecemeal fashion to distracted news consumers, who are left on their own to sort it all out. Newspapers predictably focus on outbreaks, regulatory changes, and scientific breakthroughs, seldom linking them together to demonstrate how they relate to each other and to broader cultural and economic factors. The clues are there within the stories, but they are widely scattered and difficult to pin down. Radio and television information has the capacity to arouse but is ephemeral. One is left with images and impressions and little more. CBS's 60 *Minutes* may produce a segment on filthy chickens, and NBC's *Dateline* one on *Salmonella* in eggs, but who can remember precisely what they said a day later? The result for the public: fear without facts. We are left worried about food without the hard data and solid information necessary to change behavior or challenge the system, both of which are necessary.

In the realms of science and medicine, even as the problems grow, specialists are isolated in their disciplines. Poultry producers, who know precisely what chickens are eating and how they are being reared, what becomes of their litter and how they are slaughtered and processed, seldom communicate with the medical world where clinicians deal with foodborne disease. Livestock specialists who are focused on what feed ingredients make cows produce more milk are planets away from epidemiologists who are attempting to track down the sources of human illnesses. But the lack of communication hasn't seemed to matter—until recently.

Some time ago, not long after the industrial revolution, efficiency became the standard by which all human activities were judged, triumphing in our industrial society over every other value and occasionally even over reason. The dogma of efficiency in food production, processing, and distribution could not be questioned. Both agricultural experts and epidemiologists, whose data and studies point squarely at the culprits—the intensive farming, the mass-distribution, the vast network of import and export—stare blankly if one suggests that consideration be given to changing this system, to opting instead for the small and local production of food. The great majority I have spoken to remain convinced that the system cannot be changed, as if it were a law of physics instead of a human invention. Instead, the search is on for quick fixes to a food chain out of balance, for technological cures here or there that could intervene at one or more critical points. The HACCP principle is sound, but it is being applied like a box of Band-Aids to a system in critical condition.

A visitor from another planet might well look down and observe all the factors contributing to foodborne disease and see where changes and repairs need to be made, but here on Earth there are few detached observers. Each of the participants comes equipped with an agenda and is subject to pressures that are factored into every action. Farmers want to produce food as efficiently as possible for the best return on their investment. Industry wants to produce and distribute products that appeal to consumers, have extended shelf lives and added value, and maximize profits. The regulatory agencies want to fulfill their mandates, but being subject to pressure from many different sources, they also want simply to survive—and that means taking into account the wishes of their powerful industrial constituencies as well as of those members of Congress who allocate their funding. Doctors, hospitals, HMOs, and insurance companies have the obvious goals of health and well-being for their patients, subscribers, and clients, but now their mission is increasingly conflicted by cost-cutting and profit-making pressures, which could very well have an impact on the spread of disease if diagnostic tests are performed less frequently. Scientists are eager to discover the answers, but support for research all too often depends on the goodwill and money of industry.[1] That dependence may influence both the research subject and the publication of the results. More and more research is funded not with government money or university endowments but by corporate interests that favor with grant renewals those scientists who come up with the answers the corporations are seeking. Important findings that could have an impact on food science may be withheld because they are proprietarial or because the researcher cannot find a publisher; one, two, or even three years may elapse before an important study is published, during which time rules and regulations are made and outbreaks occur that the information might have affected. The press wants a story, but it is consumed by the immediate and by the minutia and tends to miss the broader view. It may also be constrained by pressure from advertisers who look with disfavor on frightening customers and are widely known not only to complain but to withhold ads when they are irritated. Even if advertisers apply no pressure at all, editors and writers may unconsciously shape material to please. Consumers, for their part, want safe food, but they also want cheap food, as well as convenience and novelty. Demand helps to drive the system. If there is no one guilty party, neither are there any innocents.

In a perfect world these forces, acting out of individual self-interest, would reach a kind of equilibrium. But that happens only if the playing field is level. It's not. Because of the enormous resources of the now-worldwide food industry, it has the distinct advantage of being able to exert both financial and other forms of influence on lawmaking bodies and regulatory agencies. It is the eight-hundred-pound gorilla at the negotiating table when it comes time to write the rules. Against the food industry's well-organized forces, the regulatory agencies are on the defensive, or even sympathetic. Even after the western states outbreak of *E. coli* O157:H7 in 1993, when consumers were shocked into awareness and public pressure was at its peak for changes to be made in meat processing, the team of epidemiologists who reported on the outbreak, led by Dr. Patricia Griffin of the CDC, observed that the "USDA's new focus on public health and proposed reforms have been hotly contested by industry, and the final outcome is unknown."[2]

For their part, consumers are generally ignorant of the dimensions of the problem, disorganized, and equipped with few resources to apply to, say, regulatory battles in Washington.

The dice are loaded.

It is increasingly obvious from the mass of scientific material that the problem of foodborne disease is not something that can be fixed with the famous silver bullet; an intervention at one or another critical point is not going to be the final answer. The problems are complex and multifaceted, we can now see, created not by one change but by many, each contributing to the niche favored by a particular pathogen. But it can also be predicted, based on previous experience, that any attempt at a quick fix will create new problems few had anticipated. For pathogens even well-intentioned change translates into opportunity.

The other appealing end-point cure is the consumer. "If they had just prepared the food properly" is the mantra repeated endlessly by industry, the regulatory agencies, and even epidemiologists. The public can be forgiven for its confusion. A significant proportion of the recipes in even the newest cookbooks use techniques that are "improper." It seems that even the culinary experts and cookbook authors don't know what the regulatory experts expect the public to know. As for the people who have eaten their eggs sunny-side-up and become ill with salmonellosis, where precisely are the warnings that the only truly safe egg is a hard-boiled or pasteurized egg? They might be in the medical literature or CDC publications, but they certainly are not on the egg carton. They might be on brochures available from the agencies if one knows to ask, but who does? Where exactly is the public to learn that things have changed, and changed dramatically, in the foods we have consumed our entire lives? Sorting through all the explanations for why the incidence of foodborne disease is continuing to rise, one sees that the missing link is consumer information. Consumers in other countries who have learned about the microbial dangers in food have responded in ways that have forced changes in the systems. It is widespread consumer ignorance in the United States that prevents the public from demanding clean food, and the industry prefers it that way.

Clearly the USDA, with its conflicted role of promoter of agriculture and protector of food safety, is caught in a bind. To warn consumers is to tell them how badly these products are contaminated. At some point the public may think to question why the agencies responsible for food safety have let this happen.

Silencing the Activists

Agribusiness has been the driving force in attempting to maintain consumer ignorance. It has not only fought to keep warning labels off of raw meats but sought even to restrain free speech. Thirteen states have now passed "agricultural disparagement" laws designed to protect the reputation of food products, and more are pending. The model laws were written by the American Feed Industry Association in cooperation with the American Farm Bureau Federation and have been offered to states that are interested in passing them. They allow food producers to sue for public disparagement of a state's agricultural products. Food producers love them. First Amendment watchers, consumer advocates, and editors, reporters, and publishers think they are a very bad idea.

The Florida law, one of the nation's toughest, gives farmers, growers, or the industry associations that represent them the right to sue anyone who slanders their product. To say that chickens, or any other agricultural product, presents a health hazard obliges the individual or organization making the charge to back it up with scientific evidence—which sounds reasonable until one discovers that not all science satisfies the food producers.

The legislation is in reaction to the 1989 scare over Alar, say the agricultural associations that are pushing the laws. One can be sympathetic. Apple sales fell in 1989 after press attention focused on the chemical—used to improve the fruit's quality and looks—as a possible cancer threat to children. Washington State apple growers say they suffered a $130 million loss as consumers poured out apple juice and school authorities took apples off lunch trays.

Although lawsuits have been threatened under the agricultural disparagement laws, few have been brought. David Bederman, then of Emory University School of Law and now at the University of Virginia School of Law, sees that as an agribusiness strategy. "They threaten, but they don't file," he says.[3] Activist groups such as Food and Water, a Vermont organization, speculate that the laws grew out of frustration with activists who increasingly take their message to the marketplace and the consumer, thus bypassing the regulatory agencies. Food and Water says it is doing all it can to provoke a lawsuit with the hopes of challenging the statutes in court. Less bravado is seen in smaller, poorer groups, which admit they are frightened. Attorneys say that a group's members could be held personally libel for the educational information the group puts out. The members are careful with their facts and sources but now wonder, "Can I say this?"

"How would you know if something were wrong with the food if you can't talk about it?" asks Phyllis Marberger of Parents for Pesticide

Alternatives, a small Georgia group. "The United States has an awful lot of examples of food safety issues that went unnoticed until someone spoke out."[4]

If a court decides that a group can't support its claim scientifically, that decision could be costly. While some laws allow only actual monetary damages, Georgia's allows compensatory and punitive damages. Under Florida law someone found guilty of intentionally spreading information for the purpose of harming the industry can be fined triple damages. The clear purpose of these laws is to intimidate into silence those who have concerns about food safety.

Farm bureau officials across the country say farmers need protection from such attacks. The goal, says Peter Stemberg, spokesperson for the Washington State Farm Bureau, is to discourage groups and organizations from making claims in public "based on bad science or no science."[5] But the big question is, Whose science?

In the Alar case, the Environmental Protection Agency (EPA) did decide, based on scientific studies, that the chemical was a carcinogen. The manufacturer did not apply to have it reregistered, and it is no longer used. And in fact, says David Waltner-Toews of the University of Guelph in Canada, the agribusiness groups are completely misrepresenting what actually happened in the Alar episode. The EPA, he points out, announced that it would review daminozide, the chemical in Alar, in 1984; its subsequent 1985 report "estimated one extra cancer death per 10,000 individuals exposed to Alar." A ban was proposed by the agency. After industry intervened, the EPA backed down and lowered the acceptable residue level. The manufacturer, Uniroyal, was nevertheless asked to do more studies. Consumer groups protested, Ralph Nader filed a lawsuit, and boycotts were announced by supermarkets and food processors. The growers announced a voluntary ban, but Alar residues were found on apples advertised as being free of the substance. The EPA announced in February 1989 that its new studies suggested that Alar posed an unacceptable cancer risk of 3.5 per 100,000 persons. Uniroyal was asked to voluntarily withdraw it but declined to do so. It was then that the television program 60 *Minutes* focused on Alar. Although industry blamed 60 *Minutes*, Waltner-Toews says, "apple sales began declining as early as the 1984 EPA announcement, and continued to fall, with a drop (though not a statistically significant one) after the television program. In short, consumers acted in a reasonable fashion to risk messages they were receiving from scientific authorities."[6]

Despite these facts, agribusiness groups have convinced their own members and the rest of the population that the campaign against Alar was both unjustified and irresponsible.

Among journalists, the concern is also with free speech and the impact on reporting. Reporters may be left having to decide whose science is the best and most reliable before reporting on possible dangers from food, a position that Jane Kirtley of the Reporters' Committee for Freedom of the Press says they should not be put in. To avoid the question they may simply avoid the subject. The laws clearly have serious constitutional implications as well, says Kirtley. "To gag the press from reporting on ongoing scientific controversy strikes me as basically unconstitutional."[7] In fact, almost no one except the American Farm Bureau thinks the laws will stand up in court. But without a challenge they remain on the books.

The intimidation is not limited to the topic of pesticides and chemical residues by any means. Lucy Riley, news director of WSFA-TV in Montgomery, Alabama, is one of those who has felt the heat these laws can produce. When her station did an environmental story on hamburger and illness caused by *E. coli* O157:H7 bacteria, she heard "talk there might be an action against us." It wouldn't have had any effect on what they reported, she says. "We're not going to be scared or intimidated. We do stories and let the chips fall where they may." All the same, she admits, "I'm curious about the constitutionality, but I don't want to be a test case."[8]

Clearly the laws spell more trouble for food safety. Says Paul MacMasters, executive director of the Freedom Forum First Amendment Center at Vanderbilt University, "This is one more obstacle put in the way of the public getting full information about foods, and we can't afford that. Every bit of legislation put in the way of journalists who try to cover this story is detrimental to all Americans."[9]

Keeping Consumers Ignorant

Seldom has anything revealed the conflicted goals of both the food industry and the regulatory agencies more than their joint attempt to educate consumers as to the dangers of microbes on raw meats.

The USDA conducted a study in 1993 to determine the most effective way to reach the public with consumer information. The study revealed, not surprisingly, that television was the most effective, radio a close second, and pamphlets the least effective. With fascinating logic the USDA then decided to use the pamphlet approach to "warn" the public about the potential for pathogens in raw animal products. The Food Marketing Institute assisted the agency in designing the brochure. The important title of the pamphlet is printed in an almost indecipherable pale pink color that makes it unlikely that a consumer would ever pick it up.

Under pressure from consumer groups after the western states' outbreak, the USDA prepared a postcard to be distributed to children in schools to take home to parents warning them not to prepare or eat pink hamburgers. It was masterful in its design, revealing in glowing color the dual and conflicted role of the agency. On one side was a beautiful hamburger with all the trimmings—a meat promotion if there ever was one. On the back then-Secretary Mike Espy wrote,

> Like millions of Americans, I love a great hamburger. Whether it's topped with tomato, lettuce, cheese or onion; cooked on the stove or on the grill in the backyard, we all know that the hamburger is truly an American tradition. But I also know that hamburger and other meat products could contain bacteria that is harmful if not cooked or handled properly. That's why it's important to follow these simple food safety rules to protect the health of your family and so you can continue to enjoy this great American tradition.

The safe-handling information that follows is sound advice, but it's doubtful whether anyone reading the card would have any idea of the horrifying illness that can follow a safe-handling slipup. The actual consequences have been downplayed to sound like a mild cold. There is no mention of *E. coli* O157:H7, no mention of the western states outbreak, no mention of the possibility of serious illness or death. No mention that fast-food hamburgers served to children should be broken apart first to make sure they are not pink inside. The USDA announced this postcard campaign with great fanfare. What it neglected to mention in the photo-ops was that the cards would not be sent to schools automatically. School nurses would have to ask for them.

But why is the consumer the ultimate critical control point in any case? An angry Michael Osterholm at the Minnesota Health Department, while profoundly moved by the illnesses he has observed, blames parents for their children's HUS in some situations:

> Somebody had to come right out and say, because it wasn't getting said because people were tiptoeing around the issue. 'Parents, if you've got a child in a hospital laying there with HUS and you're the one who served them red hamburger, you're as responsible for the illness as if you had put them in the front seat of a car without a seat belt or a car seat and drove 90 miles an hour through red lights.[10]

He has been active in getting the safe-handling message out in his state and feels that every parent today should surely know the potentially devastating consequences of serving an undercooked hamburger or not following the exacting preparation standards real safety demands in the home kitchen. Yet in other states there have been no safe-handling campaigns at all, and parents nevertheless feel—or are made to feel—that responsibility with excruciating pain. After the death of his child from HUS, one father took his own life.

Show Me the Numbers

The lack of good solid numbers of cases of foodborne disease is another factor in delaying or preventing concerted action, but even as basic a question as to exactly how many people are getting sick becomes a political hot potato. Numbers are not neutral; they come with consequences. They can mean more funding or less for government agencies that regulate food safety or deal with the consequences; high numbers can bolster the message of consumer groups pushing for more regulation; the same numbers can translate into lost business for industry if they put off consumers or foreign trade. The obstacles to obtaining true data on foodborne disease are overwhelming. Reported numbers are most assuredly wrong, but even the very best estimates based on surveillance at hospitals and projected onto the general population may be wrong as well.

After the Schwan's ice cream outbreak, fewer than three hundred cases of infection with S. *enteritidis* related to the product had been reported to the CDC. But a careful study of the distribution patterns of the contaminated ice cream, identifying how many batches had been prepared and where they had been shipped, was combined with a meticulous survey of Schwan's customers in Minnesota to determine who had become ill. By comparing the actual number of people who had been ill with those whose disease had been reported and then applying this ratio to the ice cream shipped and eaten elsewhere, the researchers were able to estimate the magnitude of what was a national outbreak. From this they concluded that the contaminated ice cream actually caused some 224,000 cases of gastroenteritis. It was the perfect illustration of how far off the mark the official reported numbers can be. Noted *Science News*, "If CDC and state agencies missed nearly a quarter of a million cases in this heavily reported incident, imagine how difficult it is for them to detect the occasional gut-wrenching episode caused by microbes from raw poultry that were transferred to salad ingredients chopped on the same counter top 15 minutes later."[11]

The reasons for the discrepancy are obvious. Some people who are exposed and infected may not become ill—their infection may be asymptomatic. Most people with diarrheal disease treat themselves with over-the-counter medications, and most diarrhea is self-resolving—it runs its course and the patient fully recovers. Even when diarrhea is severe, some people may not seek medical care. With bloody diarrhea a victim may think it is a sign of another serious disorder, such as cancer, and avoid a doctor out of fear. Those who go to the doctor may be misdiagnosed or receive inappropriate treatment, and they are increasingly unlikely, many epidemiologists feel, to have a stool culture as clinicians respond to cost-cutting pressures, whether from insurance companies or HMOs.[12] From that perspective, the behavior of the health care system is logical. If the patient's illness is likely to be self-resolving, what is the use of spending money on a stool culture merely to satisfy curiosity? But from the perspective of public health, a goal toward which an integrated health care system should be working, such an expenditure is entirely justified because it may help tie sporadic cases into the pattern of a dispersed outbreak from a single source, an identification which could, if the contaminated product were identified and withdrawn, save both misery and money.

The problems don't end with failure to culture. When a stool culture returns negative for pathogens, the doctor and patient seldom question that conclusion. And yet there are many reasons why the test results may not be correct. The lab may not have looked for the organism responsible, or it may not have used the best methods to find it. Or the organism may no longer be present in the patient's body.

Assuming everything went perfectly—the patient sought medical help, his stools were sampled promptly and sent to the lab with the appropriate requests, the lab did everything correctly and identified the pathogen—would the illness be reported to authorities? Only recently, after much effort on the part of STOP and at the urging of the CDC and the Association of State and Territorial Epidemiologists, has *E. coli* O157:H7 been made reportable in most states. Even so, reporting is passive. The effort must come from the physician or the lab, either of whom may not remember precisely what is on the constantly changing list of reportable diseases, or they may simply forget, or neglect, to make a report. Reporting remains encouraged but voluntary; there is, as yet, no penalty for failure to report.

William Keene, an Oregon epidemiologist, created a graph to show how likely it was in 1993 that only two cases of O157 out of one hundred were actually being counted. Even for more easily recognized pathogens, vast underreporting is considered a worldwide problem. Taking one pathogen, *Campylobacter,* in the United States voluntary laboratory reporting to the CDC indicates a rate of infection of six to seven cases per one hundred thousand individuals a year, but active local surveillance reveals a much higher incidence rate in the range of thirty to sixty cases per one hundred thousand. A study in the United Kingdom—where a nationalized health system makes numbers easier to come by—revealed eighty-seven cases of *Campylobacter* infection per one hundred thousand individuals in 1990. A Dutch researcher estimated that "less than 1 percent of the real disorders caused by food [are] reported."[13] And in some developing countries there may be no reporting at all, making an accurate assessment of the extent of the problem worldwide impossible to determine.

The wonder, in fact, is that anything is reported at all, and the numbers are a credit to labs and physicians who understand the importance of public health. Although the numbers clearly do not represent the actual number of cases, there is still a benefit to reporting. Assuming that the level of reporting, as unrepresentative as it is, remains relatively constant, the CDC can observe whether a disease is increasing or decreasing in its frequency.

Despite the resistance of industry, public pressure for more action on the part of the federal agencies to clean up meat has indeed resulted in a new focus on food safety at the USDA. A separate under secretary position for food safety has now been established within the agency. And while animal health and human health staffs still work together, each is independent, to avoid conflict. Following the western states' outbreak, not only was a zero tolerance program announced for fecal matter on raw beef carcasses and warning labels mandated for raw meat at the retail level, but steps were taken to establish a program for reducing pathogens. That year $5.7 million was allocated by Congress for food safety. Consumer advocates saw the changes as real progress, but a vital preliminary step to producing safer food was clearly coming to agreement on the number of cases of foodborne disease.

To gauge the real incidence, the USDA's Food Safety and Inspection Service (FSIS) joined forces with the FDA and the CDC to collect and analyze data about diarrheal diseases—especially those caused by *E. coli* O157:H7, *Salmonella*, and, at the CDC's insistence, the neglected *Campylobacter*—by collecting numbers from five project sites in northern California, Oregon, Minnesota, Connecticut, and Atlanta. That study is now under way.

What Happened to Public Health?

Another question looms: Just how well prepared are our public health systems to deal with foodborne disease? The unfortunate answer: Not well at all. A report by the Institute of Medicine for the National Academy of Sciences in 1988 had nothing but bad news, and the situation has deteriorated since then as budget cutting at the state, and local levels has become even more acute. "This nation," the institute said bluntly, "has lost sight of its public health goals and has allowed the system of public health activities to fall into disarray."[14] The ability to act quickly in the face of an epidemic has deteriorated to the point, they said, that the health of the public is unnecessarily threatened. They deemed the situation urgent.

This disarray is the result of a lack of agreement as to what we expect from public health. It is exacerbated by political pressure that substitutes a crisis mentality and hot-button issues for steady focus on biostatistics and epidemiology to identify the health needs of the whole population. Additionally, there is an uneasy relationship between the medical profession and public health, and responsibilities are fragmented, disjointed, and sometimes duplicated between federal, state, and local agencies. The result is a lack of responsibility and accountability, inefficiency, and a generally weakened public health effort. All this at a time when we are more in need of a solid public health infrastructure than ever as we face a host of emerging diseases.

The situation is made worse by the declining number of Americans with steady access to health care. The Robert Wood Johnson Foundation in 1987 found that the proportion of people with no regular source of health care had increased by 65 percent from 1982 to 1987, and that the number of people who had health problems but refrained from seeing a physician increased by 70 percent during the same period. The problem has gotten no better. In 1985 there were 31.3 million Americans without health insurance. By 1995 that number had increased to 39.7 million.[14]

The Institute of Medicine report also found many weaknesses in public health systems, including less professional and expert competence in leadership positions, hostility to public health concepts, outdated regulations, inadequate financing, and lack of effective links between the public and private sectors. Today health departments too often become distributors of food stamps and Medicaid and Medicare funds, and public health functions, which are not well understood by the public, take a back seat. While it is hardly unique, Maine's Bureau of Health is a good example of how inadequate financial support for epidemiology—with the resulting gaps in data gathering and analysis—plays out.

Dr. Kathleen Gensheimer, Maine's state epidemiologist, works in a cramped corner of the state agency with a secretary and a nurse and no budget beyond staff salaries and overhead. The rest of the disease control division looks after immunizations, sexually transmitted diseases, TB, and AIDS, with the bulk of the support for these activities coming from federal, not state, funds. Each of these programs has more people than Gensheimer has, and yet she must contend with the multitude of other infectious disease challenges to the health of the state's residents. Her days are frantic, a combination of responding to infectious disease issues and educating the public through the media. "The state of Maine has the dubious distinction," she says, "of being one of the last states to put state dollars into the funding of the position of state epidemiologist. Maine, and northern New England in general, has not put a lot of state dollars into public health."[16]

Nevertheless, Gensheimer manages, and her manner of coping is a good example of how the CDC connection works at the state level. Because of her training there, she has no hesitation in picking up the phone and calling for help. Maine has had several outbreaks of *E. coli* O157 that needed investigation, and the CDC has lent a hand. It doesn't always. The CDC has to parcel out its limited resources to respond to a state's request for assistance. Gensheimer calls for outside help only when she feels that the greater medical community will learn something from a carefully conducted investigation. To bring out the disease detectives, the outbreak has to be especially large, of mysterious cause, or characterized by some other unusual factor. Gensheimer's activities are essentially reactive. She can tackle only the largest outbreaks or the ones involving serious illness. The others just slip by. In the summer of 1995 she noticed an increase in *Salmonella* infections but had limited resources to investigate. It was only after the Minnesota investigation that she realized that her state was part of the Schwan's outbreak.

The connection between public health and clinical medicine is equally tenuous and fragile. Gensheimer edits a newsletter to keep Maine's physicians informed of emerging infectious diseases and potential challenges to health, but physicians are even more overburdened with reading material than the rest of the population. It is difficult to get the message out. She holds conferences as well, but few health care workers can take a day off from their practices to attend. Information trickles out to the doctors on the front lines of patient care with painful slowness.

Misdiagnosed

Probable cases of foodborne disease go uncultured and unreported—and often even improperly treated—in a medical world increasingly confronted by foodborne disease with which it is not familiar. In outbreak after outbreak patients with foodborne disease are revealed to have been misdiagnosed and sometimes to have received inappropriate treatment. Diagnoses of colitis, appendicitis, and intussuption of the bowel are common, although Dr. Martin Blaser defends some of these cases by explaining that some foodborne infections have virtually identical symptoms to appendicitis, for instance, and that physicians have to make immediate decisions when the patient's condition is urgent.[17] Nevertheless, it is also clear that many obvious outbreaks, even in small towns, where they should be more obvious, are missed.

And there are real gaps in medical care, especially when physicians have no experience with an unfamiliar infection. To evaluate the preparedness of U.S. clinicians for, say, a possibly widespread outbreak of cholera, Dr. Richard Besser and his colleagues at the CDC looked at what happened to cholera patients in the United States.[18] After the major outbreak that occurred among airline passengers, the researchers found that "no passenger aboard this flight who sought care as an outpatient was treated with an appropriate oral rehydration solution." None of the facilities and pharmacies involved stocked World Health Organization oral rehydration salts solution, the preferred solution for treating cholera and other diarrheal diseases. And they found that the failure to consider the possibility of cholera and the failure to follow the most basic rehydration principles for treating diarrheal disease increased the incidence of both illness and death. The medical community in the country that "has the best health care in the world" completely missed the boat when confronted with cholera.

The WHO rehydration salts the CDC study mentions is an inexpensive powdered mixture distributed in foil packages. The cost per treatment is about fifteen cents. The preparation is made in the U.S. (and elsewhere) but is virtually unknown here, even though it has been responsible for saving the lives of hundreds of thousands of children worldwide during the fifteen years since its introduction. The treatment is simplicity itself. Dehydration is a killer in diarrheal disease. When the powder is mixed with sterile (boiled) water and administered slowly by mouth, spoonfuls at a time, it restores the lost body fluids. Why isn't the solution used in the United States? To the consternation of knowledgeable pharmacists, it is usually not even available. Perhaps we mistrust the simple and the cheap. Some pharmacists suspect that customers actually prefer the more expensive, though less effective, prepared solutions sold over the counter. Most often, however, consumers are given no choice in the matter. Certainly the WHO rehydration solution is not promoted or advertised in this country, and the small profit margin may explain the lack of drug company enthusiasm.

While the need for education about the proper treatment for diarrhea is perhaps more urgent in developing countries where medical care is scarce, it is ironic that developed countries are left out of the loop when it comes to vital education programs for parents. In combating diarrheal disease worldwide WHO and UNICEF set a goal of educating 80 percent of mothers worldwide as to the three rules for treating diarrhea in the home. (Increase fluid intake, continue feeding, and seek medical care when needed.) The effectiveness of this education was seen in Peru, where patients with cholera were often found, according to CDC reports, to have been effectively treated at home—in unfavorable contrast to professional medical care in the United States in treating the same disease.

WHO is also conducting a campaign to discourage the use of antidiarrheal medicines, especially in children, because they are not only ineffective but counterproductive and even dangerous. According to the organization, many countries have reviewed their drug policies and restricted the use of antidiarrheals in treating children with diarrhea.[19] The United States was not among the eighteen countries that did so, although the FDA has been reviewing the marketing of antidiarrheals.

Taking Action

For victims, parents of victims, and consumer advocates, the response to foodborne disease is obvious. Clean up the meat. Test for microbes at the slaughterhouse level. Use technology to design interventions that can cleanse carcasses or guarantee fully cooked hamburgers at fast-food restaurants. STOP has pushed for reporting of *E. coli* O157:H7 at the state level and new awareness and regulations at the federal level. But the first challenge of food safety advocates has been to penetrate the atmosphere of corporate denial and projection that is eerily reminiscent of the firewall the tobacco companies erected and guarded for years, maintaining against all reason a position of dumbfounded innocence—recently revealed to have been a knowing deception. An egg industry scientist, denying the huge number of outbreaks worldwide in which human illness with *Salmonella enteritidis* was associated with the consumption of eggs, cheerfully told me that it had never been proven that the outbreaks came from eggs. The beef industry reacted not very differently to the news that a virulent pathogen, *E. coli* O157:H7, had been found in hamburger and had been traced back to cows. The industry has made a point of noting every outbreak in which a food other than beef has been implicated, glossing over the obvious—that these other food vehicles may very well have been cross-contaminated with animal waste in the field, by the water supply, during transportation, at the slaughterhouse, or in the kitchen during handling.[20] The U.S. meat industry has also encouraged the notion that the meat responsible for outbreaks has been imported, although, ironically, the same inability to trace product back to its source that leaves epidemiologists confounded makes linking outbreaks to imported meat just as difficult.

That is the public posture of the meat industry. Behind the scenes the reality is quite different. The race has been on for several years now to design technological interventions at the slaughter and processing level—everything from steam-vacuum chambers and cattle-car washes to irradiation—that will reduce the threat from pathogens such as *E. coli* O157. And yet that fevered action did not begin in 1982 when the pathogen was first detected, but in 1990, increasing in intensity following the Jack-in-the-Box outbreak.

An Enlightened Industry and a European Model

It took that outbreak to make Steve Bjerklie, the former editor of *Meat and Poultry* magazine, aware of the threat posed by *E. coli* O157:H7. Evidence for the prevailing level of ignorance within the industry is that prior to the outbreak even Bjerklie, at the center of the industry information stream, had never heard of the pathogen. His first warnings were strong, and for the next two years he prodded a reluctant industry, through a series of columns and articles, to take the new bug seriously.

At one point, in a knee-jerk defense of industry, he had chided USDA Secretary Mike Espy for his efforts to put warning labels on meat. But within a year it was Bjerklie himself who was transformed as he began to understand the broad nature of the problem. When Michael Taylor, the new administrator of FSIS, in September 1994 shocked meatpackers with his announcement that *E. coli* O157:H7 would now be considered an adulterant on raw meat, Bjerklie took the part, not of industry, which would sue to try to make FSIS back down, but of Taylor. When the industry lost the lawsuit, Bjerklie showed little sympathy for the plaintiffs: "Throughout its history, the leadership of the meat and poultry industry has demonstrated a prodigious talent for mistaking its critics for its enemies," he wrote.[21]

His editorials would continue in that vein. Today he says that with a few exceptions he doesn't think the industry really understood the potential impact of bacteria and pathogens on its products.

> They didn't understand that it wasn't a food contamination issue, it was a biological issue. A piece of meat is not just a piece of meat. It is a link in a whole ecosystem—an infinitely complex ecosystem. The industry tended to treat pathogens like they try to keep metal shavings out of the meat supply, but pathogens don't work that way and are a hell of a lot more complex than metal shavings. The industry just did not get it at all.[22]

A more subtle aspect of the industry's slowness in grasping the breadth of the problem, he says, may be that the industry does not think of food animals as animals, but rather as units of production. The industry even goes so far, he points out, as to use the term "a cattle," which is not only an intentional rejection of the individual life force of the creature—a "de-animation," says Bjerklie—but a denial of the extraordinary microbiological complexity of animals. The vast import-export trade also fails to take this complexity into account when shipping meat worldwide. "A steer is not a bunch of bananas," says Bjerklie.

In August 1995 Bjerklie would take another broadside at the industry that supported his publication by challenging the supreme shibboleth in an article entitled "Who Really Has the World's Safest

Meat Supply?"[23] The industry mantra, as he called it, of "We have the safest meat and poultry supply in the world" was repeated "over and over, at convention after convention, meeting after meeting," intoned by industry leaders and government regulators and fact-finding academicians who "should have known better."

Bjerklie then systematically listed country after country with higher standards. Sweden's poultry was virtually *Salmonella*-free. The testing program for O157:H7 in the Netherlands, he wrote, made the USDA's program look like "quality control at the 'Laverne and Shirley' brewery." The Dutch also required extensive microbial sampling for other pathogens, including those in feed, and had a mandatory traceback program that U.S. industry had claimed was impossible. The European Union had passed a directive in 1988 requiring microbial checks on ground meat and meat preparations. Water chilling of poultry, the U.S. standard, was banned or being phased out in most European countries. As the United States was struggling to come up with HACCP regulations, they were already accepted practice in New Zealand and Australia. And New Zealand had strict temperature controls, which the U.S. industry was actively resisting. The EU's "much maligned" hormone ban, usually described in the United States as a trade barrier, was in fact "rooted in published concerns expressed by northern European veterinarians that hormones and antibiotics might make livestock susceptible to bacterial infection." And all meat imported into the EU in pieces larger than one hundred grams was subjected to random microbiological testing. Organic acid sprays, which U.S. meat scientists were still hoping could be used on carcasses to reduce bacterial loads, had long been discarded in Europe as ineffective, and in EU-member nations carcass sprays were illegal because they added water weight. All of these standards were tougher than those in the United States.

The EU had also established a comprehensive list of sanitary requirements for non-European meat plants that wanted to export to EU-member nations. Called the "Third Country Meat Directive," the requirements were widely protested by U.S. industry, but as the larger companies that wanted to trade with Europe put them into place, they were slowly—behind closed doors—cleaning up their act. One U.S. packinghouse consultant told Bjerklie that the EU regulations were "the best thing ever to happen to U.S. plant design."[24]

Tackling the Problem

Based on the assumption that healthy animals produce healthy human food, meat inspection has always relied on spotting and eliminating the sick ones and keeping a close eye on the slaughtering process to see that the meat is not contaminated during or after slaughter. But a growing army of microbial threats such as *Salmonella enteritidis*, *Campylobacter*, and the vicious *E. coli* O157:H7 has made the USDA rethink the inspection process. The rethinking process took three years and has been subject to much negotiation both at open meetings and behind the scenes, but unlike the conferences on *Salmonella* during the 1960s and 1970s, where people talked but little happened, there was a new element in the mix. STOP, created by angry parents of *E. coli* O157 victims, and the Safe Food Coalition, an umbrella group for a number of consumer advocacy organizations, came to the table, were listened to, and in the process changed the dynamic.

In the summer of 1996 President Clinton, surrounded by parents of victims of *E. coli* O157:H7, announced with great fanfare that with the publishing of the final HACCP rule, passage of the Act, food was going to be safer than it had been. The Safe Food Coalition called the new system a "life preserver for American children."

The white-coated troops of stern federal inspectors who eyeball, prod, and sniff meat were to be retrained and redeployed as part of a hazard analysis and critical control points (HACCP) system, the science-based method of ensuring safe food already used effectively in the processed foods industry. The system identifies the points of danger in the process, formulates how best to avoid or attack a problem, and uses numerical means to verify that the system is working. The process works well when there is a "critical control point" that can be clearly established and a surefire cure. Canned food can be assumed to be safe, for instance, if it spends a certain amount of time at a certain temperature in a retort. Inspectors check periodically to be certain that times and temperatures in the retort meet the established goals. Unfortunately, how well the HACCP system will work with raw products, which cannot be popped into retorts, is still in question.

The new system, which would be adopted gradually over the next few years, beginning after about eighteen months for larger operators and allowing smaller facilities up to three years or more to comply, signaled a shift in philosophy on the part of the U.S. Department of Agriculture. The agency would retreat from the traditional, paternalistic approach it calls "command and control" and shift toward an approach that puts more responsibility on industry to produce a clean product. Precisely how industry will do that is something it will have more leeway to determine, although all HACCP plans will have to be approved by the FSIS. Eventually it is hoped that industry will do much of its own inspecting as well. USDA inspectors will simply check to see that approved plans are in place and being followed.

Broader microbial testing will back up the new HACCP rule. Meat processors will be required to verify that their sanitation practices are effective by testing for the generic *E. coli* bacteria. The USDA will set average levels that should not be exceeded. To supplement that testing, the USDA will itself begin to monitor *Salmonella* levels on raw meat products, making certain they do not rise above certain levels. Chicken will be allowed higher levels of contamination than beef because it now is more heavily contaminated. Standards are to be raised gradually, and the level of contamination on any one carcass will not be judged but rather the percentage of contaminated carcasses, a fact that may not instill the confidence the USDA is hoping for in consumers, who cook individual chickens, not percentages. Consumers will probably be surprised to learn that products found to be contaminated with *Salmonella* will not be condemned or recalled but will continue to enter the food chain.

In fact, the plan is not very far from the idealized version of HACCP the AMI presented to the agency a few years ago—what it called a "science-based" inspection system that all but eliminates the USDA from the process. Government inspectors will, in fact—as industry hoped—move incrementally from inspecting product to chiefly inspecting paper.

While microbial testing has been included in the new system, how and where and how frequently that testing is done was left to be debated, but it is a critical issue. As is the temperature of raw meat. Everyone knows that microbes increase when a raw animal product is not kept below a certain temperature, but the industry was reluctant to have the USDA set the standards. These negotiations, too, were left for later. There are presently no requirements for keeping meat chilled during transport.

These flaws in the new HACCP plan were generally overlooked by an enthusiastic press. The president had his election-year coup; perhaps he truly believed that food was about to become safer, but whether it would or would not was still being decided out of the glare of media. Industry didn't get everything it wanted. For the moment, at least, federal inspectors will remain on the job—checking for animals with cancers, TB, worm infestations, or other diseases that should not be going into the food chain.

But consumer watchdog groups are still concerned about the new secrecy that will surround meat inspection. When the inspection of meatpacking facilities was carried out by the government, the inspectors' reports were public documents available under the Freedom of Information Act. By contrast, the inspection documents generated by HACCP systems will be industry property, unavailable to journalists and the public.[25] The meat industry maintains a level of secrecy that far exceeds that of nuclear power plants, which at least allowed visitors, and seems to have convinced regulators and consumers that this is appropriate. They are indeed private companies—but they produce very public meat. An additional concern is that the industry employees who do the inspecting will not be given safeguards under the federal whistle-blower statutes if they want to report violations or misdeeds, since they are not government employees.

While the consensus is that food safety will be much improved by the institution of HACCP, few expect it to be the definitive solution. The real evidence for success will be a decline in foodborne disease, but without firm numbers, that may be difficult to prove. Both the meat industry and the USDA say that consumers will still have to treat all raw meat as if it were potentially contaminated by cooking it thoroughly and avoiding cross-contamination in the kitchen. HACCP is a system, not a guarantee, and the consumer will, ironically, be more dependent on the good intentions of the industry than ever.

13/A Reluctant Vegetarian and the Search for Clean Food

The health of soil, plant, animal and man is one and indivisible.

LADY EVE BALFOUR,
quoted by HRH, The Prince of Wales
in his address, "The Lady Eve Balfour
Memorial Lecture," 1996

A funny thing happened to me somewhere in the late 1980s. I stopped eating hamburger. This was not a conscious decision but one that evolved gradually over time. Although the first outbreaks linked to *E. coli* O157:H7 had occurred, they weren't responsible for my reluctance. I remember reading the Associated Press report in 1982 about people becoming ill after eating fast-food hamburgers and being intrigued, but I assumed the problem had been solved and thought no more of it. More likely my growing reluctance was related to a number of subliminal sensory evaluations—something about the appearance, the smell, the feel, the taste. Whatever the reason, I bought hamburger less and less frequently and found that it stayed in the refrigerator until I had to throw it out. Or I would prepare a hamburger, take one bite, and find that I simply didn't want any more.

Making intuitive judgments about food used to be the norm. Before microbiological testing, before pasteurization, before canning and freezing and dehydrating and preserving, it was what saved us from eating bad food. That skill is rapidly being lost. In a large outbreak of foodborne disease from contaminated chocolate milk, many of those who drank the milk reported that it tasted funny but they drank it anyway. We trust our foods beyond reason, and that's a new development in human history.

I can remember living in France some thirty years ago and watching the dragon housewives who insisted on absolute top quality for everything they bought. They sniffed and prodded and poked, and they handed back with indignation anything they thought was inferior. They wanted their cheese at a precise point of ripeness, depending on when they were serving it. They looked their fish in the eye. They weren't intimidated by packaging, insisting that it be opened so that they could make these decisions. And their judgment was respected by the food merchants. Someone who sold less than top-quality produce or less than absolutely fresh fish didn't stay in business long. This was the state of food when food still mattered—to both the seller and the purchaser.

Dr. Robert Tauxe at the CDC, the well-traveled director of the CDC's Enteric Diseases Branch, reminded me of what buying hamburger in Europe was like years ago when "fresh ground" meant fresh ground: The butcher asked when precisely you were eating the ground meat, and if you planned to serve it that evening, he suggested that you come back at three or four o'clock that afternoon. The potential for microbial contamination given the multitude of cut surfaces in ground meat was always there—and that was before *E. coli* O157:H7. But fresh also had to do with taste. When last year a proud meat company scientist told me the industry had extended the shelf life of hamburger to eighteen days, my intuitive revulsion began to make sense.

Within a year or two I had stopped eating chicken for the same reasons. I was becoming a vegetarian by degrees, giving up foods I had loved because they no longer had any appeal or were actually distasteful. Only later did I understand that they were contaminated. As outbreaks of disease implicated first one and then another food, my uneasiness grew.

Avoiding meat is now, unfortunately, a reasonable decision. That more than 99 percent of chickens are found by the USDA to be contaminated with generic *E. coli* bacteria, indicating fecal contamination, should come as no surprise to those familiar with how chickens are raised and processed, but the problem is not limited to the United States by any means. In October 1996, *Which?*, a British consumer magazine, reported that one-third of the chickens it bought from supermarket chains were unfit for human consumption. That, of course, is a matter of definition. The U.S. meat industry would judge the same chickens safe if prepared "properly," but the potential consequences of spreading the bacteria these products bring into the house can no longer be ignored. And it's not being ignored at all by the manufacturers of antimicrobial sprays, disinfectants, and soaps, even antibacterial sponges, who see in the widespread distribution of contaminated food a promising new marketing opportunity.

The British Vegetarian Society reported that more people converted to vegetarianism than ever before in 1996, the year following the linking of BSE to CJD. Meat consumption all over Europe has dropped, and even in the United States vegetarianism has increased to the point where nonmeat entrées are offered at almost every decent restaurant. This was unheard of a decade ago. But of the many different reasons behind the increase in the number of those who avoid eating meat, most have nothing to do with microbes or the potential for foodborne disease. Despite the periodic revelations by the media of conditions in poultry or beef slaughterhouses, most people, to judge by how carelessly they handle meat, are unaware or unconcerned about how dirty it has become. Call it denial, call it confidence, but a large percentage of us assume we are immune to foodborne disease—until it strikes.

A More Vulnerable Population

If we are confronting new pathogens at increasing rates, then why, many people may wonder, haven't they become ill? The easy answer is that they probably have been. It was the CDC's Dr. Morris Potter who estimated that each of us has at least one episode of foodborne disease a year, whether we call it that or not.

But most (75 percent) of us come equipped with a good set of defenses, which explains why everyone isn't sick all the time and why some are less affected than others. A life threatening situation for one may be nothing more than an inconvenience for another. Saliva, stomach acid, friendly bacteria, the mucus lining of the intestines—all are there to keep out intruders. Antibiotic use can make us vulnerable to foodborne disease because the community of competitive bacteria that prevents one microbe from dominating can be disturbed, leaving the field free for a resistant pathogen—yet another reason why these lifesavers should be reserved for real medical emergencies. As sales of antacids and over-the-counter acid blockers rise, some epidemiologists are waiting to see what the effect will be on foodborne disease. The acid that troubles us may be making life miserable for bacterial intruders as well.

As Lynn Margulis and Dorion Sagan remind us, "Only today have we begun to appreciate that bacteria are normal and necessary for the human body. . . . Health is not so much a matter of destroying microorganisms as it is of restoring appropriate microbial communities."[1]

If a pathogen should overcome these barriers and break through into the bloodstream, it confronts another layer of cellular body defenses. These systems can be weakened by disease, medical treatments, and age. We can also be more vulnerable precisely because we live in more sanitary conditions and our exposure to pathogens is now infrequent and sporadic. The very sanitation that has prevented untold numbers of illnesses has ironically created more susceptible individuals. We haven't built up antibody defenses as we once did when our immediate environments presented constant microbial challenges to which the weak succumbed.

There is no ethical way around that aspect of our vulnerability. We collectively decided long ago that protecting our children and extending compassion and care to the physically fragile was more important to our culture as a whole than the possible weakening of our gene pool, and we must cope now with the not-unexpected consequence of a more vulnerable population, building ever more safety into the food supply to compensate. But we are also facing an assortment of new and especially dangerous microbes. Exposing our children intentionally to see which survive and which do not is not an ethical or moral alternative.

The other factor that will overcome our internal resources is a product so thoroughly contaminated with pathogens that our immune systems are simply overwhelmed. Too often today that is the case. Healthy young adults become desperately ill because of the sheer "load" of pathogens in a product. To assume we are immune to the problem is an illusion. Even the most knowledgeable can't avoid exposure. There is hardly an enteric disease specialist who doesn't have a personal foodborne illness experience to report.

Abandoning Responsibility and Losing Touch

When we had to gather or grow or catch it to live, our relationship to food was direct. Survival meant knowing food—knowing its cycles, where it could be found, how to store and preserve it, and how to prepare it so as to be safe and edible. Until very recently, safety was an individual responsibility. Gradually, as we developed communities and villages and then cities, we parceled out the growing or hunting of food to specialists—farmers and gatherers—and our relationship with what we ate became more and more distant. Now, when mass producers, processors, and distributors have taken over, we are positively estranged from food. We have relegated virtually all of the production and much of the preparation to others, and cooking has become for many little more than a hobby—something to do on the weekends. When our disassociation from food is coupled with the tremendous abundance and variety we find in the developed world, we are lulled into forgetting how vital it is that our nourishment be safe, wholesome, and steady. And we are paying the price.

The outbreaks of illness get the press attention, but the sporadic cases are no less important to the individuals who suffer through them. A woman tells me about her bout with *E. coli* O157:H7, an illness that required nearly two weeks of hospitalization; my dinner companion tells me of her best friend who, after infection with *Campylobacter*, was wheelchair-bound for more than six months with reactive arthritis. A thirty-five-year-old woman describes how she and a friend ate at a fast-food restaurant—the only thing they had done together for months—and both became terribly ill for more than a week. One did not see a doctor; the other one did but was not cultured. What was clearly an outbreak, probably with many other victims, remained in the shadows and was never investigated.

Foodborne disease is a thief, the stealthy hand in the pocket of our health and well-being. And like thieves' victims everywhere, we lose more than tangibles. More insidious and equally damaging is the erosion of security, the fading of that long-standing comfort and confidence we once had in the safety of food.

We would surely not have delegated so much authority for food safety if industry and government regulators had not routinely assured us that the food they jointly provided was safe, a message continually reinforced with the comforting application of sanitary packaging—a false promise when the microbes are carefully contained inside. Our confidence is further reinforced by television advertising. And yet, with cruel irony, when something goes badly wrong, the industry turns on the consumer, blaming those it has previously reassured for not being more wary.

We should be more wary. Fresh food has always had the potential to contain pathogens. Still, what is absolutely clear is that something has changed in the degree to which fresh poultry, eggs, and meat are contaminated today, and those changes are at least partly the result of an arrogant and duplicitous meat and poultry industry that is banking its future on keeping consumers ignorant and complacent. Or, as the veterinarian and academician David Waltner-Toews writes in his book *Food, Sex, and Salmonella*, "The most serious problems [with foodborne disease] invariably come back to the willingness of pure-blooded capitalists to take risks with other people's lives."[2]

What is missing in our food lives is awe and respect for this vital link to health; what must be restored is personal responsibility for determining what is safe and nutritious. We must set aside complacency and carelessness, and we must begin again with food. We must eat as if food mattered. As Waltner-Toews reminds us, eating is at least as intimate an act as sex. "When we eat," he says,

> we are selecting portions of our environment, and bringing them into intimate contact with our bodies. What sex is to interpersonal relationships, eating is to the human-environment relationship, a daily consummation of our de facto marriage to the living biosphere. And, as in sex, promiscuity in eating habits, and ignorance of eating "partners" can carry with it great risks.[3]

By these standards, eating eighteen-day-old hamburger from meat that has been mass-produced in Iowa or Colorado in huge lots from cows from several different countries would be the equivalent of having a one-night stand with a vagabond. When we consume something from the vast array of imported foods available, we are intimately connected, as Waltner-Toews reminds us, with the environmental origin of that food, like it or not. We also might want to rethink our assumption, he suggests, that we are entitled—or that it is advisable—to eat everything in the world. By the same token, imposing our foods and our culture worldwide is a cultural hegemony that may come back to haunt us.

The only real solution to the problem of emerging foodborne disease will come from reestablishing priorities: Food is part of a complex ecosystem, the health of which must be maintained at every level. Who would have thought that how a farmer houses his animals or what he feeds them or what antimicrobials he gives them might affect the level of disease in humans from a microbe that produces no apparent illness in his animals? The renderers could hardly have guessed in the 1970s that shifting from a batch to a continuous process while simultaneously abandoning the use of solvents and decreasing the rendering temperatures would eventually cause the collapse of the British beef industry. Or who would have supposed that the way food products are distributed, while creating beneficial economies of scale for both producer and consumer, might compound the consequences of contamination as well as make it difficult to spot foodborne outbreaks?

When each element in the food chain is reduced to a simplistic model, as if it were a cog in a machine that merely needs oiling or sharpening, the result shouldn't be surprising. We have only the barest appreciation for the synergistic relationships of the elements we are playing with.

What seems clear is that the cure must be as systemic as the cause, and it must involve a new consumer consciousness, a new caring about food that goes beyond the superficialities of transitory taste sensations to the very nature of food and how it is produced.

The European Revolution Versus American Complacency

The story of our estrangement from food has always been about small, seemingly insignificant acts carried out in the context of mistaken priorities. In September 1996 the Prince of Wales addressed the fiftieth anniversary gathering of the Soil Association, an organization founded by Lady Eve Balfour, a visionary in the field of sustainable farming. An organic farmer himself, Prince Charles addressed the problem directly.

> Today we are surrounded by evidence of what has happened to our farmland when husbandry-based agriculture is replaced by industrial systems and where traditional management gives way to specialisation and intensification. We see the consequences of treating animals like machines; seeking ever greater "efficiency" and even experimenting (catastrophically as we now know) with totally inappropriate alternative "fuels"—in the form of recycled animal proteins—with which to power them.

And yet, as Prince Charles noted, the paradox is that the farmers have been doing only what we asked of them. We have given them, he said,

> a remarkably narrow set of goals, and accompanying incentives to help them get there: economic performance without environmental accountability; maximum production without consideration of food quality and health; intensification without regard for animal welfare; specialisation without consideration of the maintenance of biological and cultural diversity. The signals we sent said what we wanted: cheap food and plenty of it. We can hardly blame our farmers now for their outstanding success in achieving those goals.[4]

The message to farmers must be changed. The real cost of cheap food is far greater than we imagined.

Small signs of a shift in thinking are appearing. The demand for and production and sale of organic products is slowly growing in the United States, as is the awareness of foodborne disease. The two are only partially related. But if the U.S. population as a whole remains naive and trusting about the safety of food, Europe has been utterly transformed by "mad cow disease." Suddenly the quiet organic farming movement (or what is called in Europe "biologic" or "extensive" farming) has been put on fast-forward. Sweden, Denmark, and Germany expect to have at least 10 percent of their land area in organic farming by the year 2000.

In Switzerland about 6 percent of farmers each year are switching to organic farming for competitive reasons. It is not regulation that is driving the movement, but the demands of knowledgeable consumers. Since Switzerland has had the second-worst incidence after Great Britain of bovine spongiform encephalopathy, customers there are now asking that the meat they buy be traceable, and some large food chains are obliging by putting the farm origin of each chicken on its package. Restaurants are indicating where the pork, beef, and chicken they are serving came from. Moreover, advocates of stronger regulation are standing just offstage waiting to apply pressure if these voluntary efforts begin to lag. Swiss food-buyers have been radicalized, and indeed, there has been a widespread drop in meat consumption throughout Europe.

By 1997 tantalizingly low prices and repeated assurances from government about the safety of the meat supply were reviving consumer interest even in England, and meat sales began to recover. But the old complacency and unquestioning trust are gone, replaced by disquieting suspicion and fear.

In Europe the relationship between intensive farming practices and foodborne disease is beginning to be appreciated at the highest levels of government as well. When the Economic and Social Committee of the EU reported in July 1996 on "The Bovine Spongiform Encephalopathy Crisis and Its Wide-ranging Consequences for the EU," it verified that "overproduction and the overriding need to protect food quality, the environment and animal welfare have prompted a switch to more extensive [as opposed to intensive] farming methods in the EU." Guidelines have been prepared for organic farming practices that drastically reduce the number of chickens raised in henhouses from seventy thousand or more to four thousand or less. Specifications as to how they are to be fed and treated under biologic farming are spelled out in sections of what the EU calls Common Agriculture Practices (CAP). These new organic standards are not requirements by any means, but options for farmers who want to change the way they do things. The pressures of intensive farming, the committee said, with its narrow focus on high levels of production, are threatening the ecological balance, and the attempted redress of that imbalance by the new guidelines is a "welcome trend."[5]

Although standards for organic farming have been under consideration for some time in the United States, a revolution in the way we think about food and farming is sorely needed. But there is virtually no chance—given the way industry approaches food production and its domination of the regulatory agencies and Congress—that anything of the sort will happen. In fact, "organic" has become a generic term applied incorrectly to anything not processed, rather than to those foods raised under the strict guidelines established by state organic growers' associations. "Organic" foods have come in for steady attack as a cause of foodborne disease from individuals who should, or do, know that the vast majority of cases of foodborne disease come from nonorganic foods (raw milk being the obvious exception).

The more likely approach to making food safer in the United States will be regulators and industry opting for short-term technological interventions (virtually doomed to become ineffective as pathogens adapt) within systems gone awry, rather than seeking and applying ecologically sound long-term solutions. Looking for critical control points is only useful if one realizes that there are hundreds, perhaps thousands, of such points in a process—not simply one or two. Typically, the poultry industry, for instance, looks at an obvious critical control point, such as water chilling, and instead of bagging birds or switching to air chilling, decides to add something to the water, such as chlorine, in an attempt to control bacteria. End-point "cures"—whereby manufacturers "clean up" a product just before it goes to the consumer—are especially appealing to the industry. In essence, as Waltner-Toews suggests, the institutional approach is to "send everyone to the supermarket to buy canned, packaged, or irradiated food."[6]

The weakness of this approach has cultural implications well beyond the presumed improvement to food safety. If technologically "safe" foods replace fresh foods—fresh meats in irradiated portions, eggs in pasteurized liquid form—it will mean the end of food as we know it, the discarding of centuries of food traditions. Waltner-Toews calls this "Stalin's Victory, or the Bulldozer Solution, since it promotes centralization in both the private and public sectors." Centralized technologies, he points out, take away choices and replace them with the illusion of choice.[7] It is the "have it your way" approach to burger eating: What appear to be choices are only tiny variations within a standardized offering. If fresh alternatives are eliminated, even seven brands of virtually identical frozen, pasteurized, concentrated orange juice isn't choice—it's tyranny. To be offered the choice between raw chickens, whole or cut according to every purpose, or prepared chickens in a variety of forms is no choice at all if one cannot buy safe and flavorful chickens. The food supply is thus expanding and contracting simultaneously. The centralized processed options will eventually marginalize local food production even further if the trend is not halted. If regulators have their way, there will be no legal possibility of drinking unpasteurized apple cider, even if the producer can guarantee a safe product through a healthy process. The mass-producers will have won, and food life will be significantly diminished.

To talk about the "farm to table" approach to food safety has become a meaningless cliché if the central error isn't clearly understood—that any change anywhere along that continuum must be considered with an eye to what it might produce. Even now as scientists look for ways to reduce pathogens at critical control points in food processing, have they considered the impact such innovative technologies may have? Will the steam-vacuuming of carcasses merely produce cleaner meat, as is hoped, or will it select for more virulent steam-resistant microbes? There are already signs that the processing of foods is encouraging heat-resistant microbes. The temperatures at which some pathogens can be destroyed have been increasing almost yearly.

As the food supply extends worldwide, it also becomes obvious that how we raise, import, and export food has economic and social consequences that might at first seem unrelated to food but ultimately affect it quite directly. Trade agreement pressures on small countries that eliminate protective trade barriers may well force them to accept as imports cheap, intensively produced foods. These may undercut and eventually drive their own less efficient farmers out of business. The result will be a total dependency on imported food products and a farm population out of work and short on money to buy them. Social and economic catastrophe, not to mention the total destruction of native eating patterns, will be the predictable outcome of such policies.

Many studies have shown that when ethnic populations switch to international processed-food diets, their health rapidly deteriorates. If their land is then taken over by multinational corporations taking advantage of cheap labor to mass-produce food at the maximum profit for export, as often happens, not only will the pride, national identity, and autonomy of these small nations suffer, the living conditions of those who grow and harvest and prepare our food will have a very direct effect on its safety. We ignore the social and cultural implications of intensive farming, the worldwide trade in food, the impact of trade agreements, and the health and welfare of our neighbors at our peril. What happens to them eventually happens to us.

Food Depression

How do we explain our increasing alienation from the very substance that keeps us alive? A clear sign of depression is a neglect of food, food preparation, and the drive to nourishment. Could we as a nation be in a mass depression? The signs are there, but they are disguised by abundance. We continue to eat—in mass quantities to judge from the number of overweight Americans—but eating has clearly lost its significance to the great majority of people. It is no longer a ritual to be performed with objects and manners that establish the consuming of food as something important. We eat wherever and whenever and whatever, on the street, in our cars, standing in the kitchen, waiting for the bus. Even grandmothers have forgotten how to cook, and everywhere finger foods prevail.

If we are depressed, my own guess is that the cause is the wholesale discarding of basic rituals and the replacement of every other value with that of convenience—the household term for efficiency. Our changing eating habits are really a reprioritizing, although we don't think of them that way. We have said that work or soccer or town meeting or television is more important than sitting down and eating together, and that the inherent inefficiencies of careful food preparation and sitting and eating together in a consciously attractive setting are simply not worth the trouble. The impact has been, in every sense, demoralizing.

We are not, as food consumers, entirely to blame. Everything in the culture has encouraged this shift in values. As stores abandoned nine-to-five opening hours and extended the shopping day, employees abandoned their regular habits in response. Overheard in a supermarket was a young clerk saying that she was taking her lunch break. It was 10:00 A.M. When she got off work at 3:00 P.M. she will be hungry for dinner. She will not wait for her family to eat. Our round-the-clock, ever-open society has contributed to the national snack attack. The role of mealtime has been sacrificed to economic goals.

The loss of dinner is surely a factor in undermining family relationships, a disruption that extends outward into the community in ever-widening ripples of disequilibrium. It also sets the stage for foodborne disease. When we become casual, complacent, and careless about what we eat, foodborne disease is the inevitable consequence.

And yet sometimes, paradoxically, those who have a close relationship with their food run into trouble. A battle rages now in San Francisco over the ethnic markets with live animals that are killed as the customer watches. Much of the world still demands this standard, with good reason. It is the one way to be sure that the meat one is getting is truly fresh. But animal rights activists are appalled at the sight of actual slaughter and the poor, cramped conditions they think these animals are in just prior to slaughter. Those conditions are very likely, in fact, to be much better than those at slaughterhouses, but like many people, these activists focus on what they can see rather than on what is concealed behind the walls of the slaughterhouses. What is in all likelihood the very freshest meat around is being challenged by misguided activists who should be focusing their attention on the truly tragic, stressful, and inhumane conditions under which most of the animals we eat live, are transported, then slaughtered.

Dinner Time

A Chinese definition of family is "Those who eat from the same pot." It is one that suits our times because it is inclusive and open but also establishes the centrality of food in the family. If we are to begin again with food, we must begin with dinner. Those who understand human ecology and realize that the concept of balance and equilibrium applies not just to biological but to social systems also see dinner as a touch point for revitalizing our culture.

One who has long championed the family dinner is the legendary Alice Waters, who created in her Berkeley, California, restaurant Chez Panisse a shrine for fine American cuisine, based on organic, locally grown, fresh ingredients. When these products—which taste better and *are* better than the alternative—are eaten together with family and friends, both our bodies and our spirits are restored. Waters believes that food—what we eat and how we eat—is a political issue. Food can restore communities culturally, socially, and economically when small, local producers are supported.

Fortunately, despite the horrifying decline of small farms—especially diary farms—organic growers are proliferating all over the country. Farmers' markets have been revitalized. Community-supported agriculture—subscribers paying farmers before the harvest to grow vegetables—is a growing alternative to the supermarket. Seed savers and heritage breeders are attempting to preserve old plant varieties and animal breeds in the face of the vast domination of hybridization, intense breeding, the frightening unknown of genetic engineering and now the reality of cloning. Heritage breeders are striving for a healthy diversity in the food supply even as agribusiness intentionally reduces diversity not only to create the most profitable plants and animals but also, quite clearly, to enable individual corporations to gain patentable control over what is grown so that producers are forced to buy from them. Whether cloned, engineered, or selectively bred, monocultures—and we who are increasingly forced to depend upon them—are frighteningly vulnerable to the devastating impact of disease, whether in plants, animals, or ourselves. The lesson of the Irish potato blight, which caused widespread famine because the disease spread within the one variety grown, is shrugged off with a "we can fix it" attitude. But hubris is not reality. The prospect of fixing what so clearly has gone wrong looks bleak, and yet there are faint but encouraging signs of an awakening at the federal level. The FDA's proposed ban on feeding ruminants to ruminants is a positive step forward if it can be enforced.

When we reestablish a direct relationship with food, we can take back into our own hands the power to make it safer. But we will first have to reject the powerful creed of an industrialized model. Healthy farms, healthy animals, healthy communities, healthy families, and healthy bodies may not be exemplars of efficiency, but common sense says they are the more important values for long-term survival.

Will this approach translate automatically into clean food? Actually, there is no guarantee. It is not only important what your chicken ate and how it lived, but how it was processed, how it was distributed, and how it was cooked. The path back to clean food is as complex as the explanation for the increase in foodborne disease. While the finger of guilt can be pointed clearly at industrialized models, it is not at all clear that replacing them with organic models will effect an immediate transformation. The overproduction of manure means that we are persistently spreading pathogens, through runoff and ill-advised disposal or recycling techniques, throughout the environment. It could be that we have spread the adapted pathogens so thoroughly that switching to sounder farming methods will have little effect. The achievement of clean chickens in Sweden is not a tribute to organic methods by any means, but the result of creating completely unnatural biosecure environments for poultry where they are exposed to nothing that hasn't been disinfected, sterilized, or pasteurized. The same fate may be in store for humans if we don't act now. Make no mistake: The way back to sane farming—and to sane eating—will be fraught with perplexing difficulties and worrying contradictions.

Scientific Imperialism

The challenge for those of us who want to ensure a supply of clean food should not be underestimated. Food is the single largest industry in the world, and increasingly it is in the hands of producers and distributors who operate globally, unhampered by national allegiances. The General Agreement on Tariffs and Trade (GATT), which opens the way to more trade and tolerates fewer barriers, has the potential to make food matters worse instead of better. A country that has stringent food standards will have to justify those standards scientifically; the EU, for example, will have to prove that its ban on imported beef (from the United States) that has been treated with hormones is justified because such hormone use presents a threat to human health.

That has a reassuring ring to it, and the great mass of reasonable people would consider it an appropriate standard, but we have a misplaced faith in the ability of scientific studies to determine the long-term effects of certain interventions or ingredients. Given the many and varied foods and environmental conditions we are exposed to and the long time periods between consumption and effect of such things as hormones, by the time any adverse effects appeared it would be far too late to connect them to what was consumed—and far too late to do anything about it. And with worldwide food distribution, there will be no unexposed population to use as a control group.

Many in the scientific world today are concerned about the increasing distrust of science, the turning toward mysticism and what they correctly call pseudo-science. Often today the irrational is seen as more appealing than the rational. Indeed, we seem to be divided into those who worship science and those who reject it, while, as ever, the appropriate position is somewhere in the middle. There is still no substitute for good science, but it should be understood that science can be influenced by the same cultural and economic factors that affect other institutions and systems. It can also create a closed and narrow arena where novel ideas of great merit are too quickly rejected out of hand because humans are no more open to being shaken from their complacencies today than they were four hundred years ago—especially when the new ideas challenge a dogma within which careers have been built, systems contrived, and fortunes made.

Even the speed and volume of research—leading good science to contradict itself frequently in recent years—has contributed to the confusion. The fact is, scientists don't always agree. And all the while, bad science, shoddy science, and bought-and-paid-for science abounds. With so much apparently conflicting information, it's no wonder people turn to the Ouija board for health advice.

John Gray, a fellow at Jesus College, Oxford, writing for *The Guardian*, says that science today has been elevated to the status of a religion; unlike the churches, which have lost their power, science "operates an effective system of censorship against heretics."[8] The certainty that people look for in science, says Gray, is an illusion. It did not prevent the mad cow crisis in Great Britain, despite the hints from ongoing research and warnings from various experts, because sufficient scientific evidence of the dangers of certain rendering and feeding practices was not available until the damage had been done.

What is there to be said, Gray asks, of a "society in which the feeding of cow-remains to other cows can pass almost unnoticed until disaster strikes," except that it is one "in which the very idea of nature is all but dead. No improvement in the calculus of risks is likely to protect a culture in the grip of scientific and technological hubris from further environmental catastrophes." And yet ordinary people, on discovering that diseased animals were being recycled into animal food, express appropriate shock. Nature is only dead to those in the unwitting grip of dogma, who accept without questioning the emotionless intonations of the priesthood.

Still, we continue to make incursions into areas where ordinary people raise their eyebrows. While research into the possibilities of genetic engineering is well developed, who knows what the effects will be when the genetically engineered plants are released into the natural environment? The recent cloning of sheep brought appropriate queries, yet worries were brushed away almost at once, at least in the United States, by the soothing voices of experts who emphasized the potential benefits such technologies might bring to the human condition. The potential disaster of breeding identical food animals with identical susceptibility to disease was mentioned, then, apparently, forgotten.

It would be wise, in this atmosphere, Gray says, to take a precautionary approach to environmental dangers. "We should be willing," he says, "to forgo promising technological innovations if they carry catastrophic risks, even if current scientific knowledge suggests that the probability of disaster is low." Gray suggests shifting the responsibility of possible outcomes to the manufacturers and away from the victims. He does not underestimate the radical nature of this transfer. It would involve major adjustments in industries that today base their economic survival on intensive methods. He also suggests that the evaluation of risk be reassigned, "from being the exclusive prerogative of scientific expertise to being a matter for the ethical and political judgments of lay people." These changes would require a reprioritizing of values that would be almost cataclysmic. For all our confidence in our potential power to shape our environment, it may be that such a shift is beyond us. And yet if we admit that we cannot change, that we are in the grip of economic and technological forces that we cannot control, what power, precisely, is it that we have?

There is one more reason why science should not be the sole final measure in deciding whether a food is acceptable to a population. It is simply not the standard we humans apply to food. Science would tell us that a cooked fly in our soup presents no physical danger to our bodies, or that baked manure is a perfectly safe protein ingredient in our cereal flakes. That assessment would still not make them acceptable. Common sense may not be quantifiable, but that only proves that quantifiability isn't the ultimate arbiter. We still need to use our heads. What is intuition but internalized information, and the very heads that intuitively thought that feeding rendered animal protein to vegetarian animals was a bad idea, are even now being proven right by science.

Consumer Power

But what still prevents organized consumer action in this country is the wall between the food product and real information for the average consumer. Most Americans probably don't know how their food animals are being produced, processed, and distributed. What is needed is real openness about food, and consumers should demand it. Although I no longer eat meat, I do continue to eat fish.

Living by the shore has wonderful advantages. I buy crabmeat fresh from the small plant at the dock in my village. I can walk in the office and through the glass door and watch the crab pickers in their immaculate work space as I'm buying the crabmeat. I sometimes walk back to where the crabs are boiled and can see and smell the freshness of the entire operation. Why shouldn't everyone who wants to be able to see where their food comes from? Why shouldn't we be able to view the slaughter and processing operations that produce the meat we eat? When did food production become a state secret?

As a school girl I visited a slaughter facility and followed the slaughtering of hogs from the knife stick to the bacon package. I continued to eat pork, so fears that it will turn consumers off may or may not be valid, but tours like that are a thing of the past. My requests just to interview Cargill, one of the major meat producers, about something so innocuous as their HACCP plan were flatly denied. What is the reason for this secrecy? And more to the point, why do we tolerate it? The time is right to insist on change.

Becoming a discerning food buyer is a process of training or retraining the senses as well as a matter of priority. Our eyes filter constantly, choosing only those things we want or need to see. Most afternoons I take a walk along a fire road in the national park where I live. Not long ago I became interested in mushrooms. Not to pick and eat—that is risky business for anyone other than an expert—but simply to identify. With book in hand I began to look for mushrooms and suddenly they were everywhere. There were purple fuzzy ones with pink borders, fleshy red ones, tiny yellow spikes that pressed up from mossy mounds, deadly white ones growing innocently beneath the trees. The path I had walked so many times had been overflowing with life-forms I had never even noticed. When we readjust our eyes, we see things we never saw before.

I stand before the girl at the dairy bar, and as I decide what drink to have I watch her rub the rim of the paper cup she holds around her opposite open palm, then, incredibly, she unconsciously extends her hand into the cup, turning it as she does so. She tosses it away with annoyance when I mention it. My eyes now see what I never saw before—the many violations of basic food-safety rules that go on around us daily. My local grocery store has put cooked (ready-to-eat) shrimp on the ice next to raw shrimp that undoubtedly carry pathogens. Another local store places packages of cut vegetables (presumably to inspire stir-frying) in the meat counter next to fresh beef and chicken —a practice that could easily spread contamination. Chicken drips its watery, pathogen-laden trail down the conveyor belt at the check-out counter, and I watch shoppers put fresh vegetables on the same belt. My farmstand wants to put fresh bread into a recycled bag that contained who knows what and doesn't understand my objection. My favorite takeout restaurant is totally open. I can watch my pizza being built, watch how the vegetables are cut up. That sort of openness inspires confidence.

The rebirth of the picky, demanding dragon consumer should extend to restaurant dining. An epidemiologist once told me how reluctant he was to eat the artfully arranged dinners served to him in expensive restaurants because he knew hands had been over everything. In fact, hands have been over everything you eat whether artfully arranged or not, and you had best know whether they were clean hands. A microbiologist told the *New York Observer* that he was asked to leave a restaurant after objecting to a waiter putting his finger inside the water glass he was serving. The wrong person was asked to leave.

The New Approach to Clean Food

Once the challenge for *Homo sapiens* was to find food. It may well be that the challenge of our generation will be to resist the easy availability of food—or at least to resist overconsumption and along with it the siren call of the novel, the easy, the quick, the convenient, and, above all, the seemingly cheap. While the old challenges required skills and wit coupled with strength, the new challenge will demand a thoughtfulness that goes beyond momentary satisfaction and understands the linkage between our actions in one place and their inevitable, perhaps profound consequences in another. It will require the intelligence to see clearly those forces that are driving consumption in ways that benefit only a very few at the expense of the common good, and the ability to act in a disciplined and responsible way that takes the true costs of cheap food into account. Or as Waltner-Toews puts it, the cost-benefit approach to determining how much foodborne illness can be tolerated "overlooks the fact that those who reap the greatest benefit of this food supply may not be the same people as those who are paying the cost."[9]

We have supposed that animals are here on earth solely to provide sustenance for humans, and perhaps they are, but we have gone far beyond consuming them to a willful manipulation that has strained animals to the breaking point. Turkeys have been bred with breasts so huge that they cannot maintain their balance; pigs are bred so overlong that they have trouble supporting their weight; cows, which should have a lifetime of twenty contented years, have been bred to be high milk producers, and now their milk production is being further enhanced by hormones and high doses of protein. They are treated with antibiotics to correct the diseases these stresses produce, and then, spent and exhausted and sick within a few years, they are sacrificed for hamburger. Is it any wonder that these animals, teats often dripping with infection, produce hamburger contaminated with bacteria? Even the manure these high-protein-fed animals are producing has changed. It now has much more nitrogen content than before; what that will mean to the nutritional and microbiological status of foods grown on fields fertilized with this changed material is a question for the future.[10] More recycling is not the answer to controlling the huge overproduction of manure; meat consumers should pay a manure tax to cover the cost of proper waste disposal.

We are clearly at a turning point—the BSE-CJD crisis, *E. coli* O157:H7, the other Shiga toxin–producing *E. coli*, the tremendous increase in *Salmonella enteritidis*, and all the other emerging foodborne pathogens are ample evidence of that.

If sensible agricultural practices are to take hold in the United States, the momentum will come from the bottom up, from consumers who are angry enough to take matters back into their own hands to demand access to clean food. Our belief in the "safe-food fantasy" needs to be replaced by a creative cynicism that demands to know the truth, however unpalatable. We need to ask: What was fed to the cow that produced this milk? What is the life history of this fish? Where did this lettuce come from? Was this chicken water- or air-chilled?

Wendell Berry, a Kentucky farmer and writer, has written:

> Farming cannot take place except in Nature; therefore, if Nature does not thrive, farming cannot thrive. But we know too that Nature includes us. It is not a place into which we reach from some safe standpoint outside it. We are in it and are a part of it while we use it. If it does not thrive, we cannot thrive. The appropriate measure of farming then is the world's health and our health, and this is inescapably one measure.[11]

The key word to put back into our complex relationship with food is *respect*—for the lives of animals, for how foods are grown and harvested, for the conditions of those who do the harvesting and the preparing, and finally, for ourselves and our families and what we put into our bodies. If there is something sacred about life, then it extends to all life. The point is not to reject the consumption of animal foods, for instance, but to restore mystical sacredness and profound respect to the raising and killing and consuming of animal life; to restore reverence for the earth and what it can provide us when it is carefully and respectfully nurtured and tended; and finally, to restore the self-respect that puts value back into human activities even when they produce no obvious economic return. When we reclaim this respect, not only will food be safer, but it will taste better, and so in the process we will restore joy to the sacredness not just of life but of living.

Epilogue/The Andersons' Calves

> It is this committee's considered opinion that the next major infectious agent to emerge as a threat to health in the United States may, like HIV, be a pathogen that has not previously been recognized.
>
> WAITSFIELD, VERMONT
> *Joshua Lederberg and Robert Shope, Emerging Infections,* 1992

The Mad River winds and twists down a pleasant valley in central Vermont that is banked on three sides by mountains. Like most of the state, the area has a closed-in feeling that could, depending on one's personality, feel claustrophobic or protective. One main road runs through the valley, brushing past the frame houses, sometimes cozying up to the river, sometimes stepping back, letting the fields have the honor. After a summer of frequent rains, there are broad sweeps of tall, healthy-looking corn, the rows running perpendicular to the road, creating a dizzying optical effect as you drive by. In other fields cows graze. But there are also fields overgrown with trash trees and goldenrod, going to seed, as it were.

Once the valley was virtually all farms, three hundred of them, interrupted by tiny towns with post offices and general stores. That was before the mountains just beyond began to attract skiers. Now Mad

River Glen and Sugar Bush draw crowds of visitors and swaths of cleared trails mark the highest slopes in summer like prehistoric earth markings.

The valley has changed. It is almost picture-perfect now; tourists provide both the incentive and the money to paint barns red and keep borders filled with summer flowers. They are responsible for the stylish shops and the movie theater and the restaurants whose menus with their grilled portobello mushroom sandwiches echo the latest tastes not of the hometown crowd but of the people from away. Many farms have been sold and refurbished as bed-and-breakfast inns or restored by retirees from Connecticut or Pennsylvania who don't mind the cold. The few farmers left mainly work at other things to support their families, letting their barns house their seldom-used tractors and planting mainly for their own use. Here and there a herd of dairy cows remains. Just before the crest of one steep hill a yellow warning sign flashes a light when the herd that grazes on one side of the road crosses to the barn on the other, a picturesque, almost unbelievable event for city dwellers that provides fodder for "quaint rural Vermont" stories.

The Mad River—so called because it runs backward, or north—becomes in the spring, when the snows melt, a torrent swirling down through the valley, sometimes ignoring its banks and flowing out over the fields. But in summer it is tranquil and appealing, here and there rocky and churning, in other spots widening out to a gentle flow. On a hot afternoon at the Kenneth H. Ward Memorial Access, just downriver from Moretown, the water works its appealing magic, drawing picnicking families who park their cars and clamber down the new wooden steps to a gravel bank that spreads out at the bend. The water is deep enough here for serious swimming. Boys cross the river to climb on the big rocks on the opposite bank and dive or jump into the flow with noisy abandon. The image of the river filled with children, the tranquil simplicity of it all, suits the valley. Unaware that everything is not precisely what it seems, summer visitors—a tiny proportion of the winter ski horde—can be transported momentarily to a time when farms and families and the land seemed in balance and life, if not altogether better, was simpler.

Just outside of Waitsfield a spacious white farmhouse sits close to the road with a rural practicality typical of climates where the snow is heavy. Its facade is quite elegant. The front door and the windows are topped with triangular pediments, and the corners are emphasized with handsome pilastered columns. There are mature trees on either side, and to the left a dirt driveway leads into the side yard.

The back of the house has a typical addition fronted by a long porch, and behind that the land climbs steeply to the huge old barn, a historic landmark, stained a pleasant dark brown. Through the opening where great doors once stood, a cavernous structure looms. Behind the barn is a large pen with a concrete surface, surrounded by new 8" x 8" posts and strong fence wiring.

The house and barn sit not at the crest but halfway up a long hill. To the left of the barn the land falls away sharply, and below are the fields where in summer crops are grown to supply a produce stand run by the farm's previous owner, who sold off the house, barn, and twelve acres above it to the present owner a few years ago. The house has been divided into three apartments.

In the spring of 1996 Kendal Anderson, his wife Lorey, and their two sons, Kamal and Christopher, rented one apartment. Lorey's brother Barry McAdam, his wife Elizabeth, and their eighteen-month-old son Gavin lived in another. Elizabeth was expecting a child in June. In the smaller apartment lived Kristi LaFayette, unmarried and in her twenties.

Kendal and Lorey Anderson were in the process of putting into reality a dream they'd had virtually since they had married four years before. They wanted to raise a herd of organic beef cattle, eschewing the routinely used hormones and whatever other chemicals they could avoid. Farmers around the valley seemed pleased with the idea. Anderson planned to enlist them in growing organic grains and fodder for his herd. He had a local backer.

In order to keep producing milk, cows must have calves. The bull calves are the unwanted by-product of the dairy industry, and they are quickly gotten rid of. Some, according to an expert, are just knocked on the head and taken to the woods—they aren't even worth the money to take them to the sale barn where, with other frightened, miserable, hungry calves, all looking for their mothers, they are sold. Some may be bought to raise. Others are sent to slaughter. The sales take place at night, and many go directly to be killed for veal without ever having had anything to eat. What they need to stay healthy is a precious day of their mother's colostrum, a vital substance that gives them immunity and without which they will not thrive.

In early spring the Andersons bought their first calves at the sale barn, and Kendal, who had grown up raising cattle—his father and grandfather had large herds in Nebraska—went to work feeding what would eventually be his herd. He checked to see that the ones he bought had had colostrum. The standard dry calf formula was mixed with water and the bottle-fed calves, except for one, seemed to be doing well. The questionable one belonged to Elizabeth, and she'd bought it against Kendal's advice. It was much too tiny, but she'd fallen in love. She paid only seven dollars for it, and if she had left the calf there it would have starved. Nobody wanted it, and no one would have wasted the time or money to feed it or put it down.

One of the things Kendal and Lorey wanted was for their boys to learn responsibility; each had been given set chores, which they carried out with enthusiasm. In fact, all the family members, including Lorey's sister, were involved with the effort, and the animals, as they came in, were given names. Elizabeth's tiny calf was called Daisy. There was Red Neck and Butterscotch, and as the calves accumulated, names were found for all. Anderson bought more calves.

It wasn't long before Daisy got sick. The vet took a look at her and thought it was probably pneumonia. In the age-old tradition of farm living, Elizabeth brought her into the kitchen—they'd had a thaw, but it was still bitterly cold—and various members of the family stayed up all night giving the calf electrolytes. Barry even took her on his lap on the sofa, cradling her weak head in his strong hands, stroking and encouraging. But nothing worked. Daisy died. Then Butterscotch got sick and died. It was disappointing, but calf deaths are not totally unexpected. And Daisy's had been more or less anticipated. But neither Anderson nor the vet were sure what had happened to Butterscotch. It had been sudden: She was well one moment and at death's door the next. It was strange. There was no bloody scours (calf diarrhea), no ominous hacking cough—both typical symptoms that things weren't going well.

Behind the farm and a bit to the left is an old lagoon. The foul-smelling water is held in by earth banks that have been built up around it. It is hard to know what is in there. Lorey asked someone who'd once worked on the farm whether dead animals had ever been tossed into it. He'd answered in a typical New England way: "I can't say there was, and I can't say there wasn't."

When the spring thaw began, it transformed everything. A world of snow and ice became rivulets and streams of water that ran everywhere, searching always for a way downhill. The sound of dripping was constant, and the hard soil became deep, sticky mud. A stream of water ran behind the barn and then down to the right of the house. Another stream ran down beside the barn and in front of the house. The lagoon was full to the rim with melt, and on the bank of the lagoon that stood high over the vegetable fields below, a leak began to develop. A trickle of foul water oozed and seeped through the earth bank and flowed down the hill, around the edges of the fields that in a few months would be sown with vegetables, through a culvert under the road and onward to the Mad River. On nice days Anderson let the calves out. They seemed to head right for the grass around the lagoon.

More calves died. And Barry got sick.

A sturdy man of thirty-seven, he was suddenly felled by horrible cramps so intense he couldn't keep from crying out. And severe diarrhea. When it became bloody and the cramps worsened, the family took him to the hospital. They kept him there. He was dehydrated and was given IVs to restore his fluids. He'd had a bout like this once before. It must be colitis, they told him.

Back on the farm they were losing more cows.

"The vet was going crazy," remembers Anderson. "He was treating them for first one thing then another, and nothing was working." Three more died in one day. They'd been healthy at lunch, eating, active, and then a few hours later they would start looking sick. "You could see it in their eyes."[1] The well ones were isolated from the sick ones, but nothing seemed to make a difference.

Then Elizabeth, six months pregnant, was taken sick. She had horribly severe vomiting and cramps. When the bloody diarrhea started, her mother-in-law drove her to the hospital. They kept her for two days, replenished her fluids, and sent her home, expecting her to recover momentarily. She didn't. After a week she was in the hospital again. By that time Kristi was ill as well, suffering off and on from the same symptoms. Then Lorey's two sons, Kamal and Chris, and then Kendal. All had similar symptoms: fierce diarrhea, cramps, vomiting. Kendal, like Barry and Elizabeth, had bloody diarrhea. Except for Elizabeth, they all eventually got well without medical intervention. Elizabeth was found to be positive for *Cryptosporidium*. And because she was losing weight instead of gaining it, she worried about the baby.

Looking back, Lorey remembers these days as simply insane. There were sick and dying calves to care for, as well as those still healthy. There were children to get off to school, sick family members to care for, and the overwhelming question that hung over everything: What was killing these calves? They looked for answers everywhere.

Lorey's father searched the Internet, and Kendal's father began contacting agricultural specialists he knew at midwestern universities. Lorey called the University of Iowa and other university agricultural departments to follow up on leads. The family hoped the Vermont Department of Agriculture would help as well, but calls to the department were not returned, and when the Anderson finally managed to arrange a meeting, it was not fruitful. No one seemed interested.

J. Leroy "Roy" Hadden, the Andersons' veterinarian, is well known in the small state. He is also a town selectman. He remembers clearly what happened with the Andersons' calves. "Ken called me and said a couple of cows had died," says Hadden. "They'd used another vet at first but thought I might know more about cows. I thought it was probably a fluke. All that was found in the autopsy were some lesions in the lungs. Then they started getting a death loss of one or two a day."[2]

The sick calves, who had all been vaccinated for the normal diseases, were isolated and given electrolytes and antibiotics, but it did no good. "We did a couple of autopsies," remembers Hadden, "and the symptoms were not consistent from one autopsy to the next. We would find some real hemorrhagic bowels in some and some completely normal-looking bowels in others."

Hadden sent samples to the local hospital, asking the lab to do a routine culture and look for "crypto" as well. "They reported that they had found some gram-negative rods. We didn't think it was all that significant, which, unfortunately, in retrospect, it was." Even had he known what it signified, Hadden says that his treatment—several different antibiotics—would have remained the same. Nothing he tried worked—not even a superpowerful antibiotic the FDA frowns on for use in food animals. "But at that point it was use it or shoot it and be done with it."

He also sent some tissue samples for virus isolation as well as other cultures off to Cornell to get the bigger picture. "They were very helpful, but from the beginning the frustrating thing was that it didn't matter what we did. These calves just died one after another."

Anderson dragged the dead calves out onto the cement paddock and covered them with a blue tarp. They had to be gotten rid of. He called someone he knew who picked up dead animals, and they were taken away. Anderson remembered the man asked several times precisely how long the calves had been dead. He would learn later that the man had connections to a slaughterhouse that specialized in veal—information he passed on to state officials.

Then the press got hold of the story. Kristi LaFayette's mother was concerned that the sick calves were infecting the people. She called Sam Hemingway at the *Burlington Free Press*, and at the sound of *Cryptosporidium* his antenna went up. It was a parasite that had caused more than 400,000 illnesses and 104 deaths several years before when it got into the water supply of Milwaukee, Wisconsin. His story on April 10 sounded the alarm, linking Elizabeth's apparent crypto infection to the now fifty animal deaths—one of which had also tested positive for the pathogen. But he had a difficult time talking to Kendal Anderson, he remembers, who had been warned by his father-in-law not to talk freely to the press and told by the Vermont Department of Agriculture that it might be able to help him if he didn't cause a stir.

The trouble with the theory was that crypto doesn't kill animals. Certainly not like this. Experts say it may debilitate them and make them prone to other infections, but it's found in 50–70 percent of calves and not considered serious.

The television stations got into the picture the day after Hemingway's story ran. One television station ran its story right after a segment on Britain's outbreak of "mad cow" disease, and the sensational juxtaposition caused widespread alarm. Was this BSE? It was perhaps not an unwarranted question, but of course there was no similarity at all. Bovine spongiform encephalopathy manifests itself in a slower death with completely different symptoms that begin when the animal is around five years old. Nevertheless, the Andersons suddenly found their farm surrounded by journalists. Just getting from the house to the barn meant fighting through the crowd. And more calves were dying. They had bought sixty calves over a period of weeks. Now they were down to less than a dozen, and they still didn't know what was going on.

With the public alerted and alarmed and the pressure on, the local health department came out and tested the water on the farm. The situation wasn't good. The well head was just outside the back door of the house, so low that contamination would have been easy, and more troubling was the septic system's location uphill from the well. As the thaw continued, streams of water flowed around the house, downhill from the barn. The Andersons were taking no chances in any case. They had been boiling their drinking water and sterilizing the water, bottles, and nipples they used to feed the calves. The dismal water situation still couldn't explain what was killing the Andersons' calves.

The Vermont Department of Agriculture was in siege mode. Vermont's agricultural image is pristine and squeaky clean. The department didn't need this sort of publicity and was dismissive of the problem. They reacted to the television coverage by rejecting—quite rightly—the idea that *Cryptosporidium* had killed the calves. The department spokesman, Justin Johnson, told the *Burlington Free Press*, "At this point it's not that big of a deal because there are so many options of what [could have killed them]. There's not a lot of sense in spending a lot of money to work it out."[3]

And yet, whatever was killing the calves—something clearly infectious and virulent—had the potential to spread. If the Vermont Department of Agriculture wasn't interested, residents of the area felt differently. The town buzzed with talk of the Andersons' calves.

Kendal Anderson kept expecting a visit from the state vet, who lived nearby, but by the time Sam Hutchins arrived there were only ten calves left out of sixty. Hutchins walked onto the farm and over to the blue tarp that was covering the pile of dead calves. He made certain that the calves were covered up and then, with little more than a glance around, turned to walk back to his car.[4]

Frustrated by the long days and nights, the horrifying, unending cycle of death, and the lack of attention from the state, Anderson became enraged. "Aren't you even going to look at the calves I have left?" he demanded.

Reluctantly, Hutchins entered the barn and glanced at the calves. He said, "I don't see any diarrhea," and, "They look cold."

"At that point there was a calf dying at his feet," fumes Anderson. "He never took his hands out of his pockets. And then he asked me for water to disinfect his boots."

Now there was the question of what to do with the dead calves. Anderson says that Hutchins made an agreement with him that the state would reimburse the town if it disposed of the dead animals. The state tussled with the town over when and where the calves were to be buried. According to the newspaper account, Twenty-five calves were buried at a location known to the agriculture department. Eventually Anderson, now broke, would get the bill.

If they were ever to find out definitely what was killing the calves, Anderson and his backer and partner knew they had no time to waste. It finally became clear to them that the Vermont Department of Agriculture wasn't interested in getting to the bottom of the story. Anderson's partner and a friend put the two remaining calves in the back of a van and took them to Cornell. One of the calves was sacrificed. Nothing was found. But what Cornell found in the cultures from the other animal were many "heavily encapsulated *E. coli*" bacteria, meaning that each was encased in a protective sheath. It could well be the killer.

Even with the diagnosis, Anderson and Hadden felt they were no closer to knowing the answers than before. "We knew they weren't dying from crypto, and the virus isolations were negative," remembers Hadden. "All we were left with were these encapsulated *E. coli*. The routine *E. coli* we usually see were not found. The way they died, the rapid progression, was clearly unusual. They were dehydrating even before they developed the diarrhea, and that was the frustrating thing. Fine in the morning and dead in the evening."

The only thing that Hadden knew about that could cause these problems was what he called "hot *E. coli*." With normal *E. coli*, animals get diarrhea and die from dehydration. "The hot *E. coli* moves through the animal so quickly—the animals get septic so quickly—that it doesn't matter what you do, short of putting sixty calves on an IV."

The term "hot *E. coli*" is slang, not a medical term. But it was a good description for an apparently vicious pathogen acting in ways that weren't clearly understood. The encapsulated *E. coli* that Cornell had found were negative for the virulent Shiga toxins that make some

E. coli so destructive in humans, but as the report said, that didn't mean that the strange *E. coli* (it was never serotyped) didn't produce toxins that the Cornell lab couldn't identify. Without serotyping, the mystery hit a blank wall.

Elizabeth, just seven months pregnant, was still sick. Another round of illness brought on premature labor. Her daughter, fortunately, was healthy but remained in the hospital for a month. In any case, the story wasn't over.

One day in June the landowner came to Anderson. He said the land owner had talked to Sam Hutchins, who had told him the lagoon should be drained. He asked Anderson, who had operated large equipment in the past, to empty the lagoon with a bulldozer in lieu of rent. Anderson got the equipment and broke a hole in the embankment. The foul water began flowing down the banks of the steep hill, into the brook below, past the fields where vegetables were now growing and being harvested, through the culvert under the road, across the field, and into the Mad River. It was an awful mess, and Anderson realized he couldn't continue. He pushed the earth back and closed the gap, stopping the flow with the lagoon still half-filled. But the damage was done.

Shortly thereafter, a swimmer phoned the state health department to say that pieces of what appeared to be fecal material were floating down the river. Health officials quickly became involved, and when the Mad River tested positive for coliform bacteria, it was closed to swimming for three days.

Dan Tylman, a television reporter with WPTZ-TV in Burlington, had been talking to Anderson. With the lagoon break, he had to go with the story. The public had been left in April assuming the calf and human diseases had been related to *Cryptosporidium*, a connection that the state had rightly denied. Now Anderson shared with Tylman the Cornell report—the information about the pathogenic encapsulated *E. coli*. He told Tylman on camera who had asked him to drain the lagoon. And when Tylman asked him whether the first dead calves might have made it onto the veal market, Anderson replied, "It's possible."

The Department of Agriculture quickly accused Tylman and WPTZ of "gross inaccuracies." Hours later agriculture officials backed down. They hadn't seen the report, they said. They put the deaths down to poor management. Anderson's veterinarian refuted that clearly in the WPTZ report. "The conditions were dry," said Hadden. "He did a good job of keeping the place cleaned up. They might have been a little crowded, but there was nothing out of the ordinary."

Governor Howard Dean, the state's young, attractive chief office holder, told a press conference that the possibility that the calves might have entered the veal market was laughable. But within hours WPTZ reported that the state was investigating what had happened to the first dead calves. What the investigation revealed was not clear, but the 4-D man—the term for those individuals who pick up dead, diseased, dying, or disabled animals—was found to be operating without a license. Not long afterward he received one.

State officials in Vermont are not eager to talk about this incident. They have tried to play down the deaths as nothing out of the ordinary—just a bad situation on one farm. But another Vermont veterinarian, David Webster of Bradford, can put the matter in perspective. "We don't have mortality rates like that in anything. Not that quickly. Not in anything we see out here in the field, so to speak. In daily practice that would have been just astounding."[5]

What remains now are simply questions. What was this pathogenic encapsulated *E. coli* that clearly must have produced an unknown but virulent toxin to cause such rapidly developing septicemia? Did it cause the illnesses in the people? Elizabeth McAdam had cradled a dying calf in her arms. Could she have been infected not only with *Cryptosporidium* but with another pathogen—one that no one tested for? The investigation by Vermont State Health Department had been cursory. It wasn't clear who was cultured—or for what. Or what tests were run.

And the episode would yield one more surprise. Roy Hadden, the veterinarian, admitted that he, too, had had diarrhea. And that it was bloody.

This story reveals how little we know about what is happening around us and the remarkable lack of curiosity among those who should be protecting our interests. The barriers between human and animal diseases were never as solid as we thought. The disease that killed fifty-eight calves spread no further than the Anderson farm, but that was simply luck, not effective agricultural monitoring. This pathogen might well have gotten into the food chain. If it was responsible for the illnesses in the Andersons' extended family, it could have produced more illnesses.

What is clear is that Vermont wanted the story to go away. Anderson was not from Vermont. With no long-standing association with the community, he had no one to come to his defense. No one in the Department of Agriculture knew that he had extensive experience raising cattle—and ongoing advice. He was the fall guy.

One calf remains alive—hidden away out of state. It is isolated from other animals and lonely, but the owners want to keep it just in case someone somewhere becomes interested in what caused this outbreak. The animal is still harboring and shedding the pathogen despite treatments. "The last culture we had," says Hadden, "showed that the course of antibiotics that Cornell said the bacteria was susceptible to didn't touch it. They are still growing—and growing in large amounts. Speculation? This calf is kind of a Typhoid Mary. I think it's carrying the encapsulated *E. coli.*"

Like all of the Andersons' calves, this one has a name—Dinner.

Kendal Anderson and his family have moved off the farm. New tenants have moved into their section of the house. The well head has been raised by the owner. The barn is vacant. The lagoon still looks ominous. The question as to what killed his calves is still very much on Anderson's mind, but without serotyping the *E. coli* bacteria, experts could only generalize about what the outbreak might mean. "It sounds very interesting, says David Woodward of Canada's Laboratory Center for Disease Control. "It's something that should be looked at closely."[6] Every other *E. coli* or disease expert contacted agreed that episode was alarming.

There remain many more questions than answers. Was this the first appearance of an emerging disease of the sort that Drs. Lederberg and Shope had worried about? Was it a new pathogen entirely, or an old pathogen in a new place? Was what killed the calves the pathogen that made the people ill? Had the infected animals entered the food chain? Had an opportunity to find out the answers slipped between the cracks of a defensive bureaucracy? Would this be the pathogen's only appearance or would it show up again somewhere?

Anderson's encounter with the killer bug left a vivid and unnerving impression. For him there was no doubt about one thing.

"This isn't a bad dream that will go away. It will come back."

Appendix/Eating and Cooking Safely: Short-term Defense Tactics

Although concerted effort on the part of consumers could force food producers to clean up food, that will undoubtedly take years. In the meantime, the best defense against contaminated foods is still good home cooking. It is in our homes that we have the most control over what we eat and how it is prepared. Cooking any properly prepared food[1] to an interior temperature of 160 degrees Fahrenheit is going to make it safe to eat by destroying microbes such as those described in this book (with the exception of the mad cow infectious agent—which one can hope will not be encountered). Promptly (within two hours) refrigerating foods you have cooked will keep them safe for a few days because most microbes don't grow at refrigerator temperatures. A cardinal rule of food safety is "hot foods hot and cold foods cold." Foods that are to be reheated should be brought back up to 165 degrees Fahrenheit and should be hot throughout. The exception is food produced in USDA-approved settings (such as some ham products), which may require a temperature of only 140 degrees Fahrenheit.

Because most people don't take the temperature of their foods, except perhaps when roasting meats, a good visual indicator of sufficient heat is the presence of steam. Hamburger should not be pink inside, and in the absence of a meat thermometer, the juices of poultry should run clear. The bad news is that recent studies show that hamburger may be brown without having become hot enough to kill bacteria, so be sure it is actually hot and steamy. The good news for meat eaters is that muscle meat is sterile. Steak or roast beef that has not been pierced, poked, or skewered before cooking (which can bring the pathogen into the interior of the meat) can probably be eaten less well done. Some stores, however, are using a tenderizing device that pierces the meat and appears to be designed to transport microbes into the product. Ask whether it has been used before opting to serve steak rare.

Every piece of raw animal food brought into the home should be treated as if it were contaminated. The tricky part is to avoid contaminating other foods—particularly those that will be eaten uncooked—while preparing it. Hand-washing before and during the preparation and serving of food is the second-best protection against contaminated food.

Every home should have two cutting boards, one for meats and one for vegetables, and they should not be interchanged because foods meant to be eaten raw can be contaminated from the bacteria on animal foods. The debate over plastic versus wooden cutting boards continues, but at least two new studies show less bacteria cultured from wooden boards. Apparently the cellular structure of the wood draws the microbes down into the material, while on the impervious surfaces of plastic and other materials the bacteria form biofilms that are difficult to remove. The USDA still favors plastic or glass surfaces for cutting boards but says wooden boards are fine if they are used exclusively for meat or poultry, then rinsed and air-dried.

As a general disinfecting solution in the kitchen the USDA recommends one teaspoon of liquid chlorine bleach to a quart of water. This can be used on countertops, cutting boards, and utensils.

When preparing dishes with raw meat products, have all the cooking dishes and utensils you plan to use out on the counter before you begin. Confine the meat package to the sink while opening it so that the potentially contaminated juices go down the drain. Don't touch knobs or drawer pulls or light switches with contaminated hands. Wash your hands hard for thirty seconds after handling raw meat.

All utensils used on raw animal products should be washed thoroughly in hot, soapy water before being used for anything else. Wash the counters thoroughly. Replace sponges frequently, and in between rinse them out in hot, soapy water or boil them. If you are cooking meats on the grill, be sure not to put the cooked meats back on the platter the raw meats were on. Use the same care with utensils, which could transfer bacteria from raw meats, especially hamburgers, to cooked ones. Avoid basting with marinating liquids while meat is cooking and do not use leftover marinating liquids. (Pan juices cooking with the meat are fine to baste with.)

If you think this sounds too difficult, I agree. It's why I stopped eating meat. I personally did not want something in the kitchen I had to treat this carefully.

The produce question is more complicated. The FDA is considering new regulations requiring the pasteurization of apple juice because it has been found to be responsible for outbreaks from *E. coli* O157:H7. Lettuce has been identified as a vehicle for O157 as well, and yet no one is suggesting that we sell it precooked or serve it boiled (although there are rumblings that it should be pretreated with an antibacterial solution). In some South American households vegetables are habitually washed in antibacterial solutions. There are solutions on the U.S. market, advertised in health-food magazines and sold in shops for use on vegetables. The USDA says these products are effective if the directions are followed carefully.

The idea of subjecting vegetables to yet another chemical doesn't appeal to me. The alternative is really thorough washing. I keep a separate plastic dish pan and colander for fruits and vegetables. Leafy greens get a thorough soak and several rinses; the others I scrub with a Japanese natural bristle brush, which is gentler than harsh plastics. I hope this does some good, but I don't know of any studies demonstrating that it does. I do have a good idea where my vegetables come from—at least in summer and fall. Many I grow myself, thus several possible means of contamination, such as transportation and handling, have been eliminated. I buy organically grown winter vegetables such as onions, squash, turnips, rutabagas, and potatoes from local farms and store them in my cool pantry. Each step that can be eliminated in the trail from producer to patron lowers the chances of contamination, and getting to know farmers is the first step toward building accountability into the food production process.

The question of eating out is a difficult one. No one who really knows about the spread of emerging foodborne pathogens gets much joy out of eating salads in restaurants these days, but many continue, as I do, to take the risk. Let the restaurant be your guide. I try to get to know the chefs and cooks in the restaurants I eat in and ask them, politely, about safe food-handling practices. When this is impossible, the safest choices are cooked foods served hot. Complain and return anything, such as lasagna, that comes to you with cool spots because it's been reheated in a microwave that cooks unevenly. Before ordering scrambled eggs I'd want to know whether they were pooled (they shouldn't be), and then, if told they were individually cracked, I'd make sure they were thoroughly cooked. If a waiter told me that the kitchen prepared Caesar salad from fresh eggs, I'd say "No thanks." I'd turn down lemon meringue pie unless it was made from pasteurized egg whites. And the danger of eggs makes me leery about ordering French toast because it is often not cooked in the middle. Before giving a fast-food burger to a child, break it open first and make sure it is cooked all the way through.

I avoid salad bars and hors d'oeuvre trays with prepared salads that might have sat too long. But keep in mind that heavy doses of acid (vinegar) are effective killers of pathogens, so the pickled vegetables should be fine.

Pregnant women should avoid soft cheeses, deli items, cooked meat products such as bologna, and uncooked hot dogs. Such items as vacuum-packed cooked chicken should also be avoided.

In general, it's not a bad idea to practice the conservative food traditions of times past when thoroughly cooked or pickled foods were the norm. In fact, a wonderful variety of food preparations remain safe, healthy, and delicious. We simply have to put aside the complacency and absolute trust most of us place in food providers and take responsibility for the safety of our food back into our own hands. We need to become discriminating food shoppers, veritable food dragons who reject inferior products and have the courage to send back dishes that aren't prepared properly. We should try to move incrementally closer to food production by cutting back on imported foods, buying locally, and getting to know growers whenever possible. When food producers and consumers are directly linked, it is natural to expect that more care will be taken in handling and processing. As Wendell Berry says, "Only by restoring the broken connections can we be healed."

Notes

Preface

1. Anonymous, "Regulations Breached," *Nature* 345 (1990): 652.

Chapter 1

1. Roni Rudolph, in-person and telephone interviews, 1993, 1995, 1996.

2. Much later California health officials would realize that other serious illnesses in the state were outbreak-related, and Nevada would discover that, despite its testing and reporting requirement, it too had missed related cases.

3. Patricia M. Griffin, "*Escherichia Coli* O157:H7 and Other Enterohemorrhagic *Escherichia Coli*," in *Infections of the Gastrointestinal Tract* (New York: Raven Press, 1995), p. 740.

4. Lynn Margulis and Dorion Sagan, *Microcosmos: Four Billion Years of Microbial Evolution* (New York: Simon & Schuster, 1986). (Note: The quote relating to bacteria on the skin was attributed to Margulis and Sagan by Stephen Jay Gould in his book, *Full House*. It may be from a earlier edition.)

5. Ibid., pp. 69–126.

6. Anonymous, "International Symposium on Food-based Oral Rehydration Therapy," held Nov. 12–14, 1989, Aga Khan University, Karachi, Pakistan. International Child Health Foundation Newsletter, Issue 4.

7. Conference Proceedings, "Tracking Foodborne Pathogens from Farm to Table," United States Department of Agriculture, January 9–10, 1995.

8. Jean C. Buzby and Tanya Roberts, "ERS Estimates U.S. Foodborne Disease Costs," in *Food Review Yearbook* (Washington, D.C.: U.S. Dept. of Agriculture, May–Aug. 1995), pp. 37–42. Numbers are potent political

weapons; the range of differences between the estimates may indicate not only how the numbers were compiled but a conscious or unconscious calculation by each compiling group of where its advantage lies.

9. Dr. Ruth Berkleman, Centers for Disease Control, in-person interview, Atlanta, Georgia, March 12, 1996.

10. Jean Stephenson, "When Microbes Are on the Menu," *Harvard Health Letter*, December 1994.

11. John Darnton, "The Logic of the Mad Cow Scare," *New York Times*, March 31, 1996.

12. Sen. John Warner, telephone interview, 1995.

13. C. W. Hedberg, K. L. MacDonald, and M. T. Osterholm, "Changing Epidemiology of Food-borne Disease: A Minnesota Perspective," *Clinical Infectious Diseases* 18 (1994): 671–82.

14. Kathleen Gensheimer, in-person interview, Augusta, Maine, 1993.

Chapter 2

1. T. Popovic, O. Olsvik, P. A. Blake, and K. Wachsmuth, "Cholera in the Americas: Foodborne Aspects," *Journal of Food Protection* 56, no. 9: 811–21.

2. D. L. Swerdlow and A. A. Ries, "Cholera in the Americas: Guidelines for the Clinician," *Journal of the American Medical Association* 267 (March 1992): 1495–99.

3. O. S. Levine, D. Swerdlow, L. Roberts, R. Waldman, M. Toole, and the Goma Epidemic Investigation Team, "Epidemic Cholera and Dysentery Among Rwandan Refugees in Goma, Zaire, 1994," Epidemic Intelligence Service 44th Annual Conference, March 27–31, 1995. Centers for Disease Control and Prevention, Atlanta, Georgia.

4. T. Popovic et al., "Cholera in the Americas."

5. A. DePaolo, G. M. Capers, M. L. Motes et al., "Isolation of Latin American Epidemic Strain of *Vibrio cholerae* O1 from the U.S. Gulf Coast" [letter], *The Lancet* 339 (March 1992): 624.

6. Peter Jaret, "The Disease Detectives," *National Geographic* (January 1991): 116–40.

7. R. W. Pinner, S. M. Teutsch, L. Simonsen et al., "Trends in Infectious Disease Mortality in the United States," *JAMA* 275, no. 3 (1996): 189–93.

8. D. Swerdlow et al., "Waterborne Transmission of Epidemic Cholera in Trujillo, Peru: Lessons for a Continent at Risk," *The Lancet* 340 (July 1992): 28–32. (By the time of the CDC study, the warnings against eating seviche had been heeded; only one of those interviewed had eaten it.)

9. "Update: Cholera-Western Hemisphere, 1992," *Morbidity and Mortality Weekly Report* 42, no. 5 (1992): 89–91.

10. Ibid.

11. Centers for Disease Control and Prevention, "Isolation of *Vibrio cholera* O:1 from Oysters—Mobile Bay, 1991–1992," *Morbidity and Mortality Weekly Report* 42, no. 5 (1993): 90–93.

12. World Health Organization, telephone communication, March 10, 1996.

13. William McNeill, *Plagues and Peoples* (Garden City, N.Y.: Doubleday, 1976), pp. 153–58.

14. Centers for Disease Control and Prevention, "Human Plague—India, 1994," *Morbidity and Mortality Weekly Report* 43 (1994): 689–90.

15. Mohamed Karmali, telephone interview, 1993.

16. Hans Zinsser, *Rats, Lice, and History* (New York: Blue Ribbon Books, 1934), p. 9.

17. William McNeill, *Plagues and Peoples* (Garden City, N.Y.: Doubleday, 1976), pp. 206–9.

18. Ibid., p. 20.

19. Ibid., p. 31.

20. McNeill compares the actions of biology (microparisitism) and politics (macroparisitism). They tend to act in the same way. Like excessively virulent pathogens, a political system that "overtaxes" its population, whether through oppression, conscription, or revenue demands, to the point that the population is economically, physically, or psychologically exhausted, will eventually collapse.

21. McNeill, *Plagues and Peoples*, p. 53.

22. World Health Organization, "Report of the Expert Consultation on the Attention of Persons Exposed to Rabies Transmitted by Vampire Bats," Washington, D.C., April 2–5, 1991.

23. Robert V. Tauxe, interview with the author, Atlanta, Ga., March 8, 1996.

Chapter 3

1. Sue Mobley, in-person interview, Atlanta, March 1996, and telephone interviews, 1997.

2. D. Roberts, "Factors Contributing to Outbreaks of Food Poisoning in England and Wales 1970–1979," *Journal of Hygiene* [Cambridge] 89 (1982): 491–98.

3. D. C. Rodrique, R. V. Tauxe, and B. Rowe, "International Increase in *Salmonella Enteritidis*: A New Pandemic?" *Epidemiology of Infection* 105 (1990): 21–27.

4. S. F. Altekruse and D. L. Swerdlow, "The Changing Epidemiology of Foodborne Diseases," *American Journal of Medical Sciences* 31 (1996): 23–29.

5. E. C. D. Todd, "Worldwide Surveillance of Foodborne Disease: The Need to Improve," *Journal of Food Protection* 59, no. 1 (1996): 82–92.

6. Sometimes, wild animals play a role.

7. Joshua Lederberg, Robert E. Shope, and Stanley C. O. Oaks, Jr., ed., *Emerging Infections: Microbial Threats to Health in the United States* (Washington, D.C.: National Academy Press, 1992).

8. J. L. Smith and P. M. Fratamico, "Factors Involved in the Emergence and Persistence of Foodborne Diseases," *Journal of Food Protection* 58 (1995): 696–708.

9. World Health Organization, "Worldwide Spread of Infections with *Yersinia enterocolitica*," *WHO Chronicle* 30 (1976): 494–96.

10. World Health Organization, "Consultation on Selected Emerging Food-borne Diseases," Berlin, Germany, March 20–24, 1995.

11. Sarah Muirhead, "Study shows most grow/finish pigs receive antibiotics in their diets," *Feedstuffs*, March 18, 1996. Note: The article indicated that the findings were from the National Animal Health Monitoring System's Swine '95 study, conducted by the U.S. Department of Agriculture, Animal & Plant Health Inspection Service, Veterinary Services. The antibiotics were used both for disease prevention and growth enhancement. The most frequently used antibiotics were chlortetracycline, tylosin, and bacitracin. The study also indicated that the use of antibiotics since 1990 has increased by more than 6 percent for breeding females and 27.5 percent for boars.

12. R. V. Tauxe, J. Vandepitte, G. Wauters et al., "*Yersinia enterocolitica* Infections and Pork: The Missing Link," *The Lancet* (1987): 1129–32.

13. S. M. Ostroff, G. Kapperud, L. C. Hutwagner et al., "Sources of Sporadic *Yersinia enterocolitica* Infections in Norway: A Prospective Case-Control Study," *Epidemiological Infection* 112 (1994): 133–41.

14. L. A. Lee, A. R. Gerber, D. R. Lonsway et al., "*Yersinia enterocolitica* O:3 Infections in Infants and Children, Associated with the Household Preparation of Chitterlings," *New England Journal of Medicine* 322 (1990): 984–87.

15. Susan Schoenfeld, telephone interview, 1996.

16. M. A. Tipple, L. A. Bland, J. J. Murphy et al., "Sepsis Associated with Transfusion of Red Cells Contaminated with *Yersinia enterocolitica*," *Transfusion* 30 (1990): 207–13.

17. Salsa made with "heritage" varieties of older, more acidic tomatoes should be safe.

18. J. E. Seals, J. D. Snyder, T. A. Edell, "Restaurant-Associated Type A Botulism: Transmission by Potato Salad," *American Journal of Epidemiology* 113, no. 4 (1981): 436–44.

19. K. L. McDonnell, R. F. Sponger, C. L. Hathaway, N. T. Harriet, M. Cohen, "Type A Botulism from Sauteed Onions," *JAMA* 253, no. 9 (1985): 1275–78.

20. E. C. D. Todd, "Worldwide Surveillance of Foodborne Disease: The Need to Improve," *Journal of Food Protection* 59 (1996): 84.

21. M. E. St. Louis, H. S. Shaun, M. B. Peck et al., "Botulism from

Chopped Garlic: Delayed Recognition of a Major Outbreak," *Annals of Internal Medicine* 108 (1988): 363–68.

Chapter 4

1. D. M. Ackman, R. Bills, G. Birkhead, "Fiddlehead Fern Poisoning in New York State, 1994," Epidemic Intelligence Service, 44th Annual Conference, March 27–31, 1995. Centers for Disease Control and Prevention, Atlanta, Ga.
2. Irma S. Rombauer and Marion Rombauer Becker, *The Joy of Cooking* (Indianapolis: Bobbs-Merrill, 1964), p. 512.
3. Stephen Jay Gould, *Full House: The Spread of Excellence from Plato to Darwin* (New York: Crown, 1996), p. 230.
4. Nathan Shaffer, Robert B. Wainwright, John P. Middaugh, and Robert V. Tauxe, "Botulism Among Alaska Natives: The Role of Changing Food Preparation and Consumption Practices," *Western Journal of Medicine* 153 (October 1990): 390–93.
5. Ibid., p. 391.
6. Ibid., p. 392.
7. M. Wittner, J. W. Turner, G. Jacquette, L. R. Ash, M. P. Salgo, and H. B. Tanowitz, "Eustrongylidiasis—A Parasitic Infection Acquired by Eating Sushi," *New England Journal of Medicine* 320, no. 17 (1989): 1124–26.
8. P. M. Schantz, "The Dangers of Eating Raw Fish," *The New England Journal of Medicine* 320 (1989): 1143–45.
9. J. Rabinovitz, "A Campus Appetite for à la Carte," *New York Times*, April 30, 1996.
10. Ibid.
11. R. Voelker, "Foodborne Illness Problems More than Enteric," *JAMA* 271, no. 1 (1994): 8–11.
12. M. S. Deming, R. V. Tauxe, A. Blake et al., "*Campylobactei enteritis* at a University: Transmission from Eating Chicken and from Cats," *American Journal of Epidemiology* 126 (1987): 526–34.
13. Recent studies of chemicals with mild estrogenic effects when tested alone have demonstrated that when these chemicals are combined, the effects on yeast cells genetically altered to contain human receptor proteins that respond to estrogen are magnified 160–1,600 times. W. E. Leary, "Test Developed to Weigh Impact of Hormone-like Pollutants," *New York Times*, June 7, 1996.
14. E. M. R. Critchley, P. J. Hayes, and P. E. T. Issacs, "Outbreak of Botulism in North West England and Wales, June, 1989," *The Lancet* 2 (1991): 849–53.
15. C. W. Hedberg, K. L. MacDonald, and M. T. Osterholm, "Changing Epidemiology of Food-borne Disease: A Minnesota Perspective," *Clinical Infectious Diseases* 18 (1994): 671–82.

16. Later studies by the FDA would confirm the adverse effects on rats of both the suspect batch of L-tryptophan and the suspected agent. L. A. Love et al., "Pathological and Immunological Effects of Ingesting L-tryptophan and 1,1 ' -ethylidenebis (L-tryptophan) in Lewis Rats," *Journal of Clinical Investigation* 91 (March 1993): 804–11.

17. C. W. Hedburg, K. L. MacDonald and M. T. Osterholm, "Changing Epidemiology of Food-borne Disease: A Minnesota Perspective," *Clinical Infectious Diseases* 18 (1994): 673.

18. L. M. Beuchat, "Pathogenic Microorganisms Associated with Fresh Produce," *Journal of Food Protection* 59 (1996): 204–16.

Chapter 5

1. Kim Cook, telephone interview, 1996.

2. William H. McNeill, *Plagues and Peoples* (Garden City, N.Y.: Doubleday, 1976), p. 114.

3. Evidence for this is contained in material from the International Forum on Globalization; the publication *Extra;* and my personal experience on a weekly media talk show where I was directly pressured to stop bringing up the question of the lack of media coverage of the potential impact of the trade agreements.

4. C. W. Hedberg, K. L. MacDonald, and M. T. Osterholm, "Changing Epidemiology of Food-borne Disease: A Minnesota Perspective," *Clinical Infectious Diseases* 18 (1994): 671–82.

5. Robert V. Tauxe, in-person interview, 1996.

6. Hedberg, MacDonald, and Osterholm, "Changing Epidemiology of Food-borne Disease," p. 673.

7. Craig W. Hedberg, in-person interview, Minneapolis, Minnesota, 1996.

8. L. R. Beuchat, "Pathogenic Microorganism Associated with Fresh Produce," *Journal of Food Protection* 59 (1996): 204–16.

9. Marcella Ruland, telephone interviews, 1996.

10. Jean Taylor, telephone interviews, 1996, 1997.

11. J. L. Taylor, J. Tuttle, and T. Pramukul et al., "An Outbreak of Cholera in Maryland Associated with Imported Commercial Frozen Fresh Coconut Milk," *Journal of Infectious Diseases* 167 (1993): 1330–35.

12. Ibid., p. 1334.

Chapter 6

1. This story was told on a World Wide Web site called ProMed, where physicians and health officials exchange information about emerging diseases.

2. Somerset Waters, telephone interview, 1996.

3. P. M. Griffin, R. V. Tauxe, and J. E. Koehler, "EIS Cruise Ship Primer," rev. ed. (Atlanta: Foodborne and Diarrheal Diseases Branch, Centers for Disease Control and Prevention, April 1992).

4. A predictable curve develops in every large outbreak when the cases are put into graph form. The numbers begin small, gradually accumulate to a large number that makes a bump in the graph, then taper off. This is known as the epi-curve. It is helpful to understand the pattern because one can track how rapidly cases are beginning to accumulate and, by comparing that graph to the general epi-curve, judge where in the outbreak one is at the moment and where it might go.

5. Griffin, Tauxe, and Koehler, "EIS Cruise Ship Primer," p. 3.

6. The serotypes *hartford* and *Gaminara* are very rare. Fewer than 100 S. *hartford* isolates and 50 S. *Gaminara* isolates are reported to the CDC each year, and in the past four years New Jersey had seen only six and Florida seven.

7. Kim A. Cook, "Outbreak of *Salmonella hartford* among travels to Orlando, Florida, May 1995," EPI-AID 95-62 Trip Report, Memorandum. Foodborne and Diarrheal Disease Branch, Centers for Disease Control, October 1, 1995, pp. 3–4.

8. When a culture sample comes back "negative," many people assume that the test is conclusive. But many things can go wrong in a lab test. The wrong enrichment broth can be used, or the wrong culture medium; the time and temperature may be wrong; the samples may have been mishandled. In fact, all a negative test says is that the test was negative. The CDC, of course, has developed careful techniques to maximize the potential for finding what its investigators are looking for, but other labs may not be as well informed about the appropriate techniques or as careful.

9. Kim A. Cook, "Outbreak of *Salmonella*," p. 6.

10. Ibid.

11. Kim A. Cook, telephone interview, 1996.

12. Eighteen months after the "fresh-squeezed *Salmonella* outbreak," apple cider, a New England fall tradition, would be identified as the vehicle in an outbreak from the pathogen *E. coli* O157:H7. A few weeks later Washington State would experience its own *E. coli* O157:H7 outbreak from chilled, unpasteurized juice. Commercially prepared unpasteurized juices, even when refrigerated, were rapidly looking like a thing of the past.

13. R. V. Tauxe, M. P. Formey, L. Mascola, N. T. Hargrett-Bean, P. Blake, "Salmonellosis Outbreak on Transatlantic Flights: Foodborne Illness on Aircraft, 1947–1984," *American Journal of Epidemiology* 125, no. 1 (1987): 150–57.

14. Those words were written before the massive *Cryptosporidium* outbreak from a single water source in Milwaukee in 1993, when 400,000 people were affected and 104 died.

15. Tauxe et al., "Salmonellosis Outbreak," pp. 153–55.

16. Ibid., p. 155.

Chapter 7

1. "Surveillance for Food-borne Disease Outbreaks, United States, 1988–1992," CDC Surveillance Summaries, *Morbidity and Mortality Weekly Report* 45 (1996): SS–5.

2. Michael T. Osterholm, in-person interview, Minneapolis, Minnesota, July 1996.

3. Government Accounting Office, "Food Safety and Quality: Who Does What in the Federal Government," U.S. Government Accounting Office, RCED–91–91A.

4. Government Accounting Office, "Food Safety: New Initiatives Would Fundamentally Alter the Existing System," U.S. Government Accounting Office, RCED 96–81.

5. Craig W. Hedberg, J. Korlath, J.-Y. D'Aoust et al., "A Multistate Outbreak of *Salmonella javiana* and *Salmonella oranienburg* Infections Due to Consumption of Contaminated Cheese," *JAMA* 268, no. 22 (1992): 3203–7.

6. C. W. Hedberg, K. L. MacDonald, and M. T. Osterholm, "Changing Epidemiology of Food-borne Disease: A Minnesota Perspective," *Clinical Infectious Diseases* 18 (1994): 671–82.

7. M. E. St. Louis, D. L. Morse, M. E. Potter et al., "The Emergence of Grade A Eggs as a Major Source of *Salmonella enteritidis* Infections," *JAMA* 259, no. 14 (1988): 2103–7.

8. Proceedings, National Conference on *Salmonella*, HEW, Public Health Services, March 11–13, 1964, p. 29.

9. Ibid., p. 43.

10. Ibid., p. 71.

11. Ibid., p. 76.

12. R. A. Feldman and M. A. Pollard," Uses of Epidemiology in the Control of Salmonellosis in Humans," *Proceedings: National Salmonellosis Seminar* (Washington, D.C.: USDA, January 10–11, 1978), p. 2.

13. The true implications of the change would not become obvious until well into the 1980s when "mad cow disease" appeared.

14. Confidential source, correspondence with the author, 1996.

15. P. McDonough, Cornell University College of Veterinary Medicine Diagnostic Laboratory, personal communication, September 3, 1996.

16. E. T. Mallinson, "Priority Considerations for Future *Salmonella* Central Proceedings, National *Salmonella* Seminar," U.S. Departments of Agriculture, January 10–12, 1976.

17. Confidential source, telephone interview, 1996.

18. Jeffrey A. Fisher, *The Plague Makers* (New York: Simon and Schuster, 1994), p. 90.

19. Farmers could use them for growth promotion only with a prescription from a veterinarian, and then only those that were not used to treat hu-

mans. Unfortunately, it is widely acknowledged that this safeguard is openly abused: Many veterinarians freely hand out prescriptions.

20. O. Schell, *Modern Meat* (New York: Random House, 1978), p. 107.

21. A. Rampling, R. Upson, and D. F. J. Brown, "Nitrofurantoin Resistance in Isolates of *Salmonella enteritidis* Phage Type 4 from Poultry and Humans," *Journal of Antimicrobial Chemotherapy* 25 (1990): 285–90.

22. S. D. Holmberg, M. T. Osterholm, K. A. Senger et al., "Drug-Resistant *Salmonella* from Animals Fed Antimicrobials," *New England Journal of Medicine* 311, no. 10 (1984): 617–22.

23. Ibid., p. 21

24. J. de Louvois, letter, "*Salmonella* Contamination of Eggs," *The Lancet* 342 (1993): 342–66.

25. W. Grant-Evans, "Farmers threaten to Sue Edwina Currie after she wrongly claims that millions of eggs are contaminated with *Salmonella*," *Today*, December 5, 1988.

26. Anonymous, "Currie Statement on Egg Danger Sharply Criticized, *Salmonella*, Parliament," *The Times*, December 6, 1988.

27. J. Erlichman, "Poison Cases Drop After Eggs Alarm: Fall by Third in Salmonella Rate Follows Curie Remark," *The Guardian*, December 20, 1988.

28. D. C. Rodrigue, R. V. Tavxe, and B. Rowe, "International increase of *Salmonella enteritidis:* A new pandemic?" *Epidemiology of Infection* 105 (1990): 21–27

29. M. E. St. Louis, D. L. Morse, M. Potter et al., "The Emergence of Grade A Eggs as a Major Source of *Salmonella enteritidis* infections," *JAMA* 25 (1988): 2105.

30. Ibid.

31. Kathleen Gensheimer, telephone interview, 1996.

32. T. J. Humphrey, M. Greenwood, R. J. Gilbert, B. Rowe, and P. Chapman, "The Survival of *Salmonellas* in Shell of Eggs Cooked Under Simulated Domestic Conditions," *Epidemiology Information* 103 (1989): 35–45.

33. E. D. Ebel, M. J. David, and J. Mason, "Occurrence of *Salmonella enteritis* in the U.S. Commercial Egg Industry: Report on a National Spent Hen Survey," *Avian Diseases* 36 (1992): 646–54.

34. U.S. Department of Agriculture, "Chickens Affected by *Salmonella enteritidis;* Final Rule" *Federal Register* 56 (1991): 3737.

35. *Ibid.*, p. 3730.

36. Ibid., p. 3733.

37. Michael Taylor, speech to the U.S. Animal Health Association, Reno, Nevada, November 9, 1996.

38. Centers for Disease Control and Prevention, "Outbreak of *Salmonella enteritidis* Associated with Homemade Ice Cream–Florida, 1993," *Morbidity and Mortality Weekly Report* 43, no. 36 (1994): 669–71.

39. Ibid., 669.

40. Craig W. Hedberg, in-person interview, Minneapolis, 1996.

41. T. W. Hennessy, C. W. Hedberg, L. Slutsker et al., "A National Outbreak of *Salmonella enteritidis* Infections from Ice Cream," *New England Journal of Medicine* 334, no. 20 (May 1996): 1281–86.

Chapter 8

1. U.S. Department of Agriculture, "Pathogen Reduction; Hazard Analysis and Critical Control Point (HACCP) Systems; Final Rule," *Federal Register* 61 (1996): 38806–989.

2. Ibid., p. 38867.

3. Margaret Visser, *Much Depends on Dinner* (New York: Grove Press, 1986), pp. 140–41.

4. Ibid., pp. 112–13.

5. M. J. Blaser, "How Safe Is Our Food?" *New England Journal of Medicine* 334 (1996): 1324.

6. Martin Blaser, in-person interview, New Hampshire, 1996.

7. I have noticed that even in the scientific world it is small things such as what something is named, or some other seemingly insignificant factor, often determines whether a disease, pathogen, or even a solution to a problem attracts attention, not its relative scientific importance.

8. Martin Skirrow, telephone interview, 1996.

9. Ibid.

10. R. Brower, M. J. A. Mertens, T. H. Siem, and J. Katchaki, "An Explosive Outbreak of *Campylobacter enteritis* in Soldiers," *Antonie Van Leeuwenhoek Journal of Microbiology* 45: 517–19.

11. J. Oosterrom, H. J. Beckers, L. M. van Noorle-Jansen, and M. van Schothorst, "An Outbreak of *Campylobacter* Infection in a Barrack, Probably Caused by Raw Hamburger," *Nederlands Tijdschrift voor Geneeskund* 124, no. 39 (1980): 1631–34.

12. G. R. Istre, M. J. Blaser, Pamela Shillam, and Richard S. Hopkins, "*Campylobacter enteritis* Associated with Undercooked Barbecued Chicken," *American Journal of Public Health* 74, no. 11 (1984): 1265.

13. M. J. Knill, W. G. Suckling, and A. D. Pearson, *Campylobacter: Epidemiology, Pathogenesis, and Biochemistry* (Lancaster, Eng.: MTP Press, 1982), pp. 281–84.

14. M. J. Blaser and L. B. Reller, "*Campylobacter enteritis*," *New England Journal of Medecine* 305 (1981): 1444–52.

15. J. Kaldor and B. R. Speed, "Guillain-Barré Syndrome and *Campylobacter jejuni:* A Serological Study," *British Medical Journal* 288 (1984): 1867–70.

16. A. Kuroki, T. Haruta et al., "Guillain-Barré Syndrome Associated with *Campylobacter* Infection," in *Microbial Ecology in Health and Disease*, vol. 4 (New York: Wiley, 1991).

17. Martin J. Blaser, telephone conversation, 1996.

18. Richard Behar and Michael Kramer, "Something Smells Fowl," *Time*, October 17, 1994, pp. 42–45.

19. As mentioned in chapter 7, it may be that less *Salmonella* is needed to cause infection than was thought. The amount causing illness in a recent outbreak was far less than human experiments had previously indicated. That such a basic matter is still not widely agreed upon indicates how much research remains to be done in foodborne disease.

20. W. O. James et al., "Effects of Chlorination of Chill Water on the Bacteriologic Profile of Raw Chicken Carcasses and Giblets," *Journal of the American Veterinary Medicine Association 60–63* 200, no. 1 (1992): 60–63.

21. Behar and Kramer, "Something Smells Fowl," p. 43.

22. James et al., "Effects of Chlorination of Chill Water," p. 62.

23. Behar and Kramer, "Something Smells Fowl," p. 44.

24. James et al., "Effects of Chlorination on Chill Water," p. 62.

25. Behar and Kramer, "Something Smells Fowl," p. 44.

26. N. J. Stern and R. B. Russel, "Control of *Campylobacter jejuni* in Poultry," in *Report on a WHO Consultation on the Epidemiology and Control of Campylobacteriosis*, Bilthoven, the Netherlands, April 25–27, 1994.

27. Frank Jones, telephone interviews, 1996, 1997.

28. Some experts say they can't find *Campylobacter* in litter in any case, but the Bilthaven conference report said that could be explained by the bacteria forming "viable but non-culturable" (VNC) cells, a survival strategy, or because present isolation methods are not good enough. (Because there are many different *Campylobacters*, each with different characteristics, researchers find that no one single culture medium works for all of them, an unfortunate fact that makes detection that much more complicated.) The bacteria could also be well beneath the dry top layer, down in the moist under-layers of litter, and be exposed when the chickens do what chickens instinctively do—scratch away, looking for food.

29. Eva Berndtson, Department of Food Hygiene, Faculty of Veterinary Medicine, Uppsala, Sweden, personal communication, November 13, 1996.

30. E. Berndtson and A. Engvall, "Control of *Campylobacter* in Swedish Broiler Flocks," in *Report on a Who Consultation on Epidemiology and Control of Campylobacteriosis*, Bilthoven, the Netherlands, April 25–27, 1994.

31. H. P. Endtz, G. J. Ruijs, B. van Klingerent et al., "Quinolone Resistance in *Campylobacter* Isolated from Man and Poultry Following the Introduction of Fluoroquinolones in Veterinary Medicine," *Journal of Antimicrobial Chemotherapy* 27 (1991): 199–208.

32. S. Hamlin, "Free Range? Natural? Sorting out Labels," *New York Times*, November 13, 1996.

33. Martin Blaser, personal communication.

34. Anita Rampling, telephone interview, 1996.

35. Commission of the European Communities, "Proposal for a Council

Regulation (EC) Supplementing Regulation (EEC) No. 2092–91 on Organic Production of Agricultural Products and Indications Referring Thereto on Agricultural Products and Foodstuffs to Include Livestock Production," July 26, 1996, COM(96)366 final. (Note: November 28, 1996. These proposed regulations also require that poultry have access to the open. I have seen no studies comparing the *Campylobacter* infection rate of these birds to those kept in conditions of biosecurity.)

36. Mimi Sheraton, *From My Mother's Kitchen* (New York: Harper & Row, 1977), p. 54.

37. Salvadore Iacono, telephone interview, 1996.

38. U.S. Department of Agriculture statement by Secretary Dan Glickman, "Announcing Resumption of Poultry Trade with Russia," Release No. 011446.96, March 26, 1996.

39. M. E. Potter, A. F. Kaufmann, P. A. Blake, and R. A. Feldman, "Unpasteurized Milk: The Hazards of a Health Fetish," *JAMA* 252, no. 15 (1984): 2048–52.

40. M. J. Blaser, Elizabeth Sazie, L. P. Williams, "The Influence of Immunity on Raw Milk–Associated *Campylobacter* Infection," *JAMA* 257, no. 2 (January 2, 1987): 43–46.

41. D. N. Taylor, M. T. McDermott, J. R. Little, J. G. Wells, and M. J. Blaser, "*Campylobacter enteritis* from Untreated Water in the Rocky Mountains," *Annals of Internal Medicine* 99, no. 1 (1983): 38–40.

42. Dot Pridgeon, telephone interview, 1996.

43. Charmagne Heltin, "Poultry Problems Picking Up," *USA TODAY*, August 12, 1994.

44. P. M. F. J. Koenraad, W. C. Hazeleger, T. van der Laan, R. R. Beuner, and F. M. Romboutz, "Survival of *Campylobacter* in Sewage Plants," in *Report on a Who Consultation on Epidemiology and Control of Campylobacteriosis*, Bilthoven, the Netherlands, April 25–27, 1994.

45. Figueredo Nunes, "The Issue of Extensive Agriculture as an Alternative to More Interesting Productive Systems," Comments on "Bovine Spongiform Encephalopathy (BSE) Crisis and Its Wide-ranging Consequences for the EU," prepared by the Economic and Social Committee of the EU.

46. E. B. White, *Letters of E. B. White* (New York: Perennial Library, 1989), pp. 185–86.

47. Task Force Report, "*Integrated Animal Waste Management*," Council for Agricultural Science and Technology 128 (1996): 31–33.

Chapter 9

1. Mary Heersink, in-person interview, Dothan, Alabama, and Washington, D.C.; 1993, 1994, 1995, 1996.

2. Jeffrey Tennyson, *Hamburger Heaven: The Illustrated History of the*

Hamburger (New York: Hyperion, 1993), pp. 20–24. (Note: Much of what I know about the history of the hamburger can be credited to Tennyson.)

3. Ibid., pp. 58–85.
4. Ibid., p. 80.
5. Richard Hebert, telephone interview, 1993.
6. Lee Riley, telephone interview, 1993. (Note: See also L. W. Riley, R. S. Remis, S. D. Helgerson et al., "Hemorrhagic colitis associated with a rare *Escherichia coli* serotype," *New England Journal of Medicine* 308 (1983): 681–85.)
7. Joy Wells, in-person interview, Atlanta, 1993.
8. Peggyann Hutchinson, telephone interview, 1993.
9. Lynn Turner, telephone interview, 1993.
10. Robert Remis, telephone interview, 1993.
11. Riley et al., p. 682.
12. Steve Steinberg, "Undercooked Burger Linked to Disease," *Miami Herald*, October 7, 1982.
13. Ibid.
14. Nathaniel Sheppard, Jr., "McDonald's Linked to 47 Stomach Disorder Cases," *New York Times*, October 8, 1982.
15. Edward Edelson, "A Big Mac Attack?" *New York Daily News*. October 8, 1982.
16. Edward Edelson, "McDonald's Gets Break on Ailment," *New York Daily News*, October, 9, 1982.
17. Allen Hallmark, "Illness Victims ate at McDonald's," *Medford Mail Tribune*, October 9, 1982.
18. V. C. Gasser, E. Gautier, A. Steck, R. E. Siebnmann, R. Oechslin, "Hämolytischurämmische Syndrome: Bilaterale Nierenrindennekrosen bei akuten erworbenen hamolytischen Anamien," *Schweiz Med Wochenschr* 65 (1955): 905–9.
19. Brian T. Steele, telephone interview, 1993.
20. Mohamed A. Karmali, telephone interview, 1993.
21. C. A. Ryan, R. V. Tauxe, G. W. Hosek et al., "*Escherichia coli* O157:H7 Diarrhea in an Nursing Home: Clinical, Epidemiological, and Pathological Findings," *Journal of Infectious Diseases* 154 (1986): 631–38.
22. Caroline A. Ryan, telephone interview, 1994.
23. Robert V.Tauxe, in-person interview, Atlanta, 1993.
24. A. O. Carter, A. A. Borczyk, J. A. K. Carlson et al., "A Severe Ooutbreak of *Escherichia coli* O157:H7-associated Hemorrhagic Colitis in a Nursing Home," *New England Journal of Medicine* 317 (1987): 1496–1500.
25. S. M. Ostroft, P. M. Griffin, R. V. Taue et al., "A Statewide Outbreak of Esherichia coli O157:H7 Infections in Washington State," *American Journal of Epidemiology* 132 (1990): 239–47.
26. Patricia Griffin, telephone interview, 1997.
27. In a letter from Kenneth C. Clayton, Acting Assistant Secretary for

Marketing and Inspection Services, Department of Agriculture, to MS. Kathi Smith Allen on April 21, 1993, Clayton wrote, "CDC first reported cases of hemolytic uremic syndrome, a urinary tract infection caused by the bacteria, in 1982."

28. Anonymous, "Need for USDA Beef Patty Rule Questioned; Additional Research Asked," *Food Chemical News*, January 20, 1989.

29. Robert Galler, prepared and delivered statement, Symposium on Meat Inspection, Washington, D.C., September 1993.

30. Arthur O'Connell, prepared and delivered statement, Symposium on Meat Inspection, Washington, D.C., September, 1993.

31. Wendy Feik Pinkerton, C. J. Reynold, "Washington State *E. coli* O157: H7 Outbreak," Memorandum to State Beef Council Executives, National Live Stock and Meat Board, January 19, 1993.

32. Steve Bjerklie, telephone interviews, 1996.

33. Edward Penble, "USDA Slow to Respond to Food Poisoning," *Seattle Post-Intelligencer*, January 21, 1993.

34. Ibid.

35. Dorothy Dolan, prepared and delivered statement, Symposium on Meat Inspection, Washington, D.C., September 21, 1993.

36. Diane Nole, prepared and delivered statement, Symposium on Meat Inspection, Washington, D.C., September 21, 1993.

37. Robert V. Tauxe, in-person interview, Atlanta, 1993.

38. D. Maxson, employee, Foodmaker (parent company of Jack-in-the-Box), "Patty Temperature Tracking Form: Burger Patty Observations," Store 7211, February 10, 1993.

39. Government Accountability Project, "Summary of 1990 Whistleblowing Disclosures on USDA's Proposed Streamlined Inspection System—Cattle," Governmental Accountability Project, Washington, D.C.

40. James L. Marsden, Issues Briefing "Issue: *E. coli* O157:H7," American Meat Institute, July 1992. Marsden wrote: "Dairy cattle are a major reservoir of *E. coli* O157:H7. The organism resides in their intestinal tracts, and they shed it in feces. Slaughter and milking procedures can contaminate meat and milk with *E. coli* bacteria, but because milk is generally pasteurized and meat typically is cooked prior to ingestion, the presence of *E. coli* bacteria in general is not considered a public health hazard . . . Foodborne illness cases associated with *E. coli* O157:H7 have been linked to undercooked ground beef and, to a lesser extent, improperly pasteurized or raw milk. Supermarket surveys of fresh meats and poultry revealed *E. coli* O157:H7 in 3.5 percent of ground beef, 1.5 percent of pork, 1.5 percent of poultry and 2.0 percent of lamb."

41. Charles Bartleson, personal notes, January 26, 1993.

42. H. Russell Cross, Informational Memorandum for the Chief of Staff, *E. coli* O157:H7 Outbreak," January 29, 1993.

43. Anonymous Report on "FSIS Meeting on EP case no. 7078, Edid. case no. 870290," USDA Document, August 28, 1987.

44. Meg Walker, "Meat Inspectors Fear Retaliation," *Federal Times*, March 29, 1993.

45. Ibid.

46. Ibid.

47. USDA, "Poultry Products Produced by Mechanical Deboning and Products in which Poultry Products Are Used," *Federal Register*, March 33, 1994.

48. *Los Angeles Times*, July 29, 1994.

49. Sharon LaFraniere and Gay Gugliotta, "The Promise and Puzzle of Mike Espy," *Washington Post*, September 28, 1994.

50. A copy of the anonymous letter to the president was obtained from a reliable source; it was also referenced in a *Washington Post* article, which bolsters its authenticity.

51. D. B. Laster, Central Director, Clay Center, Nebraska, "Information on Rapid Bacterial Test for Meat and Poultry," Memorandum to R. D. Plowman, Acting Assistant Secretary, USDA, and P. Jensen, Acting Assistant, USDA, August 11, 1994.

52. Michael R. Taylor, telephone interview, 1994.

53. Michael Doyle, telephone interview, 1994.

54. P. M. Griffin, B. P. Bell, P. R. Cieslak et al., "Large Outbreak of *Escherichia coli* O157:H7 Infections in the Western United States: The Big Picture," *Elsevier Science B.V.*, 1994, pp. 7–12.

55. Ibid., p. 11.

56. M. A. Rasmussen, W. C. Cray, T. A. Casey, and S. C. Whipp, "Rumen Contents as a Reservoir of Enterohemorrhagic *Escherichia coli*," *FEMS Microbiology Letters* 114 (1993): 79–84.

57. G. L. Armstrong, J. Hollingsworth, and G. Morris, Jr., "Emerging Foodborne Pathogens: *Escherichia coli* O157:H7 as a Model of Entry of a New Pathogen in the Food Supply of the Developed World," *Epidemiologic Review* 19 (1996): 1–23.

58. Ibid., pp. 15–16.

59. M. Neill, "*E. coli* O157:H7 Time Capsule: What Do We Know and When Did We Know It?" Lecture presented at the symposium on Foodborne Microbial Pathogens. Annual Meeting of the International Association of Milk, Food and Environmental Sanitarians, Atlanta, Georgia, August 2–4, 1993.

Chapter 10

1. James E. Lovelock, *Gaia: A New Look at Life on Earth* (Oxford, England: Oxford University Press, 1987).

2. Richard A. Knox, "18 Guests at Wedding Felled by Parasite," *Boston Globe*, June 22, 1996.

3. P. S. Mallard, K. F. Gensheimer et al., "An Outbreak of Cryp-

tosporidiosis from Fresh-Pressed Apple Cider," *JAMA* 272, no. 20 (1994): 1592–96.

4. American Meat Institute, *Microbial Control During Production of Ready-to-Eat Meat Products: Controlling the Incidence of* Listeria Monocytogenes (Washington, D.C.: American Meat Institute, 1987), pp. 1–13.

5. *Janet Raloff*, "Sponges and Sinks and Rags, Oh My!" *Science News* (September 1996): 172–73.

6. K. A. Glass and M. P. Doyle, "Fate of *Listeria Monocytogenes* in Processed Meat Products During Refrigerated Storage," *Applied Environmental Microbiology* 55 (1989): 1565–69.

7. B. G. Bellin, C. V. Broome, W. F. Bibb et al., "The Epidemiology of Listeriosis in the United States—1986," *American Journal of Epidemiology* 33, no. 4 (1991): 392–401.

8. J. R. Giraud, F. Denis, F. Gargot et al., "La listeriose: Incidence dans les interruptions spontanées de la grosse," *Nouvelle Press Médecin* 2 (1973): 215–18.

9. British Advisory Committee on the Microbiological Safety of Food, Working Group on Vacuum Packing and Associated Processes (London, England: HMSO, 1992).

10. Julia Langdon, "A Fatal Lack on Information," *London Telegraph*, July 12, 1995.

11. Ibid.

12. Helge M. Karcher, "Germany Fears More Deaths from *E. coli* Outbreak," *British Medical Journal* 312 (1996): 1501.

13. The other pathogens found were 13 *Salmonella*, 7 *Campylobacter*, and 4 *Shigella*.

Chapter 11

1. Anthony Bowen, telephone interview, 1996.

2. Colin Whitaker, telephone interview, 1996.

3. G. A. H. Wells, "A Novel Progressive Spongiform Encephalopathy in Cattle," *Veterinary Record*, October 31, 1987.

4. Gerald A. H. Wells, telephone interview, 1996.

5. W. J. Hadlow, "An Overview of Scrapie in the United States," *Journal of the American Veterinary Medical Association* 198, no. 10 (1991): 1676–77; J. Morse, Agricultural Research Service, Ames, Iowa, telephone interview with the author, April 8, 1996; and L. Detweiler, personal communication, March 28, 1996.

6. John W. Wilesmith, Gerald A. H. Wells, M. P. Cranwell, and J. B. M. Ryan, "Bovine Spongiform Encephalopathy: Epidemiological Studies," *Veterinary Record*, December 17, 1988, pp. 638–44.

7. J. Peder Zane, "It Ain't Just for Meat; It's for Lotion," *New York Times*, May 5, 1996.

8. The *Weekly Standard* in April 1996 ran a cartoon showing a field full of cows. One is saying, "They feed us sheep offal and then they kill and eat us and they call us mad."

9. "The British Disease," *The Economist*, March 30, 1996.

10. J. W. Wilesmith et al., "Bovine Spongiform," pp. 638–44.

11. Report of a WHO Consultation on Clinical and Neuropathological Characteristics of the New Variant and Other Human and Animal Transmissible Spongiform Encephalopathies," World Health Organization (Geneva, Switzerland, May 14–16, 1996): 2.

12. Richard Marsh, telephone interview, 1996.

13. R. F. Marsh, R. A. Bessen, S. Lehmann, G. R. Hartsough, *Journal of Genetic Virology* 72 (1991): 589–94.

14. Jessica Tuttle, EIS Officer, Enteric Diseases Branch, Memorandum, "EPI-AID Trip Report: Multistate Outbreak of *Escherichia coli* O157:H7 infections—meat Traceback," May 21, 1993. In this report, the investigation of the slaughterhouses revealed that animals identified as "downer cattle" had gone into the meat provided to suppliers, which then provided meat to Vons Meat Service Center, which, in turn, served as the western regional patty maker for Foodmaker, Incorporated, the parent company and supplier of Jack-in-the-Box restaurants. In one case, downer cattle represented 8 percent of the animals killed in an single day At the same establishment, dairy cows represented 89 percent of the total number of animals slaughtered.

15. Joel McNair, "Experts Argue over BSE Measurers," *Agri-View*, September 16, 1993.

16. Ibid.

17. APHIS, USDA, "Importation of Animal Products and By-products from Countries where BSE Exists," *Federal Register* 56, no. 235 (1991): 63835–70.

18. Ibid., p. 63837.

19. U.S. Department of Agriculture, "APHIS Questions and Answers Regarding BSE," March 25, 1996.

20. Paul Brown, "The Jury Is Still Out," *British Medical Journal* 311 (1995): 1416–18.

21. R. G. Will, J. W. Ironside, M. Zeidler et al., "A new variant of Creutzfeldt-Jakob disease in the UK," *The Lancet* 347 (1996): 921–25.

22. Richard Marsh, telephone interviews, 1996.

23. "CVM Recommends Restricting Use of Rendered Adult Sheep," *Food Chemical News* (March 15, 1993): 4.

24. Press Release, "International Center for Technology Assessment (ICTA) Files Legal Action to Prevent 'Mad Cow Disease' Epidemic in U.S.," *Center for Technology Assessment*, March 27, 1996.

25. Issue Brief, "Bovine Spongiform Encephalopathy (BSE) 1996 Update," American Veterinary Medical Association, March 27, 1996, pp. 1–2.

26. Bill Miller, "BSE-human link alleged," *Beef Today*, April 1996.

27. Irene Malbin, News Release, "Cosmetic Ingredients Not from BSE Countries," The Cosmetic, Toiletry, and Fragrance Association, April 3, 1996.

28. P. Brown, "Bovine Spongiform Encephalopathy and Creutzfeldt-Jakob Disease," *British Medical Journal* 312 (1996): 790–91.

29. Food and Drug Administration, "Substances Prohibited from Use in Animal Food or Feed; Animal Proteins Prohibited in Ruminant Feed; Proposed Rule," *Federal Register* 62, no. 2 (January 3, 1997): 552–83.

30. Commissary spokesperson, telephone interview, 1996.

31. Alan McGregor, "BSE Fears Stir the Swiss," *The Lancet* 347 (1996): 1035.

32. Much of the feed was chicken and pig feed, to which the British ban against including material from infected cattle and specified offal did not apply at the time. Now the ban extends to all animal feed.

33. Declan Butler, "Statistics sought: BSE now Europe-Wide," *Nature* 382 (1996): 4.

34. Oliver Klaffke, "Swiss Cull to Meet Fears of BSE," *Nature* 383 (1996): 289.

35. J. Collinge, C. L. Sidle, J. Meads, J. Ironside, and A. F. Hill, "Molecular Analysis of Prison Strain Variation and the Etiology of 'New Variant' CJD," *Nature* 383 (October 1996): 685–90.

36. This suggestion is my own, not that of any researchers.

37. Frank Bastian, telephone interviews, 1997.

38. Joel McNair, "BSE Voluntary Ban Having Little Effect on Producers," *Agri-View*, May 8, 1996.

39. Don Franco, telephone interviews, 1996, 1997.

40. Declan Butler, "BSE Surveillance and Saftey Measurers Still Found Wanting," *Nature* 336 (1997): 101.

Chapter 12

1. Studies of what actually goes on in animal-rearing facilities are dependent on the cooperation of private companies, which guard their methods proprietarily.

2. Patricia M. Griffin et al., "Large Outbreak of *Escherichia coli* O157:H7 Infections in the Western United States: The Big Picture," in *Recent Advances in Berocytotoxin-Producing* Escherichia coli *Infections* (New York: Elsevier, 1994), 12.

3. David Bederman, telephone interview, 1994.

4. Phyllis Marberger, telephone interview, 1994.

5. Peter Stemberg, telephone interview, 1994.

6. David Waltner-Toews, *Food, Sex and Salmonella: The Risks of Environmental Intimacy* (Toronto: NC Press Ltd., 1992), pp. 87–88.

7. Jane Kirtley, telephone interview, 1994.

8. Lucy Riley, telephone interview, 1994.

9. Paul MacMasters, telephone interview, 1994.
10. Michael Osterholm, in-person interview, July, 1996.
11. "Much Foodborne Disease Goes Unreported," *Science News* 150, no. 173 (September 14, 1996).
12. This question of whether new managed care systems do less—or perhaps even more—culturing, as some claim, will be the subject of future CDC studies, says Patricia Griffin.
13. S. Notermans, "Epidemiology and Surveillance of *Campylobacter* Infections," in *Report on a WHO Consultation on Epidemiology and Control of Campylobacteriosis*, Bilthoven, the Netherlands, April 25–27, 1994.
14. R. Remington et al., *The Future of Public Health* (Washington, D.C.: National Academy Press, 1988).
15. Robert Wood Johnson Foundation, 1987; *Sourcebook for Health and Insurance Data* (Health Insurance Association of America, 1995).
16. Kathleen Gersheimer, telephone and in-person interviews, 1993, 1994, 1995, 1996, 1997.
17. A new test for appendicitis, soon to be available, could help screen out those colitis cases caused by bacterial infection, thus avoiding unnecessary surgery.
18. R. E. Besser, D. R. Feikin, J. E. Eberhart-Phillips et al., "Diagnosis and Treatment of Cholera in the United States: Are We Prepared?" *JAMA* 272, no. 15 (October 19, 1994): 1203–5.
19. World Health Organization, "Program for Control of Diarrheal Diseases," Ninth Program Report, 1992–93.
20. Once the disease is in humans, it can be transmitted, of course, from person to person or by water contaminated by human waste.
21. Steve Bjerklie, "Know Thy Enemy," *Meat and Poultry* (February 1995): 6.
22. Steve Bjerklie, telephone interview, 1996.
23. Steve Bjerklie, "Who Really Has the World's Safest Meat Supply?" *Meat and Poultry* (August 1995): 50–54.
24. Ibid., p. 54.
25. The USDA did refuse to guarantee to industry that the papers it accumulates in an investigation of a meat plant would not be available under FOIA, so the public will have access to some reports.

Chapter 13

1. Lynn Margulis and Dorian Sagan, *Microcosmos* (New York: Simon & Schuster, 1986), p. 95.
2. David Waltner-Toews, *Food, Sex, and Salmonella,* p. 129.
3. Ibid., p. 132.
4. HRH, The Prince of Wales, "The Lady Eve Balfour Memorial Lecture," London, September 19, 1996.

5. Social and Economic Committee, "The Bovine Spongiform Encephalopathy Crisis and Its Wide-ranging Consequences for the EU," July 1996, European Union.

6. Waltner-Toews, *Food, Sex, and Salmonella*, p. 161.

7. Ibid., p. 161.

8. John Gray, "Danger Lurks in the Drive for Food Profits," *Guardian Weekly*, September 15, 1996.

9. Waltner-Toews, *Food, Sex, and Salmonella*, p. 90.

10. An outbreak of *E. coli* O157:H7 was reported on vegetables that had been grown with fresh manure, but organic farming associations point out that their standards prohibit the use of fresh manure. Manure is increasingly being recycled in intensive farming operations according to "Integrated Animal Waste Management," a report by the Council for Agricultural Science and Technology (November 1996).

11. Wendell Berry, *What Are People For* (New York: North Point Press, Farrar, Straus and Giroux, 1990), pp. 207–8.

Epilogue

1. Kendal Anderson and family, in-person interviews, Waitsfield, Vermont, 1996.

2. J. Leroy Hadden, in-person interview, Waitsfield, Vermont, 1996.

3. Quoted in anonymous, "Calf Disease," *Burlington Free Press*, April 13, 1996.

4. Recollection of Kendal Anderson. Sam Hutchins did not respond to requests for an interview, but the responses to my written questions of Justin Johnson, spokesperson for the Vermont Department of Agriculture, confirm Hutchin's impressions of how the calves appeared.

5. David Webster, telephone interview, 1996.

6. David Woodward, telephone interview, 1996.

Appendix

1. Foods that have been improperly prepared and contaminated, then left at room temperature where bacteria could grow, may contain toxins of bacteria, which take longer to kill. To destroy *C. botulinum* toxin requires boiling for ten minutes.

Glossary

bacillus – straight and rod-shaped bacteria.

bacteria – small, single-celled organisms that live in water, soil, organic matter or the bodies of animals and plants. They may be round, rod-shaped, or spiral creatures, often with the ability to move around by using flagella. Many are benign to humans, some are helpful, and some cause disease.

carrier – a person who doesn't show clinical signs of disease but can nevertheless infect others.

conjugation – a cellular contact between bacteria during which DNA is exchanged.

case-control study – done after an outbreak, it compares activities and characteristics of the ills with the wells (controls) who have been matched in certain ways. Ill individuals who attended a convention might be matched with well individuals attending the same convention, then thoroughly interviewed to see what the ill individuals did that the well individuals did not. The relative risk of one activity over another is calculated from these interviews using statistical methods.

DNA – an abbreviation for deoxyribonucleic acid. The nucleic acids are located in the nuclei of cells and the characteristic double helix form of DNA contains the molecular basis of heredity. Determining the sequence of the paired material can be used to identify individuals (microbial as well as human).

emerging pathogens – a new agent of disease, a new recognition of an agent that has previously gone undetected, or an old agent that is newly causing an increasing problem.

endemic – a constant, ongoing presence of a disease in a community.

epi-curve – the shape taken on a graph by the numbers of affected in an outbreak. The line classically begins to curve slowly, then mounts rather quickly, descends in a like manner, creating a bell shape, and then trails off. Knowing what the classic curve looks like enables epidemiologists to look at how cases are increasing in an outbreak and plot where they might be in an epidemic. The increasing numbers of cases, more closely spaced together, of V-CJD in 1996 had begun to worry epidemiologists because they looked like the beginning of an epi-curve—meaning that an epidemic might be in the process of developing.

epidemic – an unusual number of cases of disease within a given time in a specific geographical area.

epidemiology – the branch of medicine that deals with the incidence, distribution, and control of disease in a population.

epidemiologist – individuals trained in detecting and controlling disease in a population. They may or may not be medical doctors.

etiology – the cause of a disease.

foodborne disease – an illness caused by some disease-causing organism that is transmitted by food. (Some waterborne diseases may be foodborne as well.)

fecal material – waste (excrement) from the bowels.

flagella – threadlike appendages that provide one-celled organisms with the ability to move.

hemolytic uremic syndrome (HUS) – a disease that destroys red blood cells, as well as cells that cause the blood to clot, and the lining of the blood vessel walls. Kidney failure is also common.

index case – the first case of an illness in a community or in a family. It may serve as a source of infection to others.

infection – the microorganism has been deposited, has colonized, or has multiplied within the host, a process that is usually followed by an immune reaction. Infection does not necessarily mean that the patient (human or animal) shows clinical signs of the disease.

morbidity rate – number of illnesses in a given time or place.

mortality rate – number of deaths in a given time or place.

nosocomial illness – an infection that originates in the hospital environment that was not present in the patient at the time of admission.

pandemic – disease occurring over a wide range and among a significant number of the population.

pathogen – something that causes disease.

phages – (bacteriophage) a virus that infects a bacterium.

plasmids – DNA outside the chromosome that can reproduce itself.

reservoir – some part of the environment, such as an animal, soil, or a person, in which an infectious agent exists and multiplies. The reservoir may be a means of infecting others.

surveillance – in medical terms, this is the systemic collection of information in the form of data that relates to the occurrence of a particular disease.

transduction – the transfer of genetic material from one bacterial cell to another, using a bacteriophage as a carrier.

transmission – the way that an infectious agent is spread to another host.

serotype – a group of related organisms with a common set of antigens.

sequelae – an after effect of a disease, treatment, procedure, or injury.

vehicle – in foodborne disease investigations, the vehicle is the food associated with the transmission of the pathogen or disease-causing agent.

virulence – the degree to which an organism is disease-causing.

verotoxin (Shiga toxin) – an extremely toxic substance produced by around 100 serotypes of *E. coli* including *E. coli* O157:H7. It is also produced by other bacteria. It was named verotoxin by the Canadian researcher who first identified it because it killed the vero cells from green monkeys. In the U.S. it was discovered shortly afterward and was called Shiga-like toxin because it was seen to be nearly identical to the toxin produced by *Shigella*. It is now considered virtually indistinguishable and in the U.S. is called Shiga toxin. The contest between the terms verotoxin (VTEC) and shiga toxin (STEC) continues to cause confusion worldwide.

virus – a parasite that lives within another cell. It cannot reproduce itself and depends on the host cell to provide.

zoonosis – a disease or infection transmitted from animals to man.

Index